Boundary Conditions in Electromagnetics

IEEE Press
445 Hoes Lane
Piscataway, NJ 08854

Boundary Conditions in Electromagnetics

Ismo V. Lindell and Ari Sihvola
Aalto University, School of Electrical Engineering Espoo, Finland

IEEE Antennas and Propagation Society, *Sponsor*

The IEEE Press Series on Electromagnetic Wave Theory

IEEE PRESS

WILEY

Published by John Wiley & Sons, Inc., Hoboken, New Jersey.
Published simultaneously in Canada.

Library of Congress Cataloging-in-Publication Data is available.

hardback: 9781119632368

Printed in the United States of America.

V10014880_101819

Contents

Preface

Along with the Maxwell equations, boundary conditions form an essential part of defining electromagnetic problems. Mathematically, they ensure uniqueness of solution for the problem. In working with engineering problems, they offer analytic help in mathematical handling of given electromagnetic structures and synthetic help for designing new electromagnetic structures. In the former case, a given physical structure can often be approximated by boundary conditions enabling the use of simpler analytical computation schemes to find the electromagnetic response of the structure. In the latter case, certain exact boundary conditions can be used as a starting point to design (synthesize) a structure with desired electromagnetic response. This aspect is adopted throughout in the text. With the synthetic approach, the problem becomes to realize the structure in terms of suitable materials. The main topic of this book is the most general linear and local boundary conditions and the basic problem, plane-wave reflection from a planar boundary in a simple-isotropic medium.

The general boundary is approached by starting, in Chapters 1 and 2, from the simplest special cases, the Perfect Electric Conductor (PEC), the Perfect Magnetic Conductor (PMC) and their generalization, the Perfect ElectroMagnetic Conductor (PEMC). It is shown that, by applying suitable duality transformations, these boundaries can be transformed to one another. As an application, scattering of small PEMC particles is analyzed to find effective parameters of a mixture of PEMC inclusions.

In Chapter 3, the Impedance Boundary Condition (IBC) is seen to generalize the previous cases. Other special cases are obtained by splitting the impedance dyadic $\overline{\overline{\mathsf{Z}}}_t$ in its natural components. A plane wave satisfying the boundary conditions identically is called a wave matched to the boundary, an example of which is the surface wave propagating along certain impedance surfaces. In addition to finding the plane-wave reflection dyadic for the general IBC surface, and the special cases, the Soft-and-Hard (SH) boundary with its generalization (GSH boundary), the simple-isotropic boundary problem is solved for dipole excitation in terms of its image source.

In Chapter 4, boundaries defined by conditions involving components of fields normal to the boundary surface, are considered. Main attention is given to the DB boundary, at which the fields **D** and **B** have no normal components. Along with it, the one called D'B' boundary, defined by zero normal derivatives of the normal components of **D** and **B**, is analysed. Two applications, spherical DB resonator and cylindrical DB waveguide are solved for their basic modes. Finally, attention is given to the more general case of the mixed-impedance (or DB/D'B') boundary.

In Chapter 5, the most general linear and local impedance boundary (GBC boundary) is introduced as a generalization of all of the above cases except the D'B' and DB/D'B' boundaries which are not local ones. It is shown that any plane wave can be decomposed in two parts defined by polarizations depending on the parameters of the boundary and the **k** vector of the wave. It is shown that such a polarization property is retained in reflection, whence the decomposed waves do not couple at the boundary. This allows one to express the reflection dyadic as a sum of two simple parts. Conditions for waves matched to the GBC boundary are obtained from zero reflection coefficients. Special cases, like the SHDB boundary and the GSHDB boundary, not falling in the topics of the previous Chapters, are discussed with examples on matched waves in various cases. Reciprocity of the GBC boundary is discussed by analysing different special cases, and realization of the general GBC boundary in terms of the interface of a suitable bi-anisotropic medium is suggested.

As a topic different from the previous ones, based on linear boundary conditions, the class of SQL boundaries, based on sesquilinear boundary conditions, is introduced in Chapter 6. For the basic problem of plane wave incident to an isotropic SQL boundary, it is shown that there are two reflected waves, one of which is forward-reflected and, the other one, is retro-reflected. For certain parameters of the medium, only the retro-reflected wave survives, whence the SQL boundary resembles one made of corner reflectors.

Chapter 7 deals with the numerical analysis of scattering from objects defined by boundary conditions at their surfaces. The manner how electromagnetic waves are affected by such scatterers depends on their size, geometrical shape, and in particular, the character of the boundary condition of their surface. The analysis in this chapter focuses mostly on the canonical spherical geometry. Scatterers with PEC, PMC, PEMC, DB, D'B', and isotropic impedance boundaries are analysed using the full-wave Mie scattering principles. Interesting physical responses of such scatterers are discussed, such as zero-backscattering objects, unexpected front-to-back asymmetries, and sharp resonances for subwavelength impedance-boundary scatterers.

The main scope of this book is to introduce the set of most general linear and local boundary conditions and analyse the basic problems of plane-wave reflection and matching associated to a planar boundary in a simple-isotropic medium. Most of the material in this book has been first presented by these authors in various journals listed at the end of the book, and is here given a unified representation. Some of these less common boundary conditions discussed here have already found practical engineering realizations by scientists working on metamaterials and metasurfaces. They are referred to in a proper context in this book.

While the analysis applies Gibbsian 3D vector formalism, familiar to all electrical engineers, the general boundary conditions can be given a simple and natural representation in terms of 4D multiforms and dyadics, which leads to the impression of their basic physical nature. The formalism is briefly introduced in Appendix C. In Appendices A and B some useful formulas from electromagnetic analysis and Gibbsian 3D dyadics are added for convenience. Many details of the analysis are skipped and left as exercises, solutions of which are outlined in the Appendix D. The aim is as well to compact the text as to leave room for activity in self-learning.

The book is directed to anyone interested in electromagnetic theory in general. For example, those working in the fashionable engineering field of metamaterials and metasurfaces, may find the concepts discussed here useful.

Ismo V. Lindell and Ari Sihvola

Chapter 1

Introduction

1.1 Basic Equations

Maxwell Equations

Electromagnetic problems involving fields and sources are governed by the Maxwell equations, medium equations and boundary conditions. For time-harmonic fields and sources, with time dependence $\exp(j\omega t)$, the Maxwell equations, in the form of Heaviside [19] and the formalism of Gibbs [13], can be expressed as

$$\nabla \times \mathbf{E} + j\omega \mathbf{B} = -\mathbf{J}_m \tag{1.1}$$

$$\nabla \times \mathbf{H} - j\omega \mathbf{D} = \mathbf{J}_e. \tag{1.2}$$

Here, $\mathbf{J}_e$ and $\mathbf{J}_m$ denote the respective electric and magnetic current density vectors while $\mathbf{E}, \mathbf{H}, \mathbf{D}$ and $\mathbf{B}$ represent the respective electric field, magnetic field, electric flux-density and magnetic flux-density vectors. All of these vectors may have complex components.

Medium Conditions

Conditions between the field vectors in a bianisotropic medium can be expressed in the form [29, 35]

$$\begin{pmatrix} \mathbf{D} \\ \mathbf{B} \end{pmatrix} = \begin{pmatrix} \overline{\overline{\epsilon}} & \overline{\overline{\xi}} \\ \overline{\overline{\zeta}} & \overline{\overline{\mu}} \end{pmatrix} \cdot \begin{pmatrix} \mathbf{E} \\ \mathbf{H} \end{pmatrix}, \tag{1.3}$$

where $\overline{\overline{\epsilon}}, \overline{\overline{\xi}}, \overline{\overline{\zeta}}$ and $\overline{\overline{\mu}}$ are medium dyadics. In terms of a given vector basis, any dyadic can be represented in terms of a 3×3 matrix involving nine scalar components (for rules of dyadic algebra, see Appendix C).

Here we mainly consider problems involving the simple-isotropic medium ('free space, vacuum'), for which the medium conditions are reduced to

$$\mathbf{D} = \epsilon_o \mathbf{E}, \quad \mathbf{B} = \mu_o \mathbf{H}. \tag{1.4}$$

The wave impedance η_o, expressing the ratio of the electric and magnetic field magnitudes in the simple-isotropic medium, is defined by

$$\eta_o = \sqrt{\frac{\mu_o}{\epsilon_o}}. \tag{1.5}$$

Boundary Conditions

Boundary conditions form an essential part of formulating electromagnetic problems. By a boundary we mean a surface on which secondary electromagnetic sources are induced by the primary fields so that the fields beyond the surface vanish. A boundary surface is different from an interface of two media because the fields beyond the interface are not necessarily zero. From the mathematical point of view, boundary conditions are required to ensure existence and uniqueness of solutions for a particular problem. From the engineering point of view, two aspects of boundary conditions can be separated which may be called analytic and synthetic.

An *analytic* aspect of boundary conditions is encountered when a given physical problem requires mathematical analysis. Due to natural complications, a given structure must often be approximated by certain boundary conditions to find a numerical solution [21, 88]. As an example, solving radio wave propagation over ground requires that the ground be approximated by an impedance boundary or, if well-enough conducting, by the perfect electric conductor (PEC). Thus, in such a case, to analyze the effect of a given structure, we replace it by some approximate boundary conditions.

In contrast, a *synthetic* aspect emerges when we wish to realize given boundary conditions by some physical structure. As an example, when designing mobile phones with submerged antennas, the concept of perfect magnetic conductor (PMC) boundary has been suggested to solve the problem of efficient radiation [92]. In this case, the problem becomes a synthesis, how to find a physical structure to realize the PMC conditions.

In this book we are concerned about the synthetic aspect by studying different types of boundary conditions, their effect on electromagnetic fields, and possible realizations by interfaces of media defined by medium parameters. More practical physical realizations by (meta)materials are, however, beyond the topic of the book. Let us start with three most basic boundary conditions, properties of which will be considered in due course.

- Perfect Electric Conductor (PEC)

$$\mathbf{n} \times \mathbf{E} = 0 \tag{1.6}$$

- Perfect Magnetic Conductor (PMC)

$$\mathbf{n} \times \mathbf{H} = 0, \tag{1.7}$$

- Perfect Electromagnetic Conductor (PEMC) [45]

$$\mathbf{n} \times (\mathbf{H} + M\mathbf{E}) = 0 \tag{1.8}$$

The PEMC, involving a parameter M (the PEMC admittance), is a generalization of both the PEC and the PMC. For $M = 0$, the PEMC reduces to the PMC, and for $1/M = 0$, it reduces to the PEC.

1.2 Duality Transformation

A given electromagnetic problem can be transformed to another one in terms of duality transformation which does not change the geometry of the problem, but the sources, fields, medium and boundary parameters are transformed to have other values, in general. The concept of duality was evidently unknown to Maxwell, because he presented his equations in a very nonsymmetric form, in terms of 20 scalar field and potential quantities [79]. The concept was introduced by Heaviside in 1886 [19].

Duality transformation is based on the apparent symmetry of the Maxwell equations (1.1), (1.2), written more compactly as

$$\nabla \times \begin{pmatrix} \mathbf{E} \\ -\mathbf{H} \end{pmatrix} + j\omega \begin{pmatrix} 0 & 1 \\ 1 & 0 \end{pmatrix} \begin{pmatrix} \mathbf{D} \\ \mathbf{B} \end{pmatrix} = - \begin{pmatrix} \mathbf{J}_m \\ \mathbf{J}_e \end{pmatrix}. \tag{1.9}$$

In fact, for the simple change of symbols $\mathbf{E} \leftrightarrow -\mathbf{H}$, $\mathbf{B} \leftrightarrow \mathbf{D}$ and $\mathbf{J}_e \leftrightarrow \mathbf{J}_m$, the pair of equations (1.9) is invariant. More generally, the same property can be expressed in terms of the duality transformation $(\mathbf{E}, \mathbf{H}) \rightarrow (\mathbf{E}_d, \mathbf{H}_d)$, defined by [35]

$$\begin{pmatrix} \mathbf{E}_d \\ \eta_o \mathbf{H}_d \end{pmatrix} = \begin{pmatrix} A & B \\ C & D \end{pmatrix} \begin{pmatrix} \mathbf{E} \\ \eta_o \mathbf{H} \end{pmatrix}. \tag{1.10}$$

The wave impedance η_o has been included to obtain dimensionless transformation parameters $A \cdots D$. They are assumed to satisfy

$$AD - BC = 1, \tag{1.11}$$

whence the inverse transformation exists and has the form

$$\begin{pmatrix} \mathbf{E} \\ \eta_o \mathbf{H} \end{pmatrix} = \begin{pmatrix} D & -B \\ -C & A \end{pmatrix} \begin{pmatrix} \mathbf{E}_d \\ \eta_o \mathbf{H}_d \end{pmatrix}. \tag{1.12}$$

Applying (1.10) to (1.1) and (1.2), the associated transformation rules can be expressed as

$$\begin{pmatrix} \eta_o \mathbf{D}_d \\ \mathbf{B}_d \end{pmatrix} = \begin{pmatrix} D & -C \\ -B & A \end{pmatrix} \begin{pmatrix} \eta_o \mathbf{D} \\ \mathbf{B} \end{pmatrix} \tag{1.13}$$

$$\begin{pmatrix} \eta_o \mathbf{J}_{ed} \\ \mathbf{J}_{md} \end{pmatrix} = \begin{pmatrix} D & -C \\ -B & A \end{pmatrix} \begin{pmatrix} \eta_o \mathbf{J}_e \\ \mathbf{J}_m \end{pmatrix} \tag{1.14}$$

One can show that the PEMC boundary conditions (1.8) are transformed to

$$\mathbf{n} \times (\mathbf{H}_d + M_d \mathbf{E}_d) = 0, \tag{1.15}$$

with the transformed PEMC admittance satisfying

$$M_d \eta_o = -\frac{C - DM\eta_o}{A - BM\eta_o}. \tag{1.16}$$

From this it follows that both PEC and PMC boundaries are transformed to PEMC boundaries with $M_d\eta_o = -D/B$ and $M_d\eta_o = -C/A$, respectively. Also, any given PEMC boundary can be transformed to PEC and PMC boundaries when the transformation parameters are chosen to satisfy the respective restrictions $A/B = M\eta_o$ and $C/D = M\eta_o$.

In the general case, the dyadic parameters of the electromagnetic medium will be changed when the fields are subject to the duality transformation (1.10). For a bianisotropic medium defined by conditions of the form (1.3), the transformed medium dyadics can be shown to obey the relations [42]

$$\begin{pmatrix} \overline{\overline{\epsilon}}_d \\ \overline{\overline{\xi}}_d \\ \overline{\overline{\zeta}}_d \\ \overline{\overline{\mu}}_d \end{pmatrix} = \begin{pmatrix} D^2 & -CD/\eta_o & -CD/\eta_o & C^2/\eta_o^2 \\ -BD\eta_o & AD & BC & -AC/\eta_o \\ -BD\eta_o & BC & AD & -AC/\eta_o \\ B^2\eta_o^2 & -AB\eta_o & -AB\eta_o & A^2 \end{pmatrix} \begin{pmatrix} \overline{\overline{\epsilon}} \\ \overline{\overline{\xi}} \\ \overline{\overline{\zeta}} \\ \overline{\overline{\mu}} \end{pmatrix}. \tag{1.17}$$

Requiring a transformation in which the simple isotropic medium (μ_o, ϵ_o) is invariant leads to a choice of the form

$$A = D = \cos\varphi, \qquad B = -C = \sin\varphi, \tag{1.18}$$

where φ is a free transformation parameter. The resulting transformation matrix

$$\begin{pmatrix} A & B \\ C & D \end{pmatrix} = \begin{pmatrix} \cos\varphi & \sin\varphi \\ -\sin\varphi & \cos\varphi \end{pmatrix} \tag{1.19}$$

can be recognized as a 2D rotation matrix. The corresponding transformation rule for the medium dyadics (1.17) becomes in this case

$$\begin{pmatrix} \overline{\overline{\epsilon}}_d/\epsilon_o \\ \overline{\overline{\xi}}_d/\sqrt{\mu_o\epsilon_o} \\ \overline{\overline{\zeta}}_d/\sqrt{\mu_o\epsilon_o} \\ \overline{\overline{\mu}}_d/\mu_o \end{pmatrix} = \mathcal{Q}(\varphi) \begin{pmatrix} \overline{\overline{\epsilon}}/\epsilon_o \\ \overline{\overline{\xi}}/\sqrt{\mu_o\epsilon_o} \\ \overline{\overline{\zeta}}/\sqrt{\mu_o\epsilon_o} \\ \overline{\overline{\mu}}/\mu_o \end{pmatrix}, \tag{1.20}$$

with the 4×4 matrix $\mathcal{Q}(\varphi)$ defined by

$$\mathcal{Q}(\varphi) = \begin{pmatrix} \cos^2\varphi & \sin\varphi\cos\varphi & \sin\varphi\cos\varphi & \sin^2\varphi \\ -\sin\varphi\cos\varphi & \cos^2\varphi & -\sin^2\varphi & \sin\varphi\cos\varphi \\ -\sin\varphi\cos\varphi & -\sin^2\varphi & \cos^2\varphi & \sin\varphi\cos\varphi \\ \sin^2\varphi & -\sin\varphi\cos\varphi & -\sin\varphi\cos\varphi & \cos^2\varphi \end{pmatrix}. \tag{1.21}$$

For the simple-isotropic medium with $\overline{\overline{\epsilon}} = \epsilon_o \overline{\overline{\mathsf{I}}}$, $\overline{\overline{\xi}} = \overline{\overline{\zeta}} = 0$ and $\overline{\overline{\mu}} = \mu_o \overline{\overline{\mathsf{I}}}$, from (1.20) we obtain $\overline{\overline{\epsilon}}_d = \epsilon_o \overline{\overline{\mathsf{I}}}$, $\overline{\overline{\xi}}_d = \overline{\overline{\zeta}}_d = 0$ and $\overline{\overline{\mu}}_d = \mu_o \overline{\overline{\mathsf{I}}}$, as required. One can further show that the matrix $\mathcal{Q}(\varphi)$ satisfies

$$\mathcal{Q}(\varphi_1)\mathcal{Q}(\varphi_2) = \mathcal{Q}(\varphi_1 + \varphi_2), \tag{1.22}$$

$$\det \mathcal{Q}(\varphi) = 1, \tag{1.23}$$

$$\mathcal{Q}^{-1}(\varphi) = \mathcal{Q}^T(\varphi) = \mathcal{Q}(-\varphi), \tag{1.24}$$

i.e., it is an orthogonal matrix. The proof is left as an exercise.

Expressing the PEMC admittance parameter M in terms of another parameter φ as

$$M\eta_o = \tan \vartheta, \tag{1.25}$$

the transformation rule (1.16) takes the simple form

$$M_d \eta_o = \frac{M\eta_o + \tan\varphi}{1 - M\eta_o \tan\varphi} = \tan(\vartheta + \varphi). \tag{1.26}$$

Thus, for the choice $\varphi = -\vartheta$ of the duality parameter, the PEMC boundary is transformed to the PMC boundary ($M_d = 0$), while for $\varphi = \pi/2 - \vartheta$, it is transformed to the PEC boundary ($1/M_d = 0$).

1.3 Plane Waves

Basic Conditions

Let us consider time-harmonic plane waves with $\exp(j\omega t)$ dependence in a simple isotropic medium, in front of a boundary surface defined by $\mathbf{n} \cdot \mathbf{r} = 0$. For simplicity we assume that the surface is a plane, i.e., that $\mathbf{n}$ is a constant unit vector. The electric and magnetic field components of waves incident to the boundary are expressed by

$$\mathbf{E}^i(\mathbf{r}) = \mathbf{E}^i \exp(-j\mathbf{k}^i \cdot \mathbf{r}), \tag{1.27}$$

$$\mathbf{H}^i(\mathbf{r}) = \mathbf{H}^i \exp(-j\mathbf{k}^i \cdot \mathbf{r}), \tag{1.28}$$

and the reflected fields are

$$\mathbf{E}^r(\mathbf{r}) = \mathbf{E}^r \exp(-j\mathbf{k}^r \cdot \mathbf{r}), \tag{1.29}$$

$$\mathbf{H}^r(\mathbf{r}) = \mathbf{H}^r \exp(-j\mathbf{k}^r \cdot \mathbf{r}). \tag{1.30}$$

To satisfy the Maxwell equations, the two wave vectors must satisfy the dispersion equation as

$$\mathbf{k}^i \cdot \mathbf{k}^i = \mathbf{k}^r \cdot \mathbf{k}^r = k_o^2 = \omega^2 \mu_o \epsilon_o. \tag{1.31}$$

If the boundary condition is linear, the $\mathbf{k}^i$ and $\mathbf{k}^r$ vectors must have the same components tangential to the boundary surface, i.e., they can be represented as

$$\mathbf{k}^i = \mathbf{k}_t - k_n \mathbf{n}, \qquad \mathbf{k}^r = \mathbf{k}_t + k_n \mathbf{n}. \tag{1.32}$$

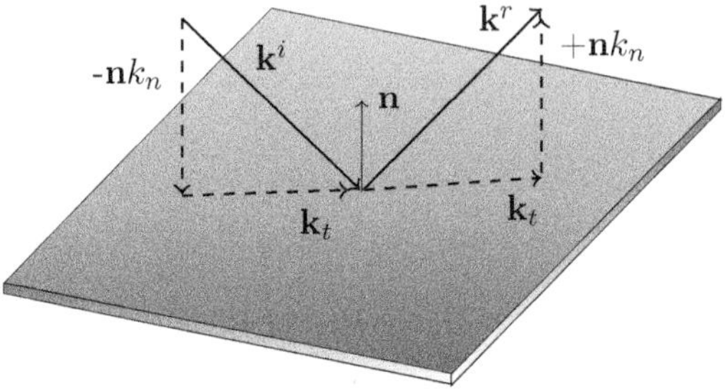

Figure 1.1: The incident and reflected plane waves have the same wave-vector components $\mathbf{k}_t$ tangential to the boundary surface.

A class of boundaries with conditions deviating from linear ones is considered in Chapter 6.

Substituting the plane-wave fields in the Maxwell equations (1.1) and (1.2), equations relating the electric and magnetic fields are obtained as

$$\mathbf{k}^i \times \mathbf{E}^i = k_o\eta_o\mathbf{H}^i, \tag{1.33}$$

$$\mathbf{k}^i \times \eta_o\mathbf{H}^i = -k_o\mathbf{E}^i, \tag{1.34}$$

and

$$\mathbf{k}^r \times \mathbf{E}^r = k_o\eta_o\mathbf{H}^r, \tag{1.35}$$

$$\mathbf{k}^r \times \eta_o\mathbf{H}^r = -k_o\mathbf{E}^r. \tag{1.36}$$

The field components normal to the boundary can be expressed in terms of their tangential components from the orthogonality conditions as

$$\mathbf{k}^i\cdot\mathbf{E}^i = 0, \quad \Rightarrow \quad \mathbf{n}\cdot\mathbf{E}^i = \frac{1}{k_n}\mathbf{k}_t\cdot\mathbf{E}^i_t, \tag{1.37}$$

$$\mathbf{k}^r\cdot\mathbf{E}^r = 0, \quad \Rightarrow \quad \mathbf{n}\cdot\mathbf{E}^r = -\frac{1}{k_n}\mathbf{k}_t\cdot\mathbf{E}^r_t, \tag{1.38}$$

valid for $k_n \neq 0$. Similar conditions are valid for the magnetic fields.

Field Relations

From (1.33) and (1.34) we obtain the following relations between the tangential components of the incident fields:

$$\mathbf{n}\cdot\mathbf{k}_t \times \mathbf{E}^i_t = \frac{k_o}{k_n}\mathbf{k}_t\cdot\eta_o\mathbf{H}^i_t, \tag{1.39}$$

$$\mathbf{n}\cdot\mathbf{k}_t \times \eta_o\mathbf{H}^i_t = -\frac{k_o}{k_n}\mathbf{k}_t\cdot\mathbf{E}^i_t. \tag{1.40}$$

Assuming also $\mathbf{k}_t \cdot \mathbf{k}_t = k_t^2 \neq 0$, the vectors $\mathbf{k}_t$ and $\mathbf{n} \times \mathbf{k}_t$ serve as a 2D basis, in terms of which we can expand the 2D unit dyadic as

$$\overline{\overline{\mathsf{I}}}_t = \overline{\overline{\mathsf{I}}} - \mathbf{nn} = \frac{1}{k_t^2}(\mathbf{k}_t\mathbf{k}_t + (\mathbf{n} \times \mathbf{k}_t)(\mathbf{n} \times \mathbf{k}_t)). \tag{1.41}$$

The relation between the tangential electric and magnetic fields can be expanded as

$$\begin{aligned} \mathbf{E}_t^i &= \frac{1}{k_t^2}(\mathbf{k}_t(\mathbf{k}_t \cdot \mathbf{E}_t^i) + (\mathbf{n} \times \mathbf{k}_t)(\mathbf{n} \times \mathbf{k}_t \cdot \mathbf{E}_t^i)) \\ &= \frac{1}{k_t^2}(-\frac{k_n}{k_o}\mathbf{k}_t(\mathbf{n} \times \mathbf{k}_t) + \frac{k_o}{k_n}(\mathbf{n} \times \mathbf{k}_t)\mathbf{k}_t) \cdot \eta_o \mathbf{H}_t^i \\ &= \frac{1}{k_o k_n}\mathbf{n} \times (\mathbf{k}_t\mathbf{k}_t + k_n^2\overline{\overline{\mathsf{I}}}_t) \cdot \eta_o \mathbf{H}_t^i. \end{aligned} \tag{1.42}$$

The dyadic in this relation is of importance in the analysis of plane waves. Denoted by

$$\overline{\overline{\mathsf{J}}}_t = \frac{1}{k_o k_n}\mathbf{n} \times (\mathbf{k}_t\mathbf{k}_t + k_n^2\overline{\overline{\mathsf{I}}}_t), \tag{1.43}$$

it can be shown to satisfy the properties

$$\mathrm{tr}\overline{\overline{\mathsf{J}}}_t = 0, \tag{1.44}$$

$$\overline{\overline{\mathsf{J}}}_t^2 = -\overline{\overline{\mathsf{I}}}_t, \tag{1.45}$$

$$\overline{\overline{\mathsf{J}}}_t^{-1} = -\overline{\overline{\mathsf{J}}}_t, \tag{1.46}$$

derivations of which are left as exercises.

The dyadic $\overline{\overline{\mathsf{J}}}_t$ allows one to write relations between the tangential electric and magnetic fields of a plane wave in compact form as

$$\mathbf{E}_t^i = \overline{\overline{\mathsf{J}}}_t \cdot \eta_o \mathbf{H}_t^i \qquad \eta_o \mathbf{H}_t^i = -\overline{\overline{\mathsf{J}}}_t \cdot \mathbf{E}_t^i. \tag{1.47}$$

For the reflected fields, similar expressions are valid when replacing k_n by $-k_n$:

$$\mathbf{E}_t^r = -\overline{\overline{\mathsf{J}}}_t \cdot \eta_o \mathbf{H}_t^r, \qquad \eta_o \mathbf{H}_t^r = \overline{\overline{\mathsf{J}}}_t \cdot \mathbf{E}_t^r. \tag{1.48}$$

The expressions (1.47) and (1.48) will be needed in the analysis of reflections from various boundaries. In this, we can also make use of the following set of relations:

$$\mathbf{k}_t \cdot \overline{\overline{\mathsf{J}}}_t = -\frac{k_n}{k_o}\mathbf{n} \times \mathbf{k}_t, \qquad \overline{\overline{\mathsf{J}}}_t \cdot \mathbf{k}_t = \frac{k_o}{k_n}\mathbf{n} \times \mathbf{k}_t, \tag{1.49}$$

$$(\mathbf{n} \times \mathbf{k}_t) \cdot \overline{\overline{\mathsf{J}}}_t = \frac{k_o}{k_n}\mathbf{k}_t, \qquad \overline{\overline{\mathsf{J}}}_t \cdot (\mathbf{n} \times \mathbf{k}_t) = -\frac{k_n}{k_o}\mathbf{k}_t, \tag{1.50}$$

and,

$$\overline{\overline{\mathsf{J}}}_t^{(2)} = \frac{1}{2}\overline{\overline{\mathsf{J}}}_t {}^\times_\times \overline{\overline{\mathsf{J}}}_t = \mathbf{nn}, \qquad \det{}_t\overline{\overline{\mathsf{J}}}_t = \mathrm{tr}\overline{\overline{\mathsf{J}}}_t^{(2)} = 1. \tag{1.51}$$

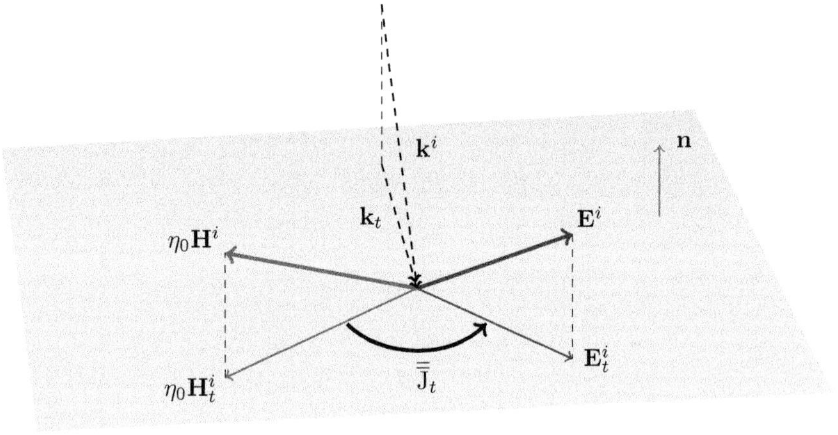

Figure 1.2: The dyadic $\overline{\overline{\mathsf{J}}}_t$ maps the tangential component of the magnetic field of an incident plane wave to the tangential component of the electric field. The vectors $\mathbf{E}^i, \mathbf{H}^i$ and $\mathbf{k}^i$ are orthogonal to one another.

The dyadic $\overline{\overline{\mathsf{J}}}_t$ and the antisymmetric dyadic $\mathbf{n}\times\overline{\overline{\mathsf{I}}}$ share the property $(\mathbf{n}\times\overline{\overline{\mathsf{I}}})^2 = \overline{\overline{\mathsf{J}}}_t^2 = -\overline{\overline{\mathsf{I}}}_t$. While $\mathbf{n}\times\overline{\overline{\mathsf{I}}}$ rotates vectors on the tangent plane by $\pi/2$, the dyadic $\overline{\overline{\mathsf{J}}}_t$ "rotates" the tangential part of the electric field to the tangential part of the magnetic field of the same plane wave. The $\overline{\overline{\mathsf{J}}}_t$ dyadic is determined by $\mathbf{k}_t$, the tangential part of the wave vector in the simple-isotropic medium.

1.4 TE/TM Decomposition

It turns out that the field vectors of a plane wave, propagating with the wave vector $\mathbf{k} = \mathbf{n}k_n + \mathbf{k}_t$, can be expressed in terms of their normal components $\mathbf{n}\cdot\mathbf{E}$ and $\mathbf{n}\cdot\mathbf{H}$ as

$$\mathbf{E} = \frac{1}{k_t^2}((k_t^2\mathbf{n} - k_n\mathbf{k}_t)(\mathbf{n}\cdot\mathbf{E}) + k_o(\mathbf{n}\times\mathbf{k})(\mathbf{n}\cdot\eta_o\mathbf{H})) \tag{1.52}$$

$$\eta_o\mathbf{H} = \frac{1}{k_t^2}((k_t^2\mathbf{n} - k_n\mathbf{k}_t)(\mathbf{n}\cdot\eta_o\mathbf{H}) - k_o(\mathbf{n}\times\mathbf{k})(\mathbf{n}\cdot\mathbf{E})). \tag{1.53}$$

Details of the derivation are left as an exercise. Applying (1.52) and (1.53), any plane wave can be decomposed in two parts as

$$\begin{pmatrix}\mathbf{E}\\ \mathbf{H}\end{pmatrix} = \begin{pmatrix}\mathbf{E}_{TE}\\ \mathbf{H}_{TE}\end{pmatrix} + \begin{pmatrix}\mathbf{E}_{TM}\\ \mathbf{H}_{TM}\end{pmatrix}, \tag{1.54}$$

with

$$\mathbf{E}_{TE} = \frac{k_o}{k_t^2}(\mathbf{n}\times\mathbf{k})(\mathbf{n}\cdot\eta_o\mathbf{H}), \quad (1.55)$$

$$\mathbf{H}_{TE} = \frac{1}{k_t^2}(k_t^2\mathbf{n} - k_n\mathbf{k}_t)(\mathbf{n}\cdot\mathbf{H}), \quad (1.56)$$

$$\mathbf{E}_{TM} = \frac{1}{k_t^2}(k_t^2\mathbf{n} - k_n\mathbf{k}_t)(\mathbf{n}\cdot\mathbf{E}), \quad (1.57)$$

$$\eta_o\mathbf{H}_{TM} = -\frac{k_o}{k_t^2}(\mathbf{n}\times\mathbf{k})(\mathbf{n}\cdot\mathbf{E}). \quad (1.58)$$

The partial fields are called tranverse electric (TE) and transverse magnetic (TM) because the two parts are restricted by the respective conditions

$$\mathbf{n}\cdot\mathbf{E}_{TE} = 0, \quad \mathbf{n}\cdot\mathbf{H}_{TM} = 0. \quad (1.59)$$

The decomposed fields satisfy the Maxwell equations separately. To check this, we can expand

$$\nabla\times\mathbf{E}_{TE} = -j\mathbf{k}\times\mathbf{E}^{TE} \quad (1.60)$$

$$= -j\frac{k_o\eta_o}{k_t^2}\mathbf{k}\times(\mathbf{n}\times\mathbf{k}_t)(\mathbf{n}\cdot\mathbf{H}) \quad (1.61)$$

$$= -j\frac{k_o\eta_o}{k_t^2}(\mathbf{n}k_t^2 - k_n\mathbf{k}_t)(\mathbf{n}\cdot\mathbf{H}) \quad (1.62)$$

$$= -jk_o\eta_o\mathbf{H}^{TE}, \quad (1.63)$$

$$\nabla\times\mathbf{H}_{TM} = -j\mathbf{k}\times\mathbf{H}^{TM} \quad (1.64)$$

$$= j\frac{k_o}{\eta_o k_t^2}\mathbf{k}\times(\mathbf{n}\times\mathbf{k}_t)(\mathbf{n}\cdot\mathbf{E}) \quad (1.65)$$

$$= j\frac{k_o}{\eta_o k_t^2}(\mathbf{n}k_t^2 - k_n\mathbf{k}_t)(\mathbf{n}\cdot\mathbf{E}) \quad (1.66)$$

$$= j\frac{k_o}{\eta_o}\mathbf{E}_{TM}. \quad (1.67)$$

The decomposition is unique for $\mathbf{k}_t \neq 0$. For $\mathbf{k}_t = 0$ the expansions (1.52) and (1.53) are not applicable. In this case the TE/TM decomposition is still possible but not unique. Actually, for $\mathbf{k}_t = 0$ the plane-wave has TEM polarization, $\mathbf{n}\cdot\mathbf{E} = \mathbf{n}\cdot\mathbf{H} = 0$.

Because the TE/TM decomposition of a plane wave does not depend on the choice of the $\mathbf{k}$ vector, it is valid for a sum of plane waves and, ultimately, for an integral of plane waves in the space of $\mathbf{k}$ vectors which may have complex values. Thus, any fields which can be expressed as an integral of plane waves can be decomposed in two independent TE and TM parts with respect to a given $\mathbf{n}$ vector.

1.5 Problems

1.1 For a given duality transformation, find a PEMC admittance which is invariant in the transformation. Apply the result to the special transformation (1.19).

1.2 Derive the transformation rule (1.17) for the medium dyadics and show that the dyadic $\overline{\overline{\xi}} - \overline{\overline{\zeta}}$ is invariant in the transformation.

1.3 Derive the transformation rule (1.20) from (1.17).

1.4 Show that the matrix $\mathcal{Q}(\varphi)$ in (1.21) is orthogonal, i.e., that it satisfies

$$\mathcal{Q}^{-1}(\varphi) = \mathcal{Q}^T(\varphi) = \mathcal{Q}(-\varphi).$$

1.5 Show that it is possible to find a duality transformation which maps two given PEMC admittances M_1 and M_2 so that $M_{1d} = -M_{2d} = M_d$ and leaves the simple-isotropic medium invariant. Find the relation between M_1, M_2 and M_d.

1.6 Derive the properties (1.45), (1.46) and (1.51) of the dyadic $\overline{\overline{\mathsf{J}}}_t$.

1.7 Derive the decomposition rules (1.52) and (1.53).

1.8 Applying (1.20), find another expression for the duality transformation of the medium dyadics in the form

$$\begin{pmatrix} \overline{\overline{\epsilon}}_d/\epsilon_o - \overline{\overline{\mu}}_d/\mu_o \\ (\overline{\overline{\xi}}_d - \overline{\overline{\zeta}}_d)/\sqrt{\mu_o\epsilon_o} \\ (\overline{\overline{\xi}}_d + \overline{\overline{\zeta}}_d)/\sqrt{\mu_o\epsilon_o} \\ \overline{\overline{\epsilon}}_d/\epsilon_o + \overline{\overline{\mu}}_d/\mu_o \end{pmatrix} = \mathcal{Q}'(\varphi) \begin{pmatrix} \overline{\overline{\epsilon}}/\epsilon_o - \overline{\overline{\mu}}/\mu_o \\ (\overline{\overline{\xi}} - \overline{\overline{\zeta}})/\sqrt{\mu_o\epsilon_o} \\ (\overline{\overline{\xi}} + \overline{\overline{\zeta}})/\sqrt{\mu_o\epsilon_o} \\ \overline{\overline{\epsilon}}/\epsilon_o + \overline{\overline{\mu}}/\mu_o \end{pmatrix}$$

in terms of a matrix $\mathcal{Q}'(\varphi)$. Which dyadics appear invariant in the transformation?

Chapter 2

Perfect Electromagnetic Conductor Boundary

2.1 PEMC Conditions

The condition (1.8), defining the perfect electromagnetic conductor (PEMC) boundary, actually represents one of the most basic conditions in electromagnetic theory. In fact, applying the four-dimensional formalism of Appendix A, the PEMC medium (also called the axion medium [20, 72, 97, 99]) is defined by the simplest possible linear relation between the electromagnetic two-forms $\mathbf{\Phi} = \mathbf{B} + \mathbf{E} \wedge \varepsilon_4$ and $\mathbf{\Psi} = \mathbf{D} - \mathbf{H} \wedge \varepsilon_4$ as

$$\mathbf{\Psi} = M\mathbf{\Phi}. \tag{2.1}$$

In terms of the spatial two-form components $\mathbf{D}, \mathbf{B}$ and spatial one-form components $\mathbf{H}, \mathbf{E}$, (2.1) equals the conditions

$$\mathbf{D} = M\mathbf{B}, \quad \mathbf{H} = -M\mathbf{E}. \tag{2.2}$$

In Gibbsian 3D formalism, the corresponding PEMC medium conditions between the field vectors[1] are

$$\begin{pmatrix} \mathbf{H} \\ \mathbf{D} \end{pmatrix} = M \begin{pmatrix} 0 & -1 \\ 1 & 0 \end{pmatrix} \begin{pmatrix} \mathbf{B} \\ \mathbf{E} \end{pmatrix}. \tag{2.3}$$

There is no natural way to express these conditions in the form (1.3) in terms of finite medium dyadics. However, they can be represented as the limit [46]

$$\begin{pmatrix} \overline{\overline{\epsilon}} & \overline{\overline{\xi}} \\ \overline{\overline{\zeta}} & \overline{\overline{\mu}} \end{pmatrix} = \lim_{q\to\infty} q \begin{pmatrix} M & 1 \\ 1 & 1/M \end{pmatrix} \overline{\overline{\mathsf{I}}}. \tag{2.4}$$

[1]Although the Gibbsian field vectors and the field one- and two-forms are here denoted by similar symbols, they obey different algebraic rules.

A slightly more natural form with the same limit is [83, 72]

$$\begin{pmatrix} \overline{\overline{\epsilon}} & \overline{\overline{\xi}} \\ \overline{\overline{\zeta}} & \overline{\overline{\mu}} \end{pmatrix} = \lim_{q\to\infty} q \begin{pmatrix} M(1+(\mu_o\epsilon_o/q^2)) & 1 \\ 1 & 1/M \end{pmatrix} \overline{\overline{\mathsf{I}}}. \tag{2.5}$$

In both of these representations all four medium dyadics $\overline{\overline{\epsilon}}\cdots\overline{\overline{\mu}}$ become ultimately infinite in magnitude. However, while in the case (2.4) the matrix has no inverse for any q, in the case of (2.5), the matrix does have an inverse for finite values of q, whence it corresponds to a more ordinary bi-anisotropic medium.

Continuity of the tangential components of **E** and **H**, and the normal components of **D** and **B**, at an interface of a PEMC medium, yields the PEMC boundary conditions,

$$\mathbf{n}\times(\mathbf{H}+M\mathbf{E}) = 0, \tag{2.6}$$

$$\mathbf{n}\cdot(\mathbf{D}-M\mathbf{B}) = 0. \tag{2.7}$$

Actually, the condition (2.7) is not necessary, because it can be shown to follow from (2.6) and the Maxwell equations. The proof is left as an exercise. Although (2.7) carries no additional information beyond (2.6), it can be useful in reducing computational effort in some practical cases.

The PEMC boundary can be realized by an interface of the PEMC medium. However, it is not obvious how to realize the PEMC medium. A more practical realization can be based on a slab of medium with suitable properties. Since the PEMC boundary is isotropic and nonreciprocal, it would be natural to apply a slab of uniaxial gyrotropic medium with axis parallel to the vector **n** [47]. In Chapter 3 such a realization (3.259), (3.260) will be considered as associated with the realization of more general impedance-boundary surfaces [51].

2.2 Eigenproblem of Dyadic $\overline{\overline{\mathsf{J}}}_t$

Assuming a field consisting of incident and reflected plane waves, the PEMC condition (2.6) can be written for the tangential field components as

$$\eta_o(\mathbf{H}_t^r+\mathbf{H}_t^i)+M\eta_o(\mathbf{E}_t^r+\mathbf{E}_t^i)=0. \tag{2.8}$$

Substituting the plane-wave relations (1.47) and (1.48), we can write

$$\overline{\overline{\mathsf{J}}}_t\cdot(\mathbf{E}_t^r-\mathbf{E}_t^i)+M\eta_o(\mathbf{E}_t^i+\mathbf{E}_t^r)=0. \tag{2.9}$$

It appears convenient to expand the tangential field components in terms of the eigenvectors of the $\overline{\overline{\mathsf{J}}}_t$ dyadic (1.43),

$$\overline{\overline{\mathsf{J}}}_t = \frac{1}{k_o k_n}\mathbf{n}\times(\mathbf{k}_t\mathbf{k}_t+k_n^2\overline{\overline{\mathsf{I}}}_t). \tag{2.10}$$

After some algebraic steps we can derive the relations

$$\overline{\overline{\mathsf{J}}}_t\cdot(k_n\mathbf{k}_t\mp jk_o\mathbf{n}\times\mathbf{k}_t)=\pm j(k_n\mathbf{k}_t\mp jk_o\mathbf{n}\times\mathbf{k}_t), \tag{2.11}$$

details of which are left as an exercise. The solutions for the eigenvalue equation

$$\overline{\overline{\mathsf{J}}}_t \cdot \mathbf{x}_{t\pm} = J_\pm \mathbf{x}_{t\pm} \tag{2.12}$$

can now be identified from (2.11) as

$$J_\pm = \pm j, \quad \mathbf{x}_{t\pm} = k_n \mathbf{k}_t \mp j k_o \mathbf{n} \times \mathbf{k}_t. \tag{2.13}$$

Applying the rule

$$\mathbf{x}_{t+}\mathbf{x}_{t-} - \mathbf{x}_{t-}\mathbf{x}_{t+} = -(\mathbf{x}_{t+} \times \mathbf{x}_{t-}) \times \overline{\overline{\mathsf{I}}} = -X\mathbf{n} \times \overline{\overline{\mathsf{I}}}, \tag{2.14}$$

$$X = \mathbf{n} \cdot (\mathbf{x}_{t+} \times \mathbf{x}_{t-}) = 2jk_n k_o k_t^2, \tag{2.15}$$

and assuming $k_n k_t^2 \neq 0$, whence $X \neq 0$, we obtain

$$\overline{\overline{\mathsf{I}}}_t = \frac{1}{X}(\mathbf{x}_{t+}\mathbf{x}_{t-} - \mathbf{x}_{t-}\mathbf{x}_{t+}) \times \mathbf{n}. \tag{2.16}$$

The eigenvectors make a 2D basis, in terms of which we can expand

$$\overline{\overline{\mathsf{J}}}_t = \overline{\overline{\mathsf{J}}}_t \cdot \overline{\overline{\mathsf{I}}}_t = \frac{j}{X}(\mathbf{x}_{t+}\mathbf{x}_{t-} + \mathbf{x}_{t-}\mathbf{x}_{t+}) \times \mathbf{n}. \tag{2.17}$$

The tangential component of the incident electric field can now be decomposed in eigenvectors as

$$\mathbf{E}_t^i = \mathbf{E}_{t+}^i + \mathbf{E}_{t-}^i = A_+ \mathbf{x}_{t+} + A_- \mathbf{x}_{t-}, \tag{2.18}$$

with

$$A_+ = \frac{1}{X}(\mathbf{x}_{t-} \times \mathbf{n}) \cdot \mathbf{E}_t^i, \quad A_- = -\frac{1}{X}(\mathbf{x}_{t+} \times \mathbf{n}) \cdot \mathbf{E}_t^i. \tag{2.19}$$

Applying $\mathbf{k}^i \cdot \mathbf{E}_\pm^i = 0$, the corresponding total incident fields can be expressed as

$$\begin{aligned} \mathbf{E}_\pm^i &= \frac{\mathbf{n}}{k_n}(\mathbf{k}_t \cdot \mathbf{E}_{t\pm}^i) + \mathbf{E}_{t\pm}^i \\ &= A_\pm(\mathbf{n}k_t^2 + k_n \mathbf{k}_t \mp jk_o \mathbf{n} \times \mathbf{k}_t). \end{aligned}$$

One can easily verify that they satisfy

$$\mathbf{E}_+^i \cdot \mathbf{E}_+^i = \mathbf{E}_-^i \cdot \mathbf{E}_-^i = 0, \tag{2.20}$$

whence the incident fields $\mathbf{E}_+^i$ and $\mathbf{E}_-^i$, whose tangential components are eigenvectors of the dyadic $\overline{\overline{\mathsf{J}}}_t$, must be circularly polarized. From (1.47) the corresponding tangential and total magnetic eigenfields satisfy

$$\eta_o \mathbf{H}_{t\pm}^i = -\overline{\overline{\mathsf{J}}}_t \cdot \mathbf{E}_{t\pm}^i = \mp j \mathbf{E}_{t\pm}^i, \tag{2.21}$$

$$\eta_o \mathbf{H}_\pm^i = \mp j \mathbf{E}_\pm^i, \tag{2.22}$$

whence also the magnetic fields $\mathbf{H}^i_+, \mathbf{H}^i_-$ are circularly polarized. Plane-wave fields reflected from the PEMC boundary can be decomposed similarly in two circularly polarized parts.

Assuming real vector of propagation $\mathbf{k}^i$, from (2.13) we have $\mathbf{x}^*_{t\pm} = \mathbf{x}_{t\mp}$, whence $\mathbf{E}^{i\,*}_+$ and $\mathbf{E}^i_-$ are multiples of the same vector. Expanding

$$\begin{aligned} \mathbf{E}^i_\pm \times \mathbf{E}^{i\,*}_\pm &= \pm j2|A_\pm|^2(\mathbf{n}k_t^2 + k_n\mathbf{k}_t) \times (k_o\mathbf{n} \times \mathbf{k}_t) \\ &= \mp j2k_o k_t^2|A_\pm|^2\mathbf{k}^i, \qquad (2.23) \\ \mathbf{E}^i_\pm \cdot \mathbf{E}^{i\,*}_\pm &= |A_\pm|^2(k_t^4 + k_n^2k_t^2 - k_o^2k_t^2) = 2k_o^2k_t^2|A_\pm|^2, \qquad (2.24) \end{aligned}$$

the polarization vectors (see Appendix B) of the two field vectors $\mathbf{E}^i_\pm$ become

$$\mathbf{p}(\mathbf{E}^i_\pm) = \frac{\mathbf{E}^i_\pm \times \mathbf{E}^{i\,*}_\pm}{j\mathbf{E}^i_\pm \cdot \mathbf{E}^{i\,*}_\pm} = \mp\mathbf{u}^i, \qquad (2.25)$$

where $\mathbf{u}^i = \mathbf{k}^i/k_o$ is a real unit vector. Because the time-harmonic vectors $\mathbf{E}^i_\pm(t)$ have right-handed rotation when looking into the directions of $\mathbf{p}(\mathbf{E}^i_\pm)$, the fields $\mathbf{E}^i_+$ and $\mathbf{E}^i_-$ have respectively left and right-handed circular polarizations when looking in the direction of propagation $\mathbf{u}^i$.

As a consequence of the property

$$\mathbf{E}^i_\pm \cdot \mathbf{E}^i_\pm = 0 \quad \Rightarrow \quad \mathbf{E}^i_\pm \cdot \mathbf{E}^{i\,*}_\mp = 0, \qquad (2.26)$$

valid for real $\mathbf{k}^i$, the Poynting vector of the incident wave becomes

$$\begin{aligned} \mathbf{S}^i &= \frac{1}{2}\mathbf{E}^i \times \mathbf{H}^{i*} = \frac{1}{2k_o\eta_o}\mathbf{E}^i \times (\mathbf{k}^i \times \mathbf{E}^{i*}) \\ &= \frac{\mathbf{k}^i}{2k_o\eta_o}|\mathbf{E}^i|^2 = \frac{\mathbf{u}^i}{2\eta_o}|\mathbf{E}^i_+ + \mathbf{E}^i_-|^2 \\ &= \frac{\mathbf{u}^i}{2\eta_o}(|\mathbf{E}^i_+|^2 + |\mathbf{E}^i_-|^2), \qquad (2.27) \end{aligned}$$

whence the incident eigenfields are power orthogonal, i.e., they carry power independently. The same applies for the reflected fields. Depending on the nature of the boundary, there may be power exchange between the incident and reflected eigenfields at the boundary. However, it turns out that, at the PEMC boundary, there is no power coupling between the eigenfields. Because of $\mathbf{n} \cdot \mathbf{k}^r = -\mathbf{n} \cdot \mathbf{k}^i$, handedness of the wave is changed in reflection, whence $\mathbf{E}^r_+$ is right handed and $\mathbf{E}^r_-$ is left handed.

2.3 Reflection from PEMC Boundary

Let us consider the problem of plane-wave reflection from the PEMC boundary defined by the boundary conditions (2.6). From (2.9) we obtain

$$(\overline{\overline{\mathsf{J}}}_t + M\eta_o\overline{\overline{\mathsf{I}}}_t) \cdot \mathbf{E}^r_t = (\overline{\overline{\mathsf{J}}}_t - M\eta_o\overline{\overline{\mathsf{I}}}_t) \cdot \mathbf{E}^i_t. \qquad (2.28)$$

Multiplying this by $(\overline{\overline{\mathsf{J}}}_t - M\eta_o\overline{\overline{\mathsf{I}}}_t)\cdot$, we obtain the relation between the tangential field components,

$$\mathbf{E}_t^r = \overline{\overline{\mathsf{R}}}_t \cdot \mathbf{E}_t^i, \tag{2.29}$$

in terms of the reflection dyadic [53, 59]

$$\begin{aligned}\overline{\overline{\mathsf{R}}}_t &= \frac{-1}{1+(M\eta_o)^2}(\overline{\overline{\mathsf{J}}}_t - M\eta_o\overline{\overline{\mathsf{I}}}_t)^2 \\ &= \frac{1-(M\eta_o)^2}{1+(M\eta_o)^2}\overline{\overline{\mathsf{I}}}_t + \frac{2M\eta_o}{1+(M\eta_o)^2}\overline{\overline{\mathsf{J}}}_t.\end{aligned} \tag{2.30}$$

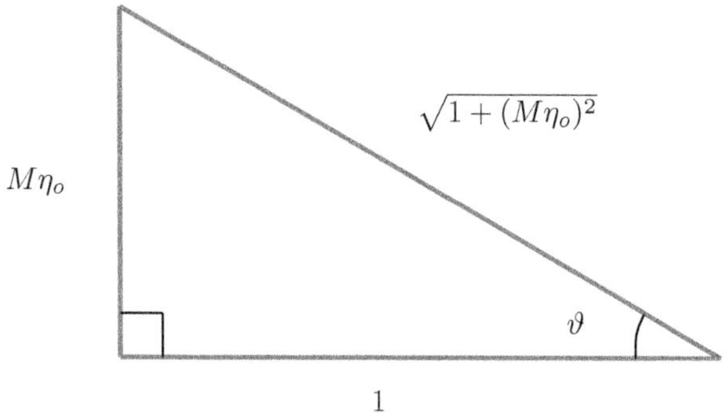

Figure 2.1: Visualization of the connection between ϑ and the PEMC admittance M.

Here we assume $M \neq j/\eta_o$ and $M \neq -j/\eta_o$. Expressing

$$M\eta_o = \tan\vartheta, \tag{2.31}$$

we obtain a compact form for the reflection dyadic,

$$\overline{\overline{\mathsf{R}}}_t = \cos 2\vartheta\ \overline{\overline{\mathsf{I}}}_t + \sin 2\vartheta\ \overline{\overline{\mathsf{J}}}_t = \exp(2\vartheta\overline{\overline{\mathsf{J}}}_t). \tag{2.32}$$

Applying properties of the $\overline{\overline{\mathsf{J}}}_t$ dyadic (see Appendix B), the reflection dyadic can be shown to satisfy

$$\mathrm{tr}\overline{\overline{\mathsf{R}}}_t = 2\cos 2\vartheta, \qquad \det{}_t\overline{\overline{\mathsf{R}}}_t = 1, \tag{2.33}$$

$$\overline{\overline{\mathsf{J}}}_t \cdot \overline{\overline{\mathsf{R}}}_t \cdot \overline{\overline{\mathsf{J}}}_t = -\overline{\overline{\mathsf{R}}}_t. \tag{2.34}$$

Applying (1.47) and (1.48), the reflection rule for the magnetic field becomes

$$\mathbf{H}_t^r = \overline{\overline{\mathsf{J}}}_t \cdot \overline{\overline{\mathsf{R}}}_t \cdot \overline{\overline{\mathsf{J}}}_t \cdot \mathbf{H}_t^i = -\overline{\overline{\mathsf{R}}}_t \cdot \mathbf{H}_t^i. \tag{2.35}$$

The reflected tangential electric field an also be expressed as

$$\mathbf{E}_t^r = \frac{1-(M\eta_o)^2}{1+(M\eta_o)^2}\mathbf{E}_t^i - \frac{2M\eta_o}{1+(M\eta_o)^2}\eta_o\mathbf{H}_t^i. \tag{2.36}$$

As is seen from (2.30), the eigenvectors of the reflection dyadic coincide with those of the dyadic $\overline{\overline{\mathsf{J}}}_t$, (2.13). Thus, the eigenproblem

$$\overline{\overline{\mathsf{R}}}_t \cdot \mathbf{x}_{t\pm} = R_\pm \mathbf{x}_{t\pm} \tag{2.37}$$

of the reflection dyadic (2.32) yields the eigenvalues

$$R_\pm = \frac{1 \pm jM\eta_o}{1 \mp jM\eta_o} = e^{\pm j2\vartheta}. \tag{2.38}$$

It is worth noting that, while the eigenvectors $\mathbf{x}_{t\pm}$ depend on the wave vector $\mathbf{k}^i$ of the incident wave, the eigenvalues do not depend on it. For the PMC ($M = 0, \vartheta = 0$) and PEC ($1/M = 0, \vartheta = \pi/2$) boundaries, the respective eigenvalues are $R_\pm = 1$ and $R_\pm = -1$.

In summary, any circularly polarized left- or right-handed incident field is reflected from the PEMC boundary with reflection coefficient either $\exp(j2\vartheta)$ or $\exp(-j2\vartheta)$, respectively. The reflected field is circularly polarized with opposite handedness.

The previous analysis appears to fail for the two special PEMC boundaries defined by

$$M = M_+ = j/\eta_o, \qquad M = M_- = -j/\eta_o. \tag{2.39}$$

In fact, from (2.38) we have the four cases

$$M = M_+, \quad \Rightarrow \quad R_+ = 0, \quad R_- = \infty, \tag{2.40}$$

$$M = M_-, \quad \Rightarrow \quad R_+ = \infty, \quad R_- = 0. \tag{2.41}$$

Thus, for $M = M_+$, we have $\mathbf{E}^r_+ = 0$ and $\mathbf{E}^i_- = 0$, while, for $M = M_-$, we have $\mathbf{E}^i_+ = 0$ and $\mathbf{E}^r_- = 0$. In both of these cases there is only one plane wave (+ or −) which satisfies the PEMC condition identically.

Considering fields of the two circularly-polarized incident and reflected plane waves, the tangential components can be expressed as

$$\mathbf{E}^i_{t\pm} = A_\pm \mathbf{x}_{t\pm}, \quad \eta_o \mathbf{H}^i_{t\pm} = -\overline{\overline{\mathsf{J}}}_t \cdot \mathbf{E}^i_{t\pm} = \mp jA_\pm \mathbf{x}_\pm, \tag{2.42}$$

$$\mathbf{E}^r_{t\pm} = B_\pm \mathbf{x}_{t\pm}, \quad \eta_o \mathbf{H}^r_{t\pm} = \overline{\overline{\mathsf{J}}}_t \cdot \mathbf{E}^r_{t\pm} = \pm jB_\pm \mathbf{x}_\pm, \tag{2.43}$$

whence the fields are related by

$$\mathbf{H}^i_{t\pm} \pm (j/\eta_o)\mathbf{E}^i_{t\pm} = 0. \tag{2.44}$$

$$\mathbf{H}^r_{t\pm} \mp (j/\eta_o)\mathbf{E}^r_{t\pm} = 0. \tag{2.45}$$

Comparing these to (2.6), the incident fields $\mathbf{E}^i_\pm, \mathbf{H}^i_\pm$ in (2.44) appear to satisfy the condition of the PEMC boundary identically for respective PEMC admittances $M = M_\pm$ with no reflected waves. Similarly, the reflected fields $\mathbf{E}^r_\pm, \mathbf{H}^r_\pm$ in (2.45) satisfy the PEMC condition for respective admittances $M = M_\mp$ without any incident waves. Actually, "incident" and "reflected" have no special meaning except that they refer to different signs of $\mathbf{n} \cdot \mathbf{k}$.

A single plane wave is called matched to a boundary if it satisfies the boundary conditions identically. Thus, for both $M = M_+$ and $M = M_-$, there exist certain circularly polarized matched waves for any $\mathbf{k}$ vector. It is left as an exercise to show that other possible matched waves associated to PEMC boundaries are lateral waves with $k_n = 0$.

For real admittance parameter M (and ϑ), (2.38) implies $|R_\pm| = 1$. In this case, the power reflected from the PEMC boundary equals the power incident to the boundary, whence the boundary is lossless. Considering the more general case, $\vartheta = \vartheta_r + j\vartheta_i$, (2.38) yields $|R_\pm|^2 = \exp(\mp 4\vartheta_i)$. In such a case, the PEMC boundary appears lossy for one of the eigenwaves and active for the other one. Thus, there is no PEMC boundary lossy for all possible waves.

2.4 Polarization Rotation

Let us consider the simple case of plane wave incident on a PEMC boundary of admittance M. The wave vector is assumed to be real as

$$\mathbf{k}^i = k_o(-\mathbf{n}\cos\theta + \mathbf{e}_2 \sin\theta), \tag{2.46}$$

where θ is the angle of incidence. The polarization of the electric field is assumed $\mathbf{E}^i = \mathbf{e}_1 E^i$, where $\mathbf{e}_1, \mathbf{e}_2$ and $\mathbf{e}_3 = \mathbf{n}$ are real orthonormal vectors. Substituting $\mathbf{k} = \mathbf{k}^i$ with $k_n = -k_o \cos\theta$ in (2.10) yields

$$\overline{\overline{\mathsf{J}}}_t = \cos\theta\ \mathbf{e}_2\mathbf{e}_1 - \frac{1}{\cos\theta}\mathbf{e}_1\mathbf{e}_2. \tag{2.47}$$

For normal incidence, $\theta = 0$, we have

$$\overline{\overline{\mathsf{J}}}_t = \mathbf{e}_2\mathbf{e}_1 - \mathbf{e}_1\mathbf{e}_2 = (\mathbf{e}_1 \times \mathbf{e}_2) \times \overline{\overline{\mathsf{I}}} = \mathbf{n} \times \overline{\overline{\mathsf{I}}}, \tag{2.48}$$

whence the reflection dyadic (2.32) becomes

$$\overline{\overline{\mathsf{R}}}_t = \cos 2\vartheta\ \overline{\overline{\mathsf{I}}}_t + \sin 2\vartheta\ \mathbf{n} \times \overline{\overline{\mathsf{I}}} = \exp(2\vartheta\mathbf{n} \times \overline{\overline{\mathsf{I}}}). \tag{2.49}$$

The reflected field

$$\mathbf{E}^r = E^i\overline{\overline{\mathsf{R}}}_t \cdot \mathbf{e}_1 = E^i\mathbf{e}', \tag{2.50}$$

has the same magnitude as the incident field. The polarization defined by the unit vector

$$\mathbf{e}' = \mathbf{e}_1 \cos 2\vartheta + \mathbf{e}_2 \sin 2\vartheta, \tag{2.51}$$

is rotated from $\mathbf{e}_1$ by the angle $\beta = 2\vartheta$, Figure 2.2.

For example, for $\vartheta = \pi/4$, or $M\eta_o = 1$, we have $\mathbf{e}' = \mathbf{e}_2$. In this case, the reflected field is completely cross-polarized, $\mathbf{E}^r \cdot \mathbf{E}^i = 0$. For general M, the reflected field contains both co-polarized and cross-polarized components. Assuming another normally incident wave polarized as $\mathbf{E}^i = E^i\mathbf{e}'$, the reflected field becomes

$$\mathbf{E}^r = E^i(\cos 2\vartheta\ \mathbf{e}' + \sin 2\vartheta\ \mathbf{n} \times \mathbf{e}') = E^i\mathbf{e}'', \tag{2.52}$$

$$\mathbf{e}'' = \mathbf{e}_1 \cos 4\vartheta + \mathbf{e}_2 \sin 4\vartheta. \tag{2.53}$$

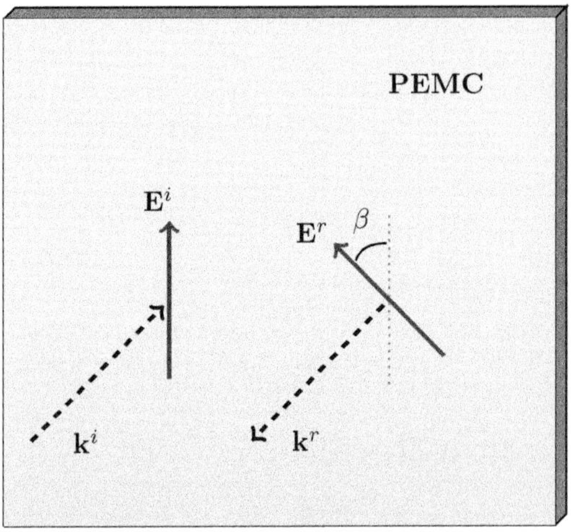

Figure 2.2: For normal incidence to the PEMC boundary, the reflected field is rotated by the angle $\beta = 2\arctan(M\eta_o) = 2\vartheta$.

The PEMC boundary is reciprocal if $\mathbf{e}'' = \mathbf{e}_1$, which requires $\sin 4\vartheta = 0$ and $\cos 4\vartheta = 1$. There are two possible cases, $\vartheta = 0$ and $\vartheta = \pi/2$, which respectively correspond to $M\eta_o = 0$ and $1/M\eta_o = 0$. In these two cases, the PEMC boundary equals the respective PEC or PMC boundary. In conclusion, except for the PEC and PMC special cases, the PEMC boundary is nonreciprocal.

2.5 Point Source and PEMC Plane

Let us consider the basic problem of an electric current source above a PEMC plane $\mathbf{n} \cdot \mathbf{r} = 0$ defined by the admittance parameter M [46]. Setting $\mathbf{n} = \mathbf{e}_3$, let us assume a vertical current element of the form

$$\mathbf{J}_e(\mathbf{r}) = \mathbf{e}_3 IL\delta(\mathbf{r} - \mathbf{e}_3 h), \tag{2.54}$$

where IL is the current moment and h is the height of the element from the PEMC ground.

To find the fields above the boundary, let us apply duality transformation so defined that the PEMC boundary is transformed to a PEC boundary, because we can apply well-known image theory. Here we must apply the transformation (1.18) which leaves the medium parameters μ_o, ϵ_o are invariant. From (1.26) it follows that the transformation parameter must be chosen to satisfy $\tan\varphi =$

$-1/M\eta_o$. The transformation rules applicable here are

$$\begin{pmatrix} \mathbf{E}_d \\ \eta_o\mathbf{H}_d \end{pmatrix} = \begin{pmatrix} \sin\vartheta & \cos\vartheta \\ -\cos\vartheta & \sin\vartheta \end{pmatrix}\begin{pmatrix} \mathbf{E} \\ \eta_o\mathbf{H} \end{pmatrix}, \tag{2.55}$$

$$\begin{pmatrix} \mathbf{E} \\ \eta_o\mathbf{H} \end{pmatrix} = \begin{pmatrix} \sin\vartheta & -\cos\vartheta \\ \cos\vartheta & \sin\vartheta \end{pmatrix}\begin{pmatrix} \mathbf{E}_d \\ \eta_o\mathbf{H}_d \end{pmatrix}, \tag{2.56}$$

$$\begin{pmatrix} \eta_o\mathbf{D}_d \\ \mathbf{B}_d \end{pmatrix} = \begin{pmatrix} \sin\vartheta & \cos\vartheta \\ -\cos\vartheta & \sin\vartheta \end{pmatrix}\begin{pmatrix} \eta_o\mathbf{D} \\ \mathbf{B} \end{pmatrix}, \tag{2.57}$$

$$\begin{pmatrix} \eta_o\mathbf{J}_{ed} \\ \mathbf{J}_{md} \end{pmatrix} = \begin{pmatrix} \sin\vartheta & \cos\vartheta \\ -\cos\vartheta & \sin\vartheta \end{pmatrix}\begin{pmatrix} \eta_o\mathbf{J}_e \\ \mathbf{J}_m \end{pmatrix}, \tag{2.58}$$

with

$$\sin\vartheta = \frac{M\eta_o}{\sqrt{1+(M\eta_o)^2}}, \quad \cos\vartheta = \frac{1}{\sqrt{1+(M\eta_o)^2}}. \tag{2.59}$$

As a check, let us write

$$\mathbf{E}_d = \frac{1}{1+(M\eta_o)^2}(M\eta_o\mathbf{E} + \eta_o\mathbf{H}) = \sin\vartheta\,\mathbf{E} + \cos\vartheta\,\eta_o\mathbf{H}, \tag{2.60}$$

whose right side has zero tangential component at the PEMC boundary, due to (2.6). Thus, we have $\mathbf{E}_{dt} = 0$ which equals the PEC condition.

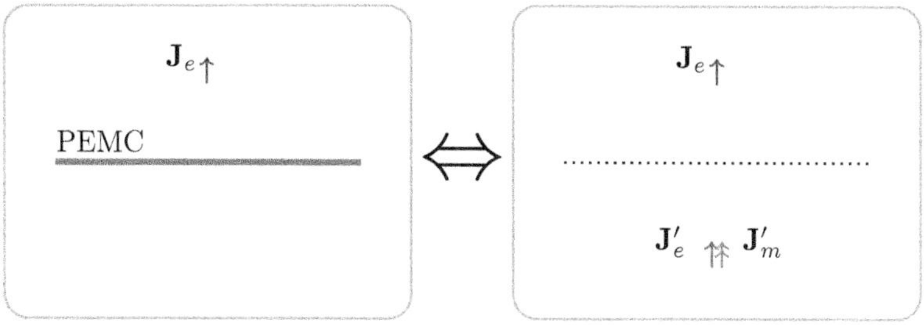

Figure 2.3: The image of an electric dipole above a PEMC boundary consists of both electric and magnetic dipoles.

The transformation requires that the electric current source (2.54) must be replaced by its dual, which consists of a combination of electric and magnetic currents as

$$\mathbf{J}_{ed}(\mathbf{r}) = \mathbf{e}_3 IL\sin\vartheta\,\delta(\mathbf{r}-\mathbf{e}_3 h), \tag{2.61}$$

$$\mathbf{J}_{md}(\mathbf{r}) = -\mathbf{e}_3 IL\cos\vartheta\,\delta(\mathbf{r}-\mathbf{e}_3 h). \tag{2.62}$$

The PEC boundary can be replaced by the image sources of both $\mathbf{J}_{ed}$ and $\mathbf{J}_{md}$ as [35]

$$\mathbf{J}'_{ed}(\mathbf{r}) = \mathbf{e}_3 IL\sin\vartheta\,\delta(\mathbf{r}+\mathbf{e}_3 h), \tag{2.63}$$

$$\mathbf{J}'_{md}(\mathbf{r}) = \mathbf{e}_3 IL\cos\vartheta\,\delta(\mathbf{r}+\mathbf{e}_3 h). \tag{2.64}$$

To obtain the image sources of the original problem, we make use of the inverse transformation,

$$\begin{pmatrix} \eta_o \mathbf{J}_e \\ \mathbf{J}_m \end{pmatrix} = \begin{pmatrix} \sin\vartheta & -\cos\vartheta \\ \cos\vartheta & \sin\vartheta \end{pmatrix} \begin{pmatrix} \eta_o \mathbf{J}_{ed} \\ \mathbf{J}_{md} \end{pmatrix}, \tag{2.65}$$

whence the image sources for the original PEMC-boundary problem are finally obtained as

$$\mathbf{J}_e'(\mathbf{r}) = -\mathbf{e}_3 IL\cos 2\vartheta\, \delta(\mathbf{r}+\mathbf{e}_3 h), \tag{2.66}$$

$$\mathbf{J}_m'(\mathbf{r}) = \mathbf{e}_3 \eta_o IL\sin 2\vartheta\, \delta(\mathbf{r}+\mathbf{e}_3 h). \tag{2.67}$$

To verify these expressions for the PEC and PMC special cases, we can respectively set $1/M = 0$ ($\vartheta = \pi/2$) and $M = 0$ ($\vartheta = 0$). In both cases the magnetic image source $\mathbf{J}_m'(\mathbf{r})$ vanishes and the electric image source equals $\mathbf{e}_3 IL\delta(\mathbf{r}+\mathbf{e}_3 h)$ and $-\mathbf{e}_3 IL\delta(\mathbf{r}+\mathbf{e}_3 h)$. For $M\eta_o = 1$ and $M\eta_o = -1$, the electric image vanishes. For such a PEMC boundary, the image of a vertical electric source is a magnetic source.

2.6 Waveguide with PEMC Walls

As another example, let us consider a rectangular waveguide with PEMC walls defined by the planes $x_1 = \mathbf{e}_1 \cdot \mathbf{r} = 0, a$ and $x_2 = \mathbf{e}_2 \cdot \mathbf{r} = 0, b$, Figure 2.4.

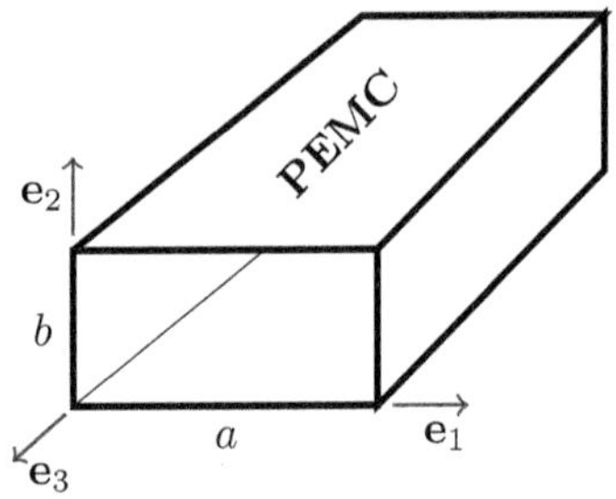

Figure 2.4: Geometry of the rectangular waveguide with PEMC walls.

PEMC waveguide can be handled through its dual PEC waveguide for which we know the modal expressions. The dominant TE_{10} mode of the PEC waveguide is known to propagate as $\exp(-j\beta x_3)$ with the propagation factor defined by [29]

$$\beta = \sqrt{k_o^2 - k_c^2}, \qquad k_c = \pi/a, \tag{2.68}$$

and the modal fields of the form

$$\mathbf{E}_d(x_1) = \mathbf{e}_2 E_d \sin k_c x_1, \tag{2.69}$$

$$\mathbf{H}_d(x_1) = \frac{jE_d}{k_o \eta_o}(\mathbf{e}_3 k_c \cos k_c x_1 + j\beta \mathbf{e}_1 \sin k_c x_1). \tag{2.70}$$

These can be understood as representing the transformed mode of a PEMC waveguide.

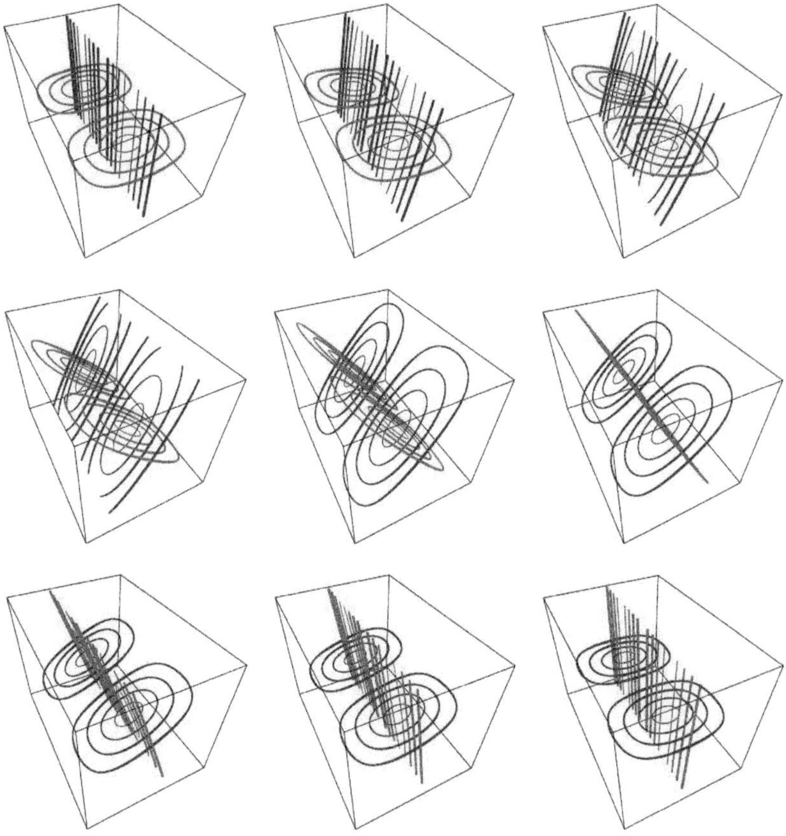

Figure 2.5: The behavior of electric-field (blue lines) and magnetic-field (red lines) patterns in a PEMC waveguide. The waveguide changes from PEC (top left) to PMC (right bottom) with PEMC parameter values $\eta_o M = \tan\vartheta$ from $\vartheta = \pi/2$ (PEC) through $\vartheta = (9-i)\pi/16$ to $\vartheta = 0$ (PMC). Note that both fields have tangential components on the walls except in the PEC and PMC cases.

The fields of the corresponding mode in the PEMC waveguide are found through the inverse transformation (2.56) to (2.69) and (2.70) as

$$\mathbf{E}(x_1) = \frac{1}{\sqrt{1+(M\eta_o)^2}}(M\eta_o\mathbf{E}_d - \eta_o\mathbf{H}_d)$$

$$= \frac{E_d}{k_o}((\mathbf{e}_1\beta\cos\vartheta + \mathbf{e}_2 k_o\sin\vartheta)\sin k_c x_1 - \mathbf{e}_3 j k_c\cos\vartheta \ \cos k_c x_1), \qquad (2.71)$$

$$\eta_o\mathbf{H}(x_1) = \frac{1}{\sqrt{1+(M\eta_o)^2}}(\mathbf{E}_d + M\eta_o^2\mathbf{H}_d)$$

$$= \frac{E_d}{k_o}((-\mathbf{e}_1\beta\sin\vartheta + \mathbf{e}_2 k_o\cos\vartheta)\sin k_c x_1 + \mathbf{e}_3 j k_c\sin\vartheta\ \cos k_c x_1). \tag{2.72}$$

Forming the product of (2.71) and (2.72) we obtain

$$\mathbf{E}\cdot\eta_o\mathbf{H} = \frac{k_c^2}{k_o^2}E_d^2\sin\vartheta\cos\vartheta. \tag{2.73}$$

From this it is seen that, unlike in the PEC waveguide, the electric and magnetic fields are not orthogonal in the corresponding mode of the general PEMC waveguide. They are orthogonal for $\sin\vartheta\cos\vartheta = 0$, i.e., for the PEC and PMC waveguides.

In Figure 2.5 the field patterns for the basic mode in a PEMC waveguide are shown for PEMC admittance M varying from PEC ($M=\infty$) to PMC ($M=0$).

2.7 Parallel-Plate PEMC Resonator

As an example of a PEMC structure which cannot be transformed to a PEC structure, let us consider a generalization of the Fabry-Perot resonator with two parallel planes $\mathbf{e}_3\cdot\mathbf{r} = x_3 = \pm d$ of different PEMC admittance, M_1 and M_2. Both planes cannot be simultaneously transformed to PEC planes making a conventional Fabry-Perot resonator. However, since it turns out that a duality transformation can be found (proof left as an exercise in Chapter 1) so that any two different admittances M_1 and M_2 can be transformed to two admittances of opposite values, $\pm M_d$, let us simplify the problem by considering the following setup:

- PEMC plane at $x_3 = d$ with admittance M,
- PEMC plane at $x_3 = -d$ with admittance $-M$.

Assuming a field consisting of TEM plane waves with $\mathbf{e}_3\cdot\mathbf{E}_t = \mathbf{e}_3\cdot\mathbf{H}_t = 0$ reflecting from both planes, we can expand the fields in terms of two circularly-polarized vectors

$$\mathbf{a}_{t+} = \mathbf{e}_1 + j\mathbf{e}_2, \qquad \mathbf{a}_{t-} = \mathbf{e}_1 - j\mathbf{e}_2, \tag{2.74}$$

as

$$\mathbf{E}_t(x_3) = \mathbf{a}_{t+}E_{t+}(x_3) + \mathbf{a}_{t-}E_{t-}(x_3), \tag{2.75}$$

$$\mathbf{H}_t(x_3) = \mathbf{a}_{t+}H_{+}(x_3) + \mathbf{a}_{t-}H_{t-}(x_3). \tag{2.76}$$

The vectors $\mathbf{a}_{t+}$ and $\mathbf{a}_{t-}$ satisfy

$$\mathbf{e}_3\times\mathbf{a}_{t\pm} = \mp j\mathbf{a}_{t\pm}, \quad \mathbf{a}_{t\pm}\cdot\mathbf{a}_{t\pm} = 0, \quad \mathbf{a}_{t\pm} = \mathbf{a}_{t\mp}^* \tag{2.77}$$

and

$$\mathbf{p}(\mathbf{a}_{t\pm}) = \frac{\mathbf{a}_{t\pm}\times\mathbf{a}_{t\pm}^*}{j\mathbf{a}_{t\pm}\cdot\mathbf{a}_{t\pm}^*} = \mathbf{e}_3\frac{\mathbf{e}_3\times\mathbf{a}_{t\pm}\cdot\mathbf{a}_{t\pm}^*}{j\mathbf{a}_{t\pm}\cdot\mathbf{a}_{t\pm}^*} = \mp\mathbf{e}_3, \tag{2.78}$$

whence $\mathbf{a}_{t+}$ is left-hand polarized, and $\mathbf{a}_{t-}$ is right-hand polarized, with respect to $\mathbf{e}_3$.

Substituted in the Maxwell equations, the fields are separated componentwise as

$$\partial_{x_3} E_\pm(x_3) = \pm k_o \eta_o H_\pm(x_3), \tag{2.79}$$
$$\partial_{x_3} \eta_o H_\pm(x_3) = \mp k_o E_\pm(x_3), \tag{2.80}$$

while the PEMC conditions at the two planes yield

$$\eta_o H_\pm(d) + M\eta_o E_\pm(d) = 0, \tag{2.81}$$
$$\eta_o H_\pm(-d) - M\eta_o E_\pm(-d) = 0. \tag{2.82}$$

Since the two polarizations are not coupled at the boundaries, we can treat them separately. From (2.79) and (2.80) we obtain

$$(\partial_{x_3}^2 + k_o^2) E_\pm(x_3) = 0, \tag{2.83}$$

the solutions of which are of the form

$$E_\pm(x_3) = A_\pm \sin k_o x_3 + B_\pm \cos k_o x_3, \tag{2.84}$$
$$\eta_o H_\pm(x_3) = \pm(A_\pm \cos k_o x_3 - B_\pm \sin k_o x_3), \tag{2.85}$$

the latter of which is obtained through (2.79).

Substituting (2.84) and (2.85) in the PEMC boundary conditions (2.81) and (2.82) yields

$$(M\eta_o \sin k_o d \pm \cos k_o d)A_\pm + (M\eta_o \cos k_o d \mp \sin k_o d)B_\pm = 0 \tag{2.86}$$
$$(M\eta_o \sin k_o d \pm \cos k_o d)A_\pm - (M\eta_o \cos k_o d \mp \sin k_o d)B_\pm = 0. \tag{2.87}$$

For $M\eta_o = \tan\vartheta$, the conditions become

$$A_\pm \cos(k_o d \mp \vartheta) = 0, \qquad B_\pm \sin(k_o d \mp \vartheta) = 0, \tag{2.88}$$

from which the possible modes are seen to split in two sets as follows.

- $A_\pm = 0$ The resonance wavenumbers are obtained from

  $$\sin(k_o d \mp \vartheta) = 0, \qquad k_o d = \pm\vartheta + n\pi, \tag{2.89}$$

  and the resonance fields are

  $$E_\pm(x_3) = B_\pm \cos k_o x_3, \tag{2.90}$$
  $$\eta_o H_\pm(x_3) = -B_\pm \cos k_o x_3, \tag{2.91}$$

  which are symmetric functions of x_3. The lowest resonances are $k_o d = \vartheta$, for E_+, H_+ and $k_o d = \pi - \vartheta$ for E_-, H_-.

- $B_\pm = 0$ The resonance wavenumbers are obtained from

$$\cos(k_o d \mp \vartheta) = 0, \qquad k_o d = \pm\vartheta + (n - 1/2)\pi. \tag{2.92}$$

and the resonance fields are

$$E_\pm(x_3) = A_\pm \sin k_o x_3, \tag{2.93}$$
$$\eta_o H_\pm(x_3) = \pm A_\pm \sin k_o x_3, \tag{2.94}$$

which are antisymmetric functions of x_3. The lowest resonances are $k_o d = \vartheta + \pi/2$ for E_+, H_+ and $k_o d = \pi/2 - \vartheta$ for E_-, H_-.

As an example, let us consider a PEMC resonator with boundary admittances $\pm M$, by assuming $M = 1/\eta_o$, or $\vartheta = \pi/4$. In this case, resonances of the two sets of modes will coincide. The lowest symmetric resonances are $k_{o+}d = \pi/4$ and $k_{o-}d = 3\pi/4$, while antisymmetric resonances are $k_{o+}d = 3\pi/4$ and $k_{o-}d = \pi/4$. For the lowest resonance $k_o d = \pi/4$, the resonator is of sub-wavelength size, $2d = \lambda/4$. The combined field in this case is

$$\mathbf{E}(x_3) = B\mathbf{a}_{t+} \cos k_o x_3 + A\mathbf{a}_{t-} \sin k_o x_3, \tag{2.95}$$
$$\eta_o \mathbf{H}(x_3) = -B\mathbf{a}_+ \sin k_o x_3 - A\mathbf{a}_- \cos k_o x_3. \tag{2.96}$$

For the choice $A = jB = E$, the combined resonance modal fields become

$$\mathbf{E}(x_3) = E(\mathbf{e}_2 e^{-jk_o x_3} - j\mathbf{e}_1 e^{jk_o x_3}), \tag{2.97}$$
$$\eta_o \mathbf{H}(x_3) = E(-\mathbf{e}_1 e^{-jk_o x_3} + j\mathbf{e}_2 e^{jk_o x_3}). \tag{2.98}$$

In this case, the combined field consists of linearly polarized plane waves reflecting back and forth between the planes. The polarization is rotated by $\pi/2$ at each reflection. When propagating in the direction $\mathbf{e}_3$, the electric field is parallel to $\mathbf{e}_2$ and when propagating in the opposite direction, it is parallel to $\mathbf{e}_1$. Such a resonator may have engineering application because a plane wave of certain polarization may exist in a compact environment. Spherical PEMC resonator has been analyzed in the literature [49].

2.8 Modeling Small PEMC Particles

The duality transformation is also useful in the analysis of small scatterers with PEMC boundary and modeling mixtures that contain PEMC particles embedded in a background medium. When a PEMC-boundary object is exposed to electromagnetic field, the cross-coupling effect of the boundary condition leads to a more complicated scattering response than in the case of pure dielectric, PEC, or PMC objects.

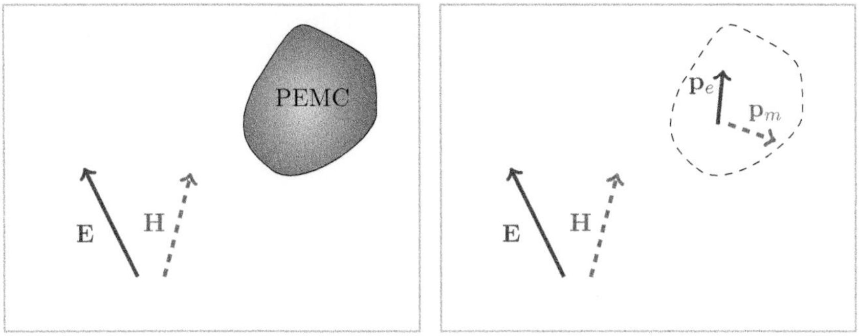

Figure 2.6: When a small scatterer with PEMC boundary condition is exposed to electric and magnetic fields, its scattered fields are those of an electric and a magnetic dipole.

Response of a Single PEMC Scatterer

For a small PEMC scatterer, the response fields are due to electric and magnetic dipoles whose magnitudes are connected to the incident fields and the PEMC parameter, in addition to the shape of the object, as depicted in Figure 2.6.

If the scattering solution for the corresponding PEC object is known, the duality transformation can be exploited for the PEMC problem [95]. The electric and magnetic dipole moments $\mathbf{p}_e, \mathbf{p}_m$ induced in a small particle are proportional to the incident electric and magnetic fields through the polarizability, and the proportionality is in general (like in the PEMC case) described by a matrix:

$$\begin{pmatrix} \mathbf{p}_e \\ \mathbf{p}_m/\eta_o \end{pmatrix} = \epsilon_o V \begin{pmatrix} \alpha_{ee} & \alpha_{em} \\ \alpha_{me} & \alpha_{mm} \end{pmatrix} \begin{pmatrix} \mathbf{E} \\ \eta_o \mathbf{H} \end{pmatrix}. \tag{2.99}$$

In this form, the polarizability matrix

$$\mathcal{A} = \begin{pmatrix} \alpha_{ee} & \alpha_{em} \\ \alpha_{me} & \alpha_{mm} \end{pmatrix} \tag{2.100}$$

is dimensionless through the normalization by the volume of the scatterer V and the necessary free-space constants $\epsilon_o, \mu_o, \eta_o$. Note that, in general, the polarizability components α_{ij} are dyadics.

If the solution of the corresponding PEC scattering problem is known, the duality transformation leads to the polarizability matrix components of the PEMC case. A small PEC object scatters as a combination of an electric and magnetic dipole:

$$\mathcal{A}_{\text{PEC}} = \begin{pmatrix} \alpha_e & 0 \\ 0 & \alpha_m \end{pmatrix}. \tag{2.101}$$

Let us consider the PEC case (with a diagonal polarizability matrix) as the duality-transformed scattering constellation. Hence the dipole moments read

$$\begin{pmatrix} \mathbf{p}_{e,d} \\ \mathbf{p}_{m,d}/\eta_o \end{pmatrix} = \epsilon_o V \mathcal{A}_{\rm PEC} \begin{pmatrix} \mathbf{E}_d \\ \eta_o \mathbf{H}_d \end{pmatrix}. \tag{2.102}$$

Using the duality relations (and noting that the fields and sources transform with the same matrix (2.55) and (2.58)), the induced moments in the PEMC case can be written as

$$\begin{pmatrix} \mathbf{p}_e \\ \mathbf{p}_m/\eta_o \end{pmatrix} = \frac{\epsilon_o V}{1+(M\eta_o)^2} \begin{pmatrix} M\eta_o & -1 \\ 1 & M\eta_o \end{pmatrix} \mathcal{A}_{\rm PEC} \begin{pmatrix} M\eta_o & 1 \\ -1 & M\eta_o \end{pmatrix} \begin{pmatrix} \mathbf{E} \\ \eta_o \mathbf{H} \end{pmatrix} \tag{2.103}$$

$$= \epsilon_o V \, \mathcal{A}_{\rm PEMC} \begin{pmatrix} \mathbf{E} \\ \eta_o \mathbf{H} \end{pmatrix}, \tag{2.104}$$

leaving us with the polarizability matrix

$$\mathcal{A}_{\rm PEMC} = \frac{1}{1+(M\eta_o)^2} \begin{pmatrix} \alpha_e (M\eta_o)^2 + \alpha_m & (\alpha_e - \alpha_m) M\eta_o \\ (\alpha_e - \alpha_m) M\eta_o & \alpha_m (M\eta_o)^2 + \alpha_e \end{pmatrix}. \tag{2.105}$$

As an example, for a PEC sphere the polarizability matrix reads

$$\mathcal{A}_{\rm PEC,sph} = \begin{pmatrix} \alpha_e & 0 \\ 0 & \alpha_m \end{pmatrix} = \begin{pmatrix} 3 & 0 \\ 0 & -3/2 \end{pmatrix}. \tag{2.106}$$

Consequently, using (2.105), we can write the polarizability matrix for a PEMC sphere as follows:

$$\mathcal{A}_{\rm PEMC,sph} = \frac{3/2}{1+(M\eta_o)^2} \begin{pmatrix} 2(M\eta_o)^2 - 1 & 3\,M\eta_o \\ 3\,M\eta_o & 2-(M\eta_o)^2 \end{pmatrix}. \tag{2.107}$$

The polarizability matrix is symmetric which is consistent with the non-reciprocal character of the PEMC medium [37].

In addition to distilling to the PEC polarizability (2.106) for $1/M = 0$, the formula (2.107) gives the polarizability matrix of the PMC sphere ($M = 0$):

$$\mathcal{A}_{\rm PMC,sph} = \begin{pmatrix} \alpha_e & 0 \\ 0 & \alpha_m \end{pmatrix} = \begin{pmatrix} -3/2 & 0 \\ 0 & 3 \end{pmatrix}. \tag{2.108}$$

Other properties of $\mathcal{A}_{\rm PEMC,sph}$ include that its eigenvalues $(3, -3/2)$ are independent of M, as also its determinant which is $-9/2$.

Figure 2.7 shows the behavior of the polarizability matrix components as functions of the PEMC parameter. The logarithmic scale of M reveals the symmetry of the components. In particular, the co-polarizability components vanish for certain values of the PEMC parameter:

$$\alpha_{ee}(M\eta_o = \pm 1/\sqrt{2}) = 0 \quad \text{and} \quad \alpha_{mm}(M\eta_o = \pm\sqrt{2}) = 0. \tag{2.109}$$

For the "most non-reciprocal" situation ($M\eta_o = \pm 1$), the electric and magnetic co-polarizabilities are equal ($\alpha_{ee} = \alpha_{mm} = 3/4$) while the cross-polarizability term is three times larger: $\alpha_{em} = \pm 9/4$.

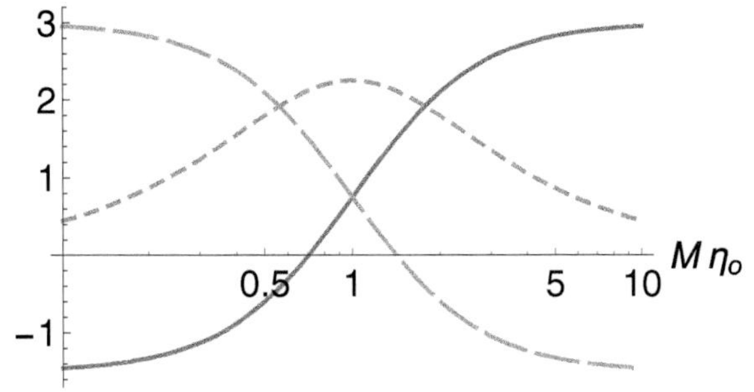

Figure 2.7: The normalized polarizability components of a PEMC sphere as functions of the PEMC parameter M. Solid blue: α_{ee}, dashed orange: α_{mm}, short-dashed green: $\alpha_{em} = \alpha_{me}$.

A real-valued M parameter corresponds to a lossless PEMC boundary condition which entails that particles with such surface impedance are also lossless. Allowing M to attain complex values, the scatterers can be active or dissipative. This is also seen in the complex-valued character of the polarizability matrix. In particular, the expression (2.107) shows that for the purely imaginary value $M\eta_o = \pm j$, all the polarizability components experience a singularity.

Mixture of PEMC Scatterers

Once the electromagnetic response of small PEMC spheres is known, it is possible to compute the effective material parameters of a mixture where such spheres (of subwavelength size) are imbedded in neutral dielectric background material as shown in Figure 2.8 [94].

According to the so-called Maxwell Garnett mixing formula [80], the effective permittivity ϵ_{eff} of a mixture where spherical inclusions of permittivity ϵ_i occupy a volume fraction p in a host medium with permittivity ϵ_e, is

$$\epsilon_{\text{eff}} = \epsilon_e + 3p\,\epsilon_e \frac{\epsilon_i - \epsilon_e}{\epsilon_i + 2\epsilon_e - p(\epsilon_i - \epsilon_e)}. \tag{2.110}$$

This, written in terms of the normalized dielectric polarizability of a spherical inclusion $\alpha = 3(\epsilon_i - \epsilon_e)/(\epsilon_i + 2\epsilon_e)$, reads

$$\frac{\epsilon_{\text{eff}}}{\epsilon_e} = 1 + \frac{p\alpha}{1 - \dfrac{p\alpha}{3}}, \tag{2.111}$$

where $p = nV$ is again the fractional volume, n being the number density of inclusions.

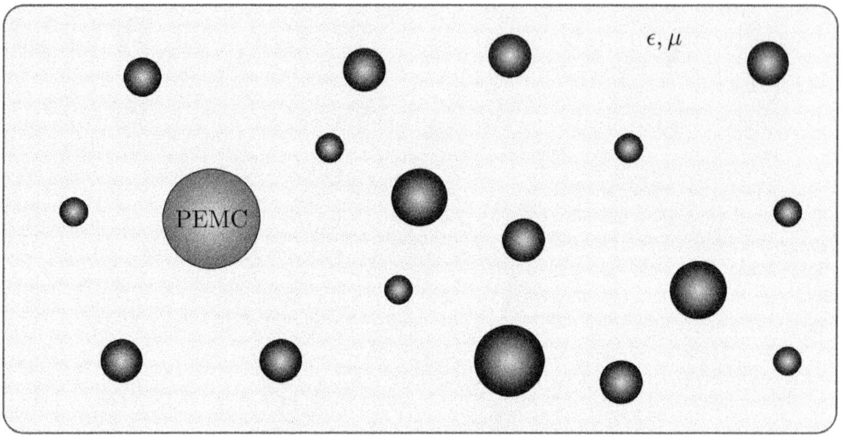

Figure 2.8: A mixture of subwavelength PEMC spheres in dielectric background becomes effectively a bi-isotropic medium.

For a mixture with PEMC inclusions of Figure 2.8, the effective relative material parameters are bi-isotropic [33]:

$$\begin{pmatrix} \mathbf{D}/\epsilon_o \\ \mathbf{B}/\sqrt{\mu_o\epsilon_o} \end{pmatrix} = \begin{pmatrix} \epsilon_r & \chi_r \\ \chi_r & \mu_r \end{pmatrix} \begin{pmatrix} \mathbf{E} \\ \eta_o\mathbf{H} \end{pmatrix} = \mathcal{M}_r \begin{pmatrix} \mathbf{E} \\ \eta_o\mathbf{H} \end{pmatrix}. \tag{2.112}$$

Here, in addition to the classical constitutive parameters, permittivity and permeability, a magneto-electric parameter is necessary in the constitutive relations: the non-reciprocal Tellegen parameter χ_r.

The generalization of the dielectric mixing principle (2.111) into the bi-isotropic domain requires the matrix form [93]

$$\mathcal{M}_{r,\text{eff}} = \mathcal{I} + p\mathcal{A}\cdot(\mathcal{I} - p\mathcal{A}/3)^{-1}, \qquad \mathcal{I} = \begin{pmatrix} 1 & 0 \\ 0 & 1 \end{pmatrix}, \tag{2.113}$$

where $\mathcal{A}$ is the polarizability of a single inclusion, in the case of Figure 2.8 given by Eq. (2.107). The effective relative parameters of the bi-isotropic constitutive matrix $\mathcal{M}_{r,\text{eff}}$ are

$$\epsilon_{r,\text{eff}} = 1 + 3p\frac{2(M\eta_o)^2 - 1 + p(1 + (M\eta_o)^2)}{(1-p)(2+p)(1+(M\eta_o)^2)}, \tag{2.114}$$

$$\chi_{r,\text{eff}} = 3p\frac{3M\eta_o}{(1-p)(2+p)(1+(M\eta_o)^2)}, \tag{2.115}$$

$$\mu_{r,\text{eff}} = 1 + 3p\frac{2 - (M\eta_o)^2 + p(1 + (M\eta_o)^2)}{(1-p)(2+p)(1+(M\eta_o)^2)}. \tag{2.116}$$

If the inclusions are PEC spheres ($1/M = 0$), these parameters simplify to

$$\epsilon_{r,\text{eff}} = 1 + \frac{3p}{1-p}, \quad \mu_{r,\text{eff}} = 1 - \frac{3p}{2+p}, \quad \chi_{r,\text{eff}} = 0, \tag{2.117}$$

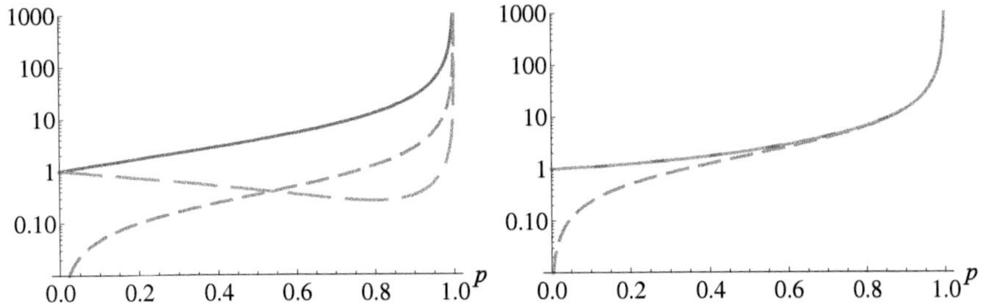

Figure 2.9: The effective parameters of a mixture of Figure 2.8 as functions of the volume fraction p of the PEMC inclusions. Solid blue: $\epsilon_{r,\text{eff}}$, dashed orange: $\mu_{r,\text{eff}}$, short-dashed green: $\chi_{r,\text{eff}}$. Left: close-to-PEC ($M\eta_o = 10$), right: $M\eta_o = 1$ where the blue and orange lines fall on each other.

and the dual case (PMC inclusions, $M = 0$) leads to the electric–magnetic interchange of the effective parameters in (2.117). This representation satisfies the limits of $p \to 1$ which correspond to the models for solid PEC ($\epsilon = \infty$, $\mu = 0$) and PMC ($\epsilon = 0$, $\mu = \infty$).

Figure 2.9 shows the behavior of the three effective parameters of the mixture as functions of the volume fraction of the PEMC inclusions p. For a close-to-PEC case ($M\eta_o = 10$) the effective permittivity is the largest of the parameters, but for $p \to 1$ (solid PEMC material), all parameters grow without limit as is known [45]. For the special case of $M\eta_o = 1$, the effective relative permittivity and permeability are equal over the whole p-range.

On the other hand, as function of M parameter, Figure 2.10 illustrates the effective parameters for a case when PEMC spheres occupy a volume fraction of $p = 0.5$ in free space, showing the facts that ϵ_{eff} and μ_{eff} are even and χ_{eff} odd functions of the PEMC parameter. Despite the variation of the polarizability components with respect to M, the eigenvalues $\lambda_{1,2}$ of the effective parameter matrix (2.113) are independent of the strength of the PEMC parameter:

$$\lambda_1 = 1 + \frac{3p}{1-p} \quad \text{and} \quad \lambda_2 = 1 - \frac{3p}{2+p}, \tag{2.118}$$

which obviously match the effective permittivity and permeability values of the PMC (or PEC) mixture (2.117).

2.9 Problems

2.1 Show by applying the representation (2.4) that a plane wave in the PEMC medium has no dispersion equation, i.e., that it is satisfied for any $\mathbf{k}$ vector.

2.2 Derive the eigenvector relation (2.11) for the $\overline{\overline{\mathsf{J}}}_t$ dyadic.

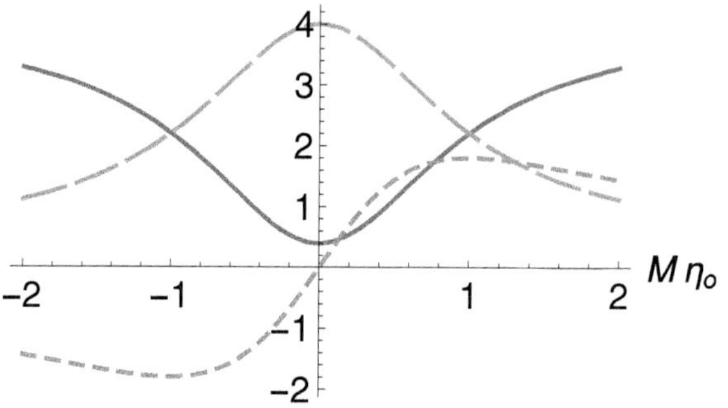

Figure 2.10: The effective parameters of a mixture of Figure 2.8 as functions of the PEMC parameter. Solid blue: $\epsilon_{r,\text{eff}}$, dashed orange: $\mu_{r,\text{eff}}$, short-dashed green: $\chi_{r,\text{eff}}$. The volume fraction of PEMC inclusions is $p = 0.5$.

2.3 Show that the PEMC boundary condition (2.7) follows from the PEMC condition (2.6) and the Maxwell equations. Assume that the boundary is represented by a function $f(\mathbf{r}) = 0$, whence the normal unit vector is of the form $\mathbf{n} = \nabla f(\mathbf{r})/|\nabla f(\mathbf{r})|$.

2.4 Derive the eigenfunction expansion (2.17) by expanding the dyadic in terms of its eigenvectors as

$$\overline{\overline{\mathsf{J}}}_t = \mathbf{x}_{t+}\mathbf{a}_{t+} + \mathbf{x}_{t-}\mathbf{a}_{t-}.$$

2.5 Show that the dyadics $\overline{\overline{\mathsf{J}}}_t \pm j\overline{\overline{\mathsf{I}}}_t$ do not have two-dimensional inverses.

2.6 Figure 2.7 shows that for some PEMC parameter values, the co- and cross-polarizability components may be equal. Find the conditions for this to happen. How large are the components when this condition is met?

2.7 According to Equation (2.117), the relative effective permittivity $\epsilon_{r,\text{eff}}$ of a mixture with PEC spheres increases monotonically from unity to infinity while the (relative effective) permeability $\mu_{r,\text{eff}}$ decreases to zero. However, for PEMC mixtures both parameters grow without limit for $p \to 0$. As Figure 2.9 shows, there is a minimum for $\mu_{r,\text{eff}}$ when the PEMC parameter is close to PEC ($M\eta_o = 10$). Find the volume fraction p_{min} for which this happens for a given $M\eta_o$. What is the minimum value for $\mu_{r,\text{eff}}$?

2.8 If the PEMC parameter of a boundary is complex, the scatterer with such a surface impedance is no longer lossless. However, certain conditions hold for the polarizability components of a PEMC sphere with complex-valued M parameter. Show that the following is valid

$$\text{Im}\{\alpha_{ee}\} = -\text{Im}\{\alpha_{mm}\}$$

for any complex valued PEMC parameter for the co-polarizability components of a PEMC sphere.

2.9 The refractive index for a bi-isotropic Tellegen material with parameters ϵ, μ and χ is $n = \sqrt{\epsilon\mu - \chi^2}$ [33]. For the homogenized PEMC mixture of Figure 2.8 with effective parameters (2.114)–(2.116), compute the refractive index as function of $M\eta_o$ and volume fraction p.

2.10 Check that (2.71) and (2.72) yield the PEC waveguide fields for $M \to \infty$. Find the fields of the corresponding PMC waveguide.

2.11 Find conditions for matched waves at the PEMC boundary.

2.12 Derive the conditions (2.33) of the reflection dyadic $\overline{\overline{\mathsf{R}}}_t$.

Chapter 3

Impedance Boundary

3.1 Basic Conditions

The class of impedance boundaries is defined by linear conditions between tangential electric and magnetic field components at the boundary surface. One of the common representations for the impedance-boundary condition (IBC) is [35]

$$\mathbf{E}_t = \overline{\overline{\mathsf{Z}}}_t \cdot (\mathbf{n} \times \mathbf{H}_t), \tag{3.1}$$

where $\overline{\overline{\mathsf{Z}}}_t$ is the two-dimensional boundary-impedance dyadic satisfying $\mathbf{n} \cdot \overline{\overline{\mathsf{Z}}}_t = \overline{\overline{\mathsf{Z}}}_t \cdot \mathbf{n} = 0$. Denoting the electric surface current at the boundary by

$$\mathbf{J}_{es} = \mathbf{n} \times \mathbf{H}_t, \tag{3.2}$$

the boundary condition (3.1) can be interpreted as a two-dimensional version of Ohm's law,

$$\mathbf{E}_t = \overline{\overline{\mathsf{Z}}}_t \cdot \mathbf{J}_{es}. \tag{3.3}$$

Denoting the magnetic surface current by

$$\mathbf{J}_{ms} = -\mathbf{n} \times \mathbf{E}_t, \tag{3.4}$$

another form for the impedance-boundary condition (3.1) is

$$\mathbf{H}_t = \overline{\overline{\mathsf{Y}}}_t \cdot \mathbf{J}_{ms} = -\overline{\overline{\mathsf{Y}}}_t \cdot (\mathbf{n} \times \mathbf{E}_t), \tag{3.5}$$

where $\overline{\overline{\mathsf{Y}}}_t$ is the boundary-admittance dyadic [35] .

The impedance-boundary condition equivalent to its simple form with $\overline{\overline{\mathsf{Z}}}_t = Z_s \overline{\overline{\mathsf{I}}}_t$ has been called by the name Leontovich condition in the past, because it was introduced by M. A. Leontovich in a paper of 1944 [32]. However, as has been pointed out in [84], it had been introduced already in 1940 in a book by A. N. Shchukin [91].

The conditions (3.5) and (3.1) are equivalent in defining the impedance boundary, provided the dyadics $\overline{\overline{\mathsf{Z}}}_t$ and $\overline{\overline{\mathsf{Y}}}_t$ are related in a proper manner. The relation can be found from

$$\mathbf{E}_t = (\overline{\overline{\mathsf{Z}}}_t \times \mathbf{n}) \cdot \mathbf{H}_t = -(\overline{\overline{\mathsf{Z}}}_t \times \mathbf{n}) \cdot (\overline{\overline{\mathsf{Y}}}_t \times \mathbf{n}) \cdot \mathbf{E}_t. \tag{3.6}$$

Because this must be valid for any $\mathbf{E}_t$, we obtain

$$(\overline{\overline{\mathsf{Z}}}_t \times \mathbf{n}) \cdot (\overline{\overline{\mathsf{Y}}}_t \times \mathbf{n}) = -\overline{\overline{\mathsf{Z}}}_t \cdot (\overline{\overline{\mathsf{Y}}}_t {}^\times_\times \mathbf{nn}) = -\overline{\overline{\mathsf{I}}}_t, \tag{3.7}$$

or,

$$\overline{\overline{\mathsf{Z}}}_t^{-1} = \overline{\overline{\mathsf{Y}}}_t {}^\times_\times \mathbf{nn} = \mathbf{nn} {}^\times_\times \overline{\overline{\mathsf{Y}}}_t. \tag{3.8}$$

Applying the expansion rule for the inverse of two-dimensional dyadics (see Appendix C),

$$\overline{\overline{\mathsf{A}}}_t^{-1} = \frac{\overline{\overline{\mathsf{A}}}_t^T {}^\times_\times \mathbf{nn}}{\mathrm{det}_t \overline{\overline{\mathsf{A}}}_t}, \tag{3.9}$$

$$\mathrm{det}_t \overline{\overline{\mathsf{A}}}_t = \mathrm{tr}\overline{\overline{\mathsf{A}}}_t^{(2)} = \frac{1}{2}\mathbf{nn} : \overline{\overline{\mathsf{A}}}_t {}^\times_\times \overline{\overline{\mathsf{A}}}_t \neq 0, \tag{3.10}$$

where det_t denotes the two-dimensional determinant of the two-dimensional dyadic $\overline{\overline{\mathsf{A}}}_t$, the relation (3.8) can be written in the form

$$\overline{\overline{\mathsf{Z}}}_t = \frac{1}{\mathrm{det}_t \overline{\overline{\mathsf{Y}}}_t} \overline{\overline{\mathsf{Y}}}_t^T. \tag{3.11}$$

Because this implies

$$\mathrm{det}_t \overline{\overline{\mathsf{Y}}}_t \, \mathrm{det}_t \overline{\overline{\mathsf{Z}}}_t = 1, \tag{3.12}$$

the relation can be inverted as

$$\overline{\overline{\mathsf{Y}}}_t = \frac{1}{\mathrm{det}_t \overline{\overline{\mathsf{Z}}}_t} \overline{\overline{\mathsf{Z}}}_t^T. \tag{3.13}$$

The PEMC boundary is an example of the impedance boundary,

$$\mathbf{n} \times (\mathbf{H} + M\mathbf{E}) = 0, \quad \Rightarrow \quad \mathbf{E}_t = \frac{1}{M}\mathbf{n} \times (\mathbf{n} \times \mathbf{H}). \tag{3.14}$$

In this case, the impedance and admittance dyadics are antisymmetric,

$$\overline{\overline{\mathsf{Z}}}_t = \frac{1}{M}\mathbf{n} \times \overline{\overline{\mathsf{I}}}_t, \quad \mathrm{det}_t \overline{\overline{\mathsf{Z}}}_t = 1/M^2. \tag{3.15}$$

$$\overline{\overline{\mathsf{Y}}}_t = M\mathbf{n} \times \overline{\overline{\mathsf{I}}}_t, \quad \mathrm{det}_t \overline{\overline{\mathsf{Y}}}_t = M^2. \tag{3.16}$$

The converse of this is also true: any impedance boundary with an antisymmetric impedance or admittance dyadic is an example of the PEMC boundary.

The representation (3.1) or (3.5) is not convenient in cases when some of the impedance or admittance parameters may not have finite values. For example, for the PMC boundary, the representation (3.1) would require an infinite magnitude for $\overline{\overline{\mathsf{Z}}}_t$. To avoid this, one can replace the impedance-boundary conditions (3.1) or (3.5)by

$$\mathbf{a}_{1t} \cdot \mathbf{E}_t + \mathbf{b}_{1t} \cdot \eta_o \mathbf{H}_t = 0 \tag{3.17}$$

$$\mathbf{a}_{2t} \cdot \mathbf{E}_t + \mathbf{b}_{2t} \cdot \eta_o \mathbf{H}_t = 0, \tag{3.18}$$

where we assume that the dimensionless tangential vectors satisfy $\mathbf{a}_{1t} \times \mathbf{a}_{2t} \neq 0$ and $\mathbf{b}_{1t} \times \mathbf{b}_{2t} \neq 0$. Relation between (3.1) and (3.17), (3.18) can be represented by

$$\overline{\overline{\mathsf{Z}}}_t = \frac{\eta_o}{\mathbf{n} \cdot \mathbf{a}_{1t} \times \mathbf{a}_{2t}}((\mathbf{n} \times \mathbf{a}_{2t})(\mathbf{n} \times \mathbf{b}_{1t}) - (\mathbf{n} \times \mathbf{a}_{1t})(\mathbf{n} \times \mathbf{b}_{2t})). \tag{3.19}$$

To define a given impedance boundary, the form (3.17), (3.18) is not, however, unique. In fact, expressing the conditions as

$$\begin{pmatrix} \mathbf{a}_{1t} & \mathbf{b}_{1t} \\ \mathbf{a}_{2t} & \mathbf{b}_{2t} \end{pmatrix} \cdot \begin{pmatrix} \mathbf{E}_t \\ \eta_o \mathbf{H}_t \end{pmatrix} = \begin{pmatrix} 0 \\ 0 \end{pmatrix}, \tag{3.20}$$

the matrix of four tangential vectors can be multiplied by any matrix of four scalars with nonzero determinant without changing the definition of the boundary,

$$\begin{pmatrix} C_1 & D_1 \\ C_2 & D_2 \end{pmatrix} \begin{pmatrix} \mathbf{a}_{1t} & \mathbf{b}_{1t} \\ \mathbf{a}_{2t} & \mathbf{b}_{2t} \end{pmatrix} \cdot \begin{pmatrix} \mathbf{E}_t \\ \eta_o \mathbf{H}_t \end{pmatrix} = \begin{pmatrix} 0 \\ 0 \end{pmatrix}. \tag{3.21}$$

In this, the set of four vectors is replaced by another one equivalent to the original set. The definition can be made unique, e.g., by choosing the scalars so that

$$C_1 \mathbf{a}_{1t} + D_1 \mathbf{a}_{2t} = \mathbf{e}_1, \qquad C_2 \mathbf{a}_{1t} + D_2 \mathbf{a}_{2t} = \mathbf{e}_2, \tag{3.22}$$

where $\mathbf{e}_1, \mathbf{e}_2, \mathbf{n}$ is a given orthonormal basis of 2D vectors. In such a case the conditions (3.17) and (3.18) are equivalent to conditions of the form

$$\mathbf{e}_1 \cdot \mathbf{E} + \mathbf{c}_{1t} \cdot \eta_o \mathbf{H} = 0, \tag{3.23}$$

$$\mathbf{e}_2 \cdot \mathbf{E} + \mathbf{c}_{2t} \cdot \eta_o \mathbf{H} = 0, \tag{3.24}$$

whence the impedance boundary is defined in terms of two tangential vectors $\mathbf{c}_{1t}, \mathbf{c}_{2t}$.

The number of parameters defining the impedance boundary is thus reduced from 8 to $2(\mathbf{c}_{1t}) + 2(\mathbf{c}_{2t}) = 4$, which equals the number of the impedance parameters Z_{ij} of (3.1).

It is well known that an impedance boundary is reciprocal when the impedance dyadic is symmetric [35], whence the impedance dyadic must satisfy

$$\overline{\overline{\mathsf{Z}}}_t^T = \overline{\overline{\mathsf{Z}}}_t. \tag{3.25}$$

In the representation (3.17), (3.18), the condition of reciprocity requires

$$\mathbf{a}_{1t} \times \mathbf{b}_{2t} = \mathbf{a}_{2t} \times \mathbf{b}_{1t}. \tag{3.26}$$

Also, the impedance boundary is lossless if the impedance dyadic is antihermitian [35], i.e., if it satisfies

$$\overline{\overline{\mathsf{Z}}}_t^T = -\overline{\overline{\mathsf{Z}}}_t^*. \tag{3.27}$$

Thus, a simple-isotropic boundary with $\overline{\overline{\mathsf{Z}}}_t = Z_s \overline{\overline{\mathsf{I}}}_t$ is inherently reciprocal. It is lossless when Z_s is imaginary [50].

The impedance boundary is lossy when the power absorbed is positive, which requires that

$$\Re(-\frac{1}{2}\mathbf{n}\cdot\mathbf{E}_t\times\mathbf{H}_t^*) = \frac{1}{4}(\mathbf{n}\times\mathbf{H})\cdot(\overline{\overline{\mathsf{Z}}}_t^* + \overline{\overline{\mathsf{Z}}}_t^T)\cdot(\mathbf{n}\times\mathbf{H})^* > 0 \tag{3.28}$$

must be valid for all possible fields $\mathbf{H}_t$ (note that $-\mathbf{n}$ points towards the boundary). This requires that the Hermitian part of the impedance dyadic $\overline{\overline{\mathsf{Z}}}_t$ be positive definite, i.e., the dyadic must have positive real symmetric part and negative imaginary antisymmetric part. For example, a simple-isotropic boundary is lossy when the real part of Z_s is positive. Because for antisymmetric $\overline{\overline{\mathsf{Z}}}_t$, the dyadic $\overline{\overline{\mathsf{Z}}}_t^* + \overline{\overline{\mathsf{Z}}}_t^T$ is always imaginary, the PEMC boundary cannot be lossy.

3.2 Subclasses of Impedance Boundaries

Basic subclasses of impedance boundaries can be defined as based on the 2D dyadic eigenproblem

$$\mathbf{nn}{}^\times_\times\overline{\overline{\mathsf{Z}}}_t = \overline{\overline{\mathsf{Z}}}_t{}^\times_\times\mathbf{nn} = \lambda\overline{\overline{\mathsf{Z}}}_t. \tag{3.29}$$

Applying the dyadic rule

$$\mathbf{nn}{}^\times_\times(\mathbf{nn}{}^\times_\times\overline{\overline{\mathsf{Z}}}_t) = \overline{\overline{\mathsf{Z}}}_t = \lambda^2\overline{\overline{\mathsf{Z}}}_t, \tag{3.30}$$

the two eigenvalues become $\lambda_\pm = \pm 1$, whence the corresponding eigendyadics $\overline{\overline{\mathsf{Z}}}_{t\pm}$ satisfy

$$\mathbf{nn}{}^\times_\times\overline{\overline{\mathsf{Z}}}_{t+} = \overline{\overline{\mathsf{Z}}}_{t+}, \qquad \mathbf{nn}{}^\times_\times\overline{\overline{\mathsf{Z}}}_{t-} = -\overline{\overline{\mathsf{Z}}}_{t-}. \tag{3.31}$$

Any given impedance dyadic $\overline{\overline{\mathsf{Z}}}_t$ can be decomposed as

$$\overline{\overline{\mathsf{Z}}}_t = \overline{\overline{\mathsf{Z}}}_{t+} + \overline{\overline{\mathsf{Z}}}_{t-}, \tag{3.32}$$

with

$$\overline{\overline{\mathsf{Z}}}_{t+} = \frac{1}{2}(\overline{\overline{\mathsf{Z}}}_t + \overline{\overline{\mathsf{Z}}}_t{}^\times_\times\mathbf{nn}), \tag{3.33}$$

$$\overline{\overline{\mathsf{Z}}}_{t-} = \frac{1}{2}(\overline{\overline{\mathsf{Z}}}_t - \overline{\overline{\mathsf{Z}}}_t{}^\times_\times\mathbf{nn}). \tag{3.34}$$

Because of the rule

$$\mathrm{tr}(\overline{\overline{\mathsf{A}}}_t{}^\times_\times\mathbf{nn}) = \mathrm{tr}\overline{\overline{\mathsf{A}}}_t, \tag{3.35}$$

we have

$$\mathrm{tr}\overline{\overline{\mathsf{Z}}}_{t-} = 0, \tag{3.36}$$

whence the possible eigendyadics $\overline{\overline{\mathsf{Z}}}_{t-}$ are trace free. Applying the rule (See Appendix C)

$$\overline{\overline{\mathsf{A}}}_t{}^\times_\times\mathbf{nn} = (\mathrm{tr}\overline{\overline{\mathsf{A}}}_t)\overline{\overline{\mathsf{I}}}_t - \overline{\overline{\mathsf{A}}}_t^T, \tag{3.37}$$

valid for any 2D dyadic $\overline{\overline{\mathsf{A}}}_t$, we obtain

$$\overline{\overline{\mathsf{Z}}}_{t-}^T = -\overline{\overline{\mathsf{Z}}}_{t-}{}^\times_\times\mathbf{nn} = \overline{\overline{\mathsf{Z}}}_{t-}. \tag{3.38}$$

Thus, the eigendyadics $\overline{\overline{\mathsf{Z}}}_{t-}$ are both symmetric and trace free.

Substituting $\overline{\overline{\mathsf{A}}}_t = \overline{\overline{\mathsf{Z}}}_{t+}$ in (3.37) yields

$$\overline{\overline{\mathsf{Z}}}_{t+} = (\mathrm{tr}\overline{\overline{\mathsf{Z}}}_{t+})\overline{\overline{\mathsf{I}}}_t - \overline{\overline{\mathsf{Z}}}_{t+}^T, \tag{3.39}$$

or,

$$\overline{\overline{\mathsf{Z}}}_{t+} = \frac{1}{2}(\mathrm{tr}\overline{\overline{\mathsf{Z}}}_{t+})\overline{\overline{\mathsf{I}}}_t + \frac{1}{2}(\overline{\overline{\mathsf{Z}}}_{t+} - \overline{\overline{\mathsf{Z}}}_{t+}^T), \tag{3.40}$$

whence any eigendyadic $\overline{\overline{\mathsf{Z}}}_{t+}$ can be expressed as a multiple of the unit dyadic $\overline{\overline{\mathsf{I}}}_t$ and an antisymmetric dyadic, which is a multiple of $\mathbf{n}\times\overline{\overline{\mathsf{I}}}$.

To summarize, any impedance dyadic can be decomposed in three parts,

$$\overline{\overline{\mathsf{Z}}}_t = Z_s\overline{\overline{\mathsf{I}}}_t + Z_n\mathbf{n}\times\overline{\overline{\mathsf{I}}} + \overline{\overline{\mathsf{Z}}}_{ta}, \tag{3.41}$$

- The simple-isotropic part, $Z_s\overline{\overline{\mathsf{I}}}_t$
- The antisymmetric (PEMC) part, $Z_n\mathbf{n}\times\overline{\overline{\mathsf{I}}}$

  Together these two parts make the eigendyadic $\overline{\overline{\mathsf{Z}}}_+$. It is isotropic, because it requires no special direction on the boundary surface.

- The anisotropic part, $\overline{\overline{\mathsf{Z}}}_{ta}$, which equals the eigendyadic $\overline{\overline{\mathsf{Z}}}_-$.

  This is a symmetric and trace-free 2D dyadic and can be represented in any orthonormal 2D vector basis as

$$\overline{\overline{\mathsf{Z}}}_{ta} = Z_1(\mathbf{e}_1\mathbf{e}_1 - \mathbf{e}_2\mathbf{e}_2) + Z_2(\mathbf{e}_1\mathbf{e}_2 + \mathbf{e}_2\mathbf{e}_1). \tag{3.42}$$

  One can show that $\mathbf{n}\times\overline{\overline{\mathsf{Z}}}_{ta}$ and $\overline{\overline{\mathsf{Z}}}_{ta}\times\mathbf{n}$ are anisotropic whenever $\overline{\overline{\mathsf{Z}}}_{ta}$ is anisotropic.

An impedance boundary is *isotropic* if it contains no component $\overline{\overline{\mathsf{Z}}}_{ta}$. For $\overline{\overline{\mathsf{Z}}}_t = Z_s\overline{\overline{\mathsf{I}}}_t$ it is a *simple-isotropic* boundary. If $\overline{\overline{\mathsf{Z}}}_t$ contains a nonzero component $\overline{\overline{\mathsf{Z}}}_{ta}$, it is *anisotropic*. If $\overline{\overline{\mathsf{Z}}}_t = \overline{\overline{\mathsf{Z}}}_{ta}$, it can be called *perfectly anisotropic* [52].

From (3.25) it follows that an impedance boundary is reciprocal if it does not contain the antisymmetric part: $\overline{\overline{\mathsf{Z}}}_t = Z_s\overline{\overline{\mathsf{I}}}_t + \overline{\overline{\mathsf{Z}}}_{ta}$. From (3.27) it follows that an isotropic impedance boundary is lossless if Z_s is imaginary and Z_n is real. A perfectly anisotropic impedance dyadic is reciprocal. It is also lossless if $\overline{\overline{\mathsf{Z}}}_{ta}$ is an imaginary dyadic.

In the representation (3.17), (3.18), the impedance dyadic takes the form (3.19). In this case, the decomposition (3.41) contains the terms

$$Z_s = \frac{\eta_o}{2\mathbf{n}\cdot\mathbf{a}_{1t}\times\mathbf{a}_{2t}}(\mathbf{a}_{2t}\cdot\mathbf{b}_{1t} - \mathbf{a}_{1t}\cdot\mathbf{b}_{2t}), \tag{3.43}$$

$$Z_n = \frac{\eta_o}{2\mathbf{n}\cdot\mathbf{a}_{1t}\times\mathbf{a}_{2t}}\mathbf{n}\cdot(\mathbf{b}_{1t}\times\mathbf{a}_{2t} - \mathbf{b}_{2t}\times\mathbf{a}_{1t}), \tag{3.44}$$

$$\overline{\overline{\mathsf{Z}}}_{ta} = \frac{\eta_o}{2\mathbf{n}\cdot\mathbf{a}_{1t}\times\mathbf{a}_{2t}}(\mathbf{a}_{1t}\mathbf{b}_{2t} + \mathbf{b}_{2t}\mathbf{a}_{1t} - \mathbf{a}_{2t}\mathbf{b}_{1t} - \mathbf{b}_{1t}\mathbf{a}_{2t} + (\mathbf{a}_{2t}\cdot\mathbf{b}_{1t} - \mathbf{a}_{1t}\cdot\mathbf{b}_{2t})\overline{\overline{\mathsf{I}}}_t). \tag{3.45}$$

Here we have assumed $\mathbf{a}_{1t}\times\mathbf{a}_{2t} \neq 0$.

3.3 Reflection from Impedance Boundary

Considering plane-wave reflection from the general impedance boundary defined by the surface-impedance dyadic $\overline{\overline{\mathsf{Z}}}_t$, the relation between the incident and reflected fields can be written, applying (1.47) and (1.48), as

$$\begin{aligned}\mathbf{E}_t^r + \mathbf{E}_t^i &= \overline{\overline{\mathsf{Z}}}_t\cdot(\mathbf{n}\times(\mathbf{H}_t^r + \mathbf{H}_t^i)) \\ &= \frac{1}{\eta_o}\overline{\overline{\mathsf{Z}}}_t\cdot(\mathbf{n}\times\overline{\overline{\mathsf{J}}}_t)\cdot(\mathbf{E}_t^r - \mathbf{E}_t^i).\end{aligned} \tag{3.46}$$

The reflected field is obtained from

$$\mathbf{E}_t^r = \overline{\overline{\mathsf{R}}}_t\cdot\mathbf{E}_t^i, \tag{3.47}$$

in terms of the reflection dyadic

$$\begin{aligned}\overline{\overline{\mathsf{R}}}_t &= (\overline{\overline{\mathsf{Z}}}_t\cdot\mathbf{n}\times\overline{\overline{\mathsf{J}}}_t - \eta_o\overline{\overline{\mathsf{I}}}_t)^{-1}\cdot(\overline{\overline{\mathsf{Z}}}_t\cdot\mathbf{n}\times\overline{\overline{\mathsf{J}}}_t + \eta_o\overline{\overline{\mathsf{I}}}_t) \\ &= \overline{\overline{\mathsf{I}}}_t + 2\eta_o(\overline{\overline{\mathsf{Z}}}_t\cdot\mathbf{n}\times\overline{\overline{\mathsf{J}}}_t - \eta_o\overline{\overline{\mathsf{I}}}_t)^{-1}.\end{aligned} \tag{3.48}$$

Applying the rule (3.9) for the 2D inverse of a 2D dyadic and the property (see Appendix B)

$$\overline{\overline{\mathsf{J}}}_t{}^{\times}_{\times}\mathbf{n}\mathbf{n} = -\overline{\overline{\mathsf{J}}}_t^T, \tag{3.49}$$

after some steps, the reflection dyadic (3.48) can be expressed in the form

$$\overline{\overline{\mathsf{R}}}_t = (1 - \frac{2\eta_o^2}{\Delta})\overline{\overline{\mathsf{I}}}_t + \frac{2\eta_o}{\Delta}\overline{\overline{\mathsf{J}}}_t\cdot\overline{\overline{\mathsf{Z}}}_t^T\times\mathbf{n}, \tag{3.50}$$

with

$$\Delta = \mathrm{det}_t\overline{\overline{\mathsf{Z}}}_t + \eta_o^2 + \frac{\eta_o}{k_o k_n}(\overline{\overline{\mathsf{Z}}}_t : \mathbf{k}_t\mathbf{k}_t + k_n^2\mathrm{tr}\overline{\overline{\mathsf{Z}}}_t). \tag{3.51}$$

Details of the derivation are left as an exercise.

Eigenproblem

Let us consider the eigenproblem of the reflection dyadic $\overline{\overline{\mathsf{R}}}_t$,

$$\overline{\overline{\mathsf{R}}}_t \cdot \mathbf{x}_t = R\mathbf{x}_t. \tag{3.52}$$

From (3.50) it is obvious that the eigenvectors $\mathbf{x}_t$ are shared by the eigenproblem

$$(\overline{\mathsf{J}}_t \cdot \overline{\overline{\mathsf{Z}}}_t^T \times \mathbf{n}) \cdot \mathbf{x}_t = \lambda \mathbf{x}_t, \tag{3.53}$$

with

$$\lambda = \eta_o + \frac{\Delta}{2\eta_o}(R-1). \tag{3.54}$$

The eigenvalues λ satisfy the equation

$$\mathrm{det}_t(\overline{\overline{\mathsf{Z}}}_t^T \times \mathbf{n} + \lambda\overline{\mathsf{J}}_t) = \mathrm{tr}(\overline{\overline{\mathsf{Z}}}_t^T \times \mathbf{n} + \lambda\overline{\mathsf{J}}_t)^{(2)} = 0, \tag{3.55}$$

which is of the quadratic form,

$$\mathrm{tr}(\overline{\overline{\mathsf{Z}}}_t^T \times \mathbf{n})^{(2)} + \lambda\mathrm{tr}((\overline{\overline{\mathsf{Z}}}_t^T \times \mathbf{n})^{\times}_{\times}\overline{\mathsf{J}}_t) + \lambda^2\mathrm{tr}\overline{\mathsf{J}}_t^{(2)} = 0. \tag{3.56}$$

Substituting the expressions

$$\mathrm{tr}(\overline{\overline{\mathsf{Z}}}_t^T \times \mathbf{n})^{(2)} = \mathrm{tr}(\overline{\overline{\mathsf{Z}}}_t^{(2)T} \cdot \mathbf{nn}) = \mathrm{det}_t\overline{\overline{\mathsf{Z}}}_t, \tag{3.57}$$

$$\begin{aligned} \mathrm{tr}((\overline{\overline{\mathsf{Z}}}_t^T \times \mathbf{n})^{\times}_{\times}\overline{\mathsf{J}}_t) &= (\overline{\overline{\mathsf{Z}}}_t^T \times \mathbf{n}) : (\overline{\mathsf{J}}_t{}^{\times}_{\times}\mathbf{nn}) \\ &= \frac{1}{k_o k_n}(\overline{\overline{\mathsf{Z}}}_t : \mathbf{k}_t\mathbf{k}_t + k_n^2\mathrm{tr}\overline{\overline{\mathsf{Z}}}_t), \end{aligned} \tag{3.58}$$

$$\mathrm{tr}\overline{\mathsf{J}}_t^{(2)} = \mathrm{tr}(\mathbf{nn}) = 1, \tag{3.59}$$

the eigenvalue equation becomes

$$\lambda^2 + \lambda\frac{1}{k_o k_n}(\overline{\overline{\mathsf{Z}}}_t : \mathbf{k}_t\mathbf{k}_t + k_n^2\mathrm{tr}\overline{\overline{\mathsf{Z}}}_t) + \mathrm{det}_t\overline{\overline{\mathsf{Z}}}_t = 0. \tag{3.60}$$

Its solutions will be considered in sections considering special cases of the impedance boundary. The eigenvector corresponding to an eigenvalue λ can be obtained from

$$\mathbf{x}_t = \mathbf{a}_t \cdot \overline{\overline{\mathsf{B}}}_t(\lambda) \times \mathbf{n}, \qquad \overline{\overline{\mathsf{B}}}_t(\lambda) = \overline{\overline{\mathsf{Z}}}_t^T \times \mathbf{n} + \lambda\overline{\mathsf{J}}_t, \tag{3.61}$$

for any vector $\mathbf{a}_t$ yielding a nonzero vector $\mathbf{x}_t$.

PEMC special case

To verify the expression for the reflection dyadic (3.50), let us consider the special case of PEMC boundary. Substituting $\overline{\overline{\mathsf{Z}}}_t = (1/M)\mathbf{n} \times \overline{\mathsf{I}}$ in (3.51), we obtain

$$\Delta = \mathrm{det}_t\overline{\overline{\mathsf{Z}}}_t + \eta_o^2 = \frac{1}{M^2} + \eta_o^2, \tag{3.62}$$

which inserted in (3.50) yields

$$\overline{\overline{\mathsf{R}}}_t = \frac{1-(M\eta_o)^2}{1+(M\eta_o)^2}\overline{\overline{\mathsf{I}}}_t + \frac{2M\eta_o}{1+(M\eta_o)^2}\overline{\overline{\mathsf{J}}}_t. \tag{3.63}$$

This expression coincides with (2.30). The eigenvalues are obtained from (3.60) as

$$\lambda^2 + \mathrm{det}_t\overline{\overline{\mathsf{Z}}}_t = \lambda^2 + \frac{1}{M^2} = 0 \quad \Rightarrow \quad \lambda_\pm = \pm j\frac{1}{M}, \tag{3.64}$$

and, from (3.54), as

$$R_\pm = \frac{1-(M\eta_o)^2}{1+(M\eta_o)^2} \pm j\frac{2M\eta_o}{1+(M\eta_o)^2}. \tag{3.65}$$

Setting $M\eta_o = \tan\vartheta$ yields the compact form (2.38),

$$R_\pm = \cos 2\vartheta \pm j\sin 2\vartheta = \exp(\pm j2\vartheta). \tag{3.66}$$

The PMC and PEC special cases $R_+ = +1$ and $R_- = -1$ are respectively obtained for $\vartheta = 0$ and $\vartheta = \pi/2$.

3.4 Matched Waves

Plane waves which satisfy the boundary conditions identically are called waves matched to the boundary. Familiar examples of matched waves include surface waves and leaky waves attached to certain boundaries. Any incident wave for which the reflecting wave vanishes, appears matched to the boundary. This case corresponds to a zero of the reflection coefficient and is associated to a certain wave vector $\mathbf{k}^i$. Also, if for another wave vector a reflection coefficient becomes infinite in magnitude, it corresponds to a reflected wave without any incident wave, in which case the "reflected" wave can be considered a plane wave matched to the boundary.

Let us find conditions for a plane wave to be matched to an impedance boundary. Substituting the plane-wave field relation

$$k_o\eta_o\mathbf{H} = \mathbf{k}\times\mathbf{E} \tag{3.67}$$

in the boundary condition (3.1), we obtain

$$k_o\eta_o\mathbf{E}_t = \overline{\overline{\mathsf{Z}}}_t\cdot(\mathbf{n}\times(\mathbf{k}\times\mathbf{E})) = (\overline{\overline{\mathsf{Z}}}_t\cdot\mathbf{k})(\mathbf{n}\cdot\mathbf{E}) - (\mathbf{n}\cdot\mathbf{k})\overline{\overline{\mathsf{Z}}}_t\cdot\mathbf{E}_t. \tag{3.68}$$

Applying $\mathbf{k}\cdot\mathbf{E} = 0$ and assuming $\mathbf{n}\cdot\mathbf{k} \neq 0$, the condition can be expressed as

$$\overline{\overline{\mathsf{D}}}_t(\mathbf{k})\cdot\mathbf{E}_t = 0, \tag{3.69}$$

with

$$\overline{\overline{\mathsf{D}}}_t(\mathbf{k}) = k_o\eta_o(\mathbf{n}\cdot\mathbf{k})\overline{\overline{\mathsf{I}}}_t + \overline{\overline{\mathsf{Z}}}_t\cdot\mathbf{k}_t\mathbf{k}_t + (\mathbf{n}\cdot\mathbf{k})^2\overline{\overline{\mathsf{Z}}}_t. \tag{3.70}$$

The condition (3.69) restricting the $\mathbf{k}$ vector and the field vector $\mathbf{E}_t \neq 0$, for the wave to be matched to the boundary, is valid when the $\mathbf{k}$ vector satisfies

$$\mathrm{det}_t \overline{\overline{\mathsf{D}}}_t(\mathbf{k}) = 0. \tag{3.71}$$

Because of $\mathbf{k}\cdot\mathbf{k} = k_o^2$, (3.71) actually defines an equation for $\mathbf{k}_t$. It is similar to the dispersion equation satisfied by the wave vector of a plane wave propagating in a linear medium and, thus, can be called the dispersion equation of a wave matched to an impedance boundary. For $\mathbf{k}_t = k_t\mathbf{u}_t$ with $\mathbf{u}_t\cdot\mathbf{u}_t = 1$, the dispersion equation in the form $k_t = f(\mathbf{u}_t)$ defines two-dimensional dispersion curves on the boundary surface corresponding to the dispersion surfaces of waves in linear media. Of course, $f(\mathbf{u}_t)$ may have complex values, whence the curves have real and imaginary parts.

Inserting (3.70) in (3.71), and denoting $\mathbf{n}\cdot\mathbf{k} = k_n$, yields an explicit form for the dispersion equation,

$$k_o k_n(\eta_o^2 + \mathrm{det}_t\overline{\overline{\mathsf{Z}}}_t) + \eta_o(\overline{\overline{\mathsf{Z}}}_t : \mathbf{k}_t\mathbf{k}_t + k_n^2\mathrm{tr}\overline{\overline{\mathsf{Z}}}_t) = 0. \tag{3.72}$$

Details of the derivation are left as a topic of an exercise. For the simple-isotropic boundary, $\overline{\overline{\mathsf{Z}}}_t = Z_s\overline{\overline{\mathsf{I}}}_t$, (3.72) is reduced to

$$(k_n Z_s + k_o\eta_o)(k_o Z_s + k_n\eta_o) = 0. \tag{3.73}$$

3.5 Simple-Isotropic Impedance Boundary

The simple-isotropic boundary is defined by

$$\overline{\overline{\mathsf{Z}}}_t = Z_s\overline{\overline{\mathsf{I}}}_t, \tag{3.74}$$

or, in the representation (3.17) and (3.18) with $\mathbf{b}_{1t} = Z_s\mathbf{n}\times\mathbf{a}_{1t}$ and $\mathbf{b}_{2t} = Z_s\mathbf{n}\times\mathbf{a}_{2t}$. The reflection dyadic (3.48) can be expanded as

$$\begin{aligned}
\overline{\overline{\mathsf{R}}}_t &= \overline{\overline{\mathsf{I}}}_t + 2((Z_s/\eta_o)\mathbf{n}\times\overline{\overline{\mathsf{J}}}_t - \overline{\overline{\mathsf{I}}}_t)^{-1} \\
&= \overline{\overline{\mathsf{I}}}_t - 2k_o k_n\eta_o(Z_s\mathbf{k}_t\mathbf{k}_t + k_n(k_n Z_s + k_o\eta_o)\overline{\overline{\mathsf{I}}}_t)^{-1} \\
&= \frac{k_n Z_s - k_o\eta_o}{k_n Z_s + k_o\eta_o}\overline{\overline{\mathsf{I}}}_t + \frac{2\eta_o Z_s}{(k_n Z_s + k_o\eta_o)(k_o Z_s + k_n\eta_o)}\mathbf{k}_t\mathbf{k}_t \\
&= \frac{k_o Z_s - k_n\eta_o}{k_o Z_s + k_n\eta_o}\frac{\mathbf{k}_t\mathbf{k}_t}{\mathbf{k}_t\cdot\mathbf{k}_t} + \frac{k_n Z_s - k_o\eta_o}{k_n Z_s + k_o\eta_o}\frac{(\mathbf{n}\times\mathbf{k}_t)(\mathbf{n}\times\mathbf{k}_t)}{\mathbf{k}_t\cdot\mathbf{k}_t},
\end{aligned} \tag{3.75}$$

As a check, for $Z_s = 0$ and $Z_s = \infty$, (3.75) yields the respective PEC and PMC reflection dyadics $\overline{\overline{\mathsf{R}}}_t = -\overline{\overline{\mathsf{I}}}_t$ and $\overline{\overline{\mathsf{R}}}_t = \overline{\overline{\mathsf{I}}}_t$. It is left as an exercise to show that (3.75) can also be obtained by starting from the form (3.50).

Because of the representations (1.52) and (1.53), the field incident in a simple isotropic medium to the boundary can be decomposed in TE and TM parts as

$$\mathbf{E}^i = \mathbf{E}^i_{\mathrm{TE}} + \mathbf{E}^i_{\mathrm{TM}}, \tag{3.76}$$

satisfying

$$\mathbf{n}\cdot\mathbf{E}^i_{\rm TE}=0, \quad \Rightarrow \quad \mathbf{k}_t\cdot\mathbf{E}^i_{\rm TE}=0, \tag{3.77}$$

$$\mathbf{n}\cdot\mathbf{H}^i_{\rm TM}=0, \quad \Rightarrow \quad (\mathbf{n}\times\mathbf{k}_t)\cdot\mathbf{E}^i_{\rm TM}=0. \tag{3.78}$$

Thus, the partial tangential fields are polarized as

$$\mathbf{E}^i_{TEt}=A^i_{TE}\mathbf{n}\times\mathbf{k}_t, \quad \mathbf{E}^i_{TMt}=A^i_{TM}\mathbf{k}_t, \tag{3.79}$$

and they satisfy

$$\mathbf{E}^i_{TEt}\cdot\mathbf{E}^i_{TMt}=0. \tag{3.80}$$

For real $\mathbf{k}^i_t$ they also satisfy

$$\mathbf{E}^i_{TEt}\cdot\mathbf{E}^{i*}_{TMt}=0. \tag{3.81}$$

Applying (3.75), the reflected field is decomposed similarly in TE and TM parts as

$$\begin{aligned}\mathbf{E}^r &= \overline{\overline{\mathsf{R}}}_{\rm TE}\cdot\mathbf{E}^i_{\rm TE}+\overline{\overline{\mathsf{R}}}_{\rm TM}\cdot\mathbf{E}^i_{\rm TM}\\ &= R_{TE}\mathbf{E}^i_{TE}+R_{TM}\mathbf{E}^i_{TM}.\end{aligned}$$

The two reflection dyadic components can be identified as

$$\begin{aligned}\overline{\overline{\mathsf{R}}}_{\rm TE} &= R_{TE}\frac{(\mathbf{n}\times\mathbf{k}_t)(\mathbf{n}\times\mathbf{k}_t)}{\mathbf{k}_t\cdot\mathbf{k}_t},\\ \overline{\overline{\mathsf{R}}}_{\rm TM} &= R_{TM}\frac{\mathbf{k}_t\mathbf{k}_t}{\mathbf{k}_t\cdot\mathbf{k}_t},\end{aligned} \tag{3.82}$$

with

$$R_{TE} = \frac{k_nZ_s-k_o\eta_o}{k_nZ_s+k_o\eta_o}, \tag{3.83}$$

$$R_{TM} = \frac{k_oZ_s-k_n\eta_o}{k_oZ_s+k_n\eta_o}. \tag{3.84}$$

It is left as an exercise to verify that, for imaginary Z_s, the fields satisfy

$$|\mathbf{E}^r_t|^2=|\mathbf{E}^i_t|^2, \tag{3.85}$$

whence the boundary is lossless.

Matched Waves

An incident wave can be interpreted as a matched wave when the corresponding reflected wave vanishes, i.e., for vanishing reflection coefficient. From (3.83) and (3.84) we have two possibilities,

$$k_n=k_o(\eta_o/Z_s), \quad \Rightarrow \quad \mathbf{E}^r_{TE}=0, \tag{3.86}$$

$$k_n = k_o(Z_s/\eta_o), \quad \Rightarrow \quad \mathbf{E}^r_{TM} = 0. \tag{3.87}$$

The respective wave vectors are

$$\mathbf{k}^i_{TE} = -\mathbf{n}k_o(\eta_o/Z_s) + \mathbf{k}_t, \tag{3.88}$$

$$\mathbf{k}^i_{TM} = -\mathbf{n}k_o(Z_s/\eta_o) + \mathbf{k}_t. \tag{3.89}$$

Reflected waves can also serve as matched waves when the corresponding incident waves vanish. This occurs for reflection coefficients becoming infinite in magnitude. The two possibilities are obtained from (3.83) and (3.84) as

$$k_n = -k_o(\eta_o/Z_s), \quad \Rightarrow \quad \mathbf{E}^i_{TE} = 0, \tag{3.90}$$

$$k_n = -k_o(Z_s/\eta_o), \quad \Rightarrow \quad \mathbf{E}^i_{TM} = 0. \tag{3.91}$$

The respective wave vectors are now

$$\mathbf{k}^r_{TE} = -\mathbf{n}k_o(\eta_o/Z_s) + \mathbf{k}_t, \tag{3.92}$$

$$\mathbf{k}^r_{TM} = -\mathbf{n}k_o(Z_s/\eta_o) + \mathbf{k}_t. \tag{3.93}$$

Comparing (3.88) and (3.92) on one hand, and (3.89) and (3.93) on the other hand, the wave vectors of the incident and reflected matched waves are seen to coincide. In other words, the same matched wave is obtained by considering reflection coefficient being zero or infinite in magnitude. Actually, the single equation (3.73) covers both incident- and reflected-wave representations of the matched wave.

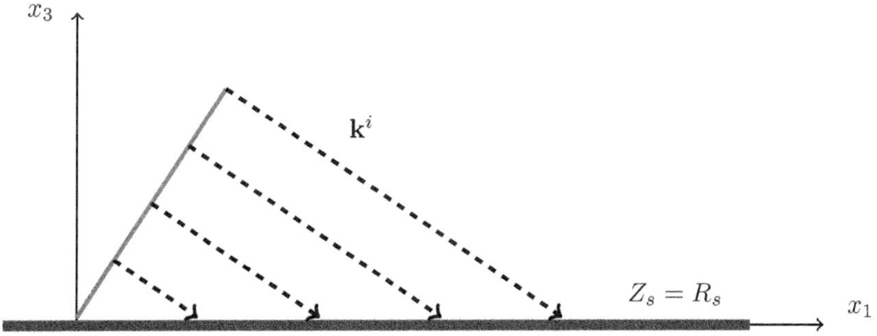

Figure 3.1: An incident wave matched to a boundary satisfies the boundary condition for a certain wave vector $\mathbf{k}^i$ and does not give rise to any reflected wave. $\mathbf{k}^i$ is different for the TE and TM polarizations.

Because of the dependence $\exp(-jk_n\mathbf{n}\cdot\mathbf{r})$ above the boundary, for $\Im\{k_n\} < 0$ the matched wave is a surface wave decaying exponentially away from the surface. For $\Im\{k_n\} > 0$ it is a leaky wave. From (3.88) and (3.92) we see that, for a capacitive surface impedance, $\Im\{Z_s\} < 0$, a matched TE wave is a surface

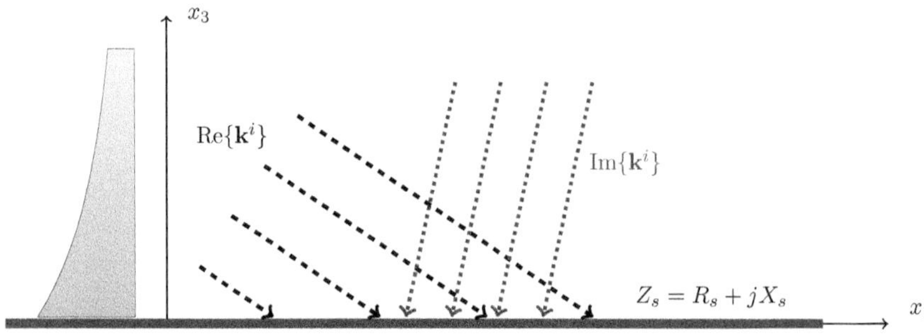

Figure 3.2: For a complex-valued impedance boundary and certain complex wave vector $\mathbf{k}^i$ the matched wave can be a surface wave with exponential decay away from the boundary. For $X_i > 0$ this occurs for the TM wave.

wave, while from (3.89), (3.93), a matched TM wave appears as a leaky wave. The opposite is the case for an inductive surface impedance $\Im\{Z_s\} > 0$.

It is known that a well-conducting ground can be approximated by an inductive impedance surface, whence the possible surface wave, called the Zenneck wave, is a TM wave [8, 23, 100].

Image of a Current Element

Let us consider the problem of vertical electric current element (dipole) at height h above a planar simple-isotropic impedance boundary $\mathbf{n}\cdot\mathbf{r} = \mathbf{e}_3\cdot\mathbf{r} = 0$ (see Figure 3.3),

$$\mathbf{J}_e(\mathbf{r}) = \mathbf{e}_3 J_3(x_3)\delta(\boldsymbol{\rho}) = \mathbf{e}_3 IL\delta(x_3 - h)\delta(\boldsymbol{\rho}), \tag{3.94}$$

with

$$\boldsymbol{\rho} = \overline{\overline{\mathsf{I}}}_t\cdot\mathbf{r} = \rho(\mathbf{e}_1\cos\varphi + \mathbf{e}_2\sin\varphi). \tag{3.95}$$

Due to rotational symmetry, a vertical electric current gives rise to a TM polarized field $\mathbf{H} = \mathbf{e}_\varphi H_\varphi$, while $\mathbf{E}_t$ has only the radial component $\mathbf{e}_\rho E_\rho$. Because the impedance boundary does not destroy the symmetry, one can imagine that the reflected field can be created by an image source, which is an electric line current of the form

$$\mathbf{J}^r(\mathbf{r}) = \mathbf{e}_3 ILf(x_3)\delta(\boldsymbol{\rho}). \tag{3.96}$$

The function $f(x_3)$ must vanish in the region above the boundary.

To find the function $f(x_3)$, let us apply the 2D Fourier transformation

$$\mathbf{F}(\mathbf{K}, x_3) = \int \mathbf{f}(\boldsymbol{\rho}, x_3)\exp(j\mathbf{K}\cdot\boldsymbol{\rho})dS_{\boldsymbol{\rho}}, \tag{3.97}$$

$$\mathbf{f}(\boldsymbol{\rho}, x_3) = \frac{1}{(2\pi)^2}\int \mathbf{F}(\mathbf{K}, x_3)\exp(-j\mathbf{K}\cdot\boldsymbol{\rho})dS_{\mathbf{K}}, \tag{3.98}$$

where $\mathbf{e}_3 \cdot \mathbf{K} = 0$, to the Maxwell equations as

$$j\mathbf{K} \times \mathbf{E} + \mathbf{e}_3 \times \partial_{x_3}\mathbf{E} + j\omega\mu_o\mathbf{H} = 0, \tag{3.99}$$

$$j\mathbf{K} \times \mathbf{H} + \mathbf{e}_3 \times \partial_{x_3}\mathbf{H} - j\omega\epsilon_o\mathbf{E} = \mathbf{e}_3 J_3(x_3). \tag{3.100}$$

Multiplying the equations successively by $\mathbf{e}_3\cdot$, $\mathbf{K}\cdot$ and $(\mathbf{e}_3 \times \mathbf{K})\cdot$, the resulting six scalar equations can be split in two non-interacting groups. One of them contains no source term, whence the corresponding field components must vanish,

$$\mathbf{e}_3 \cdot \mathbf{H} = 0, \quad \mathbf{K} \cdot \mathbf{H} = 0, \quad (\mathbf{e}_3 \times \mathbf{K}) \cdot \mathbf{E} = 0. \tag{3.101}$$

Let us denote the remaining field components by

$$I = \mathbf{e}_3 \times \mathbf{K} \cdot \mathbf{H}_t, \qquad U = \mathbf{K} \cdot \mathbf{E}_t, \qquad E_3 = \mathbf{e}_3 \cdot \mathbf{E}, \tag{3.102}$$

because I and U are analogous to respective current and voltage quantities.

Assuming $\mathbf{K} \cdot \mathbf{K} = K^2 \neq 0$, after eliminating E_3, the Maxwell equations are reduced to

$$\partial_{x_3} I(x_3) + j\frac{k_o}{\eta_o}U(x_3) = 0 \tag{3.103}$$

$$\partial_{x_3} U(x_3) + j\frac{\beta^2\eta_o}{k_o}I(x_3) = -\frac{K^2\eta_o}{k_o^2}J_3(x_3), \tag{3.104}$$

with

$$\beta = \sqrt{k_o^2 - K^2}. \tag{3.105}$$

(3.103) and (3.104) have the form of transmission-line equations. The second-order equations become

$$(\partial_{x_3}^2 + \beta^2)U(x_3) = -\frac{K^2\eta_o}{k_o^2}\partial_{x_3}J_3(x_3), \tag{3.106}$$

$$(\partial_{x_3}^2 + \beta^2)I(x_3) = j\frac{K^2}{k_o}J_3(x_3). \tag{3.107}$$

In the region $0 < x_3 < h$, the voltage and current consist of incident and reflected waves as

$$U(x_3) = U^i e^{j\beta x_3} + U^r e^{-j\beta x_3}, \tag{3.108}$$

$$I(x_3) = I^i e^{j\beta x_3} + I^r e^{-j\beta x_3}, \tag{3.109}$$

whose amplitudes obey relations obtained from (3.103) and (3.104),

$$U^i = -\frac{\beta}{k_o}\eta_o I^i, \qquad U^r = \frac{\beta}{k_o}\eta_o I^r. \tag{3.110}$$

At the source of the incident wave, $J_3^i(x_3) = IL\delta(x_3 - h)$, $I^i(x_3)$ is continuous, while $U^i(x_3)$ changes sign. Thus, the original current source acts as a series voltage source in a transmission line.

Invoking the property of the one-dimensional Green function

$$(\partial_x^2 + \beta^2)\frac{e^{-j\beta|x|}}{2j\beta} = -\delta(x), \tag{3.111}$$

to (3.107) and applying (3.103), the wave incident to the boundary can be expressed as

$$I^i(x_3) = -\frac{ILK^2}{2k_o\beta}e^{-j\beta|x_3-h|} \tag{3.112}$$

$$U^i(x_3) = -\frac{\eta_o}{jk_o}\partial_{x_3}I^i(x_3) = -\frac{ILK^2\eta_o}{2k_o^2}e^{-j\beta|x_3-h|}. \tag{3.113}$$

The impedance boundary acts as a terminating impedance for the transmission line, defined by the condition

$$\mathbf{E}_t - Z_s\mathbf{e}_3 \times \mathbf{H} = 0 \quad \Rightarrow \quad U + Z_sI = 0. \tag{3.114}$$

Substituting $U = U^i + U^r$ and $I = I^i + I^r$ and applying (3.110), we obtain

$$Z_s(I^i + I^r) = \frac{\beta\eta_o}{k_o}(I^i - I^r), \tag{3.115}$$

whence the reflected amplitude at the boundary can be expressed as

$$\begin{aligned} I^r &= \frac{\beta\eta_o - k_oZ_s}{\beta\eta_o + k_oZ_s}I^i = I^i - \frac{2k_oZ_s}{\beta\eta_o + k_oZ_s}I^i \\ &= -\frac{ILK^2}{2k_o\beta}e^{-j\beta h} + \frac{2k_oZ_s}{\beta\eta_o + k_oZ_s}\frac{ILK^2}{2k_o\beta}e^{-j\beta h}. \end{aligned} \tag{3.116}$$

Thus, the reflected current wave has the form

$$I^r(x_3) = I_a^r(x_3) + I_b^r(x_3), \tag{3.117}$$

with

$$I_a^r(x_3) = -\frac{ILK^2}{2k_o\beta}e^{-j\beta(x_3+h)}, \tag{3.118}$$

$$I_b^r(x_3) = \frac{2k_oZ_s}{\beta\eta_o + k_oZ_s}\frac{ILK^2}{2k_o\beta}e^{-j\beta(x_3+h)}. \tag{3.119}$$

The image of the original source $J_3^i(x_3)$ is the apparent source of the reflected wave. Comparing with (3.107), it can be obtained through the operation

$$J_3^r(x_3) = -j\frac{k_o}{K^2}(\partial_{x_3}^2 + \beta^2)I^r(x_3). \tag{3.120}$$

Inserting the first term of the reflected wave (3.118) in (3.120), where we can replace $(x_3 + h)$ by $|x_3 + h|$, and applying (3.111), yields

$$J_{3a}^r(x_3) = -IL(\partial_{x_3}^2 + \beta^2)\frac{e^{-j\beta|x_3+h|}}{2j\beta} = IL\delta(x_3 + h), \tag{3.121}$$

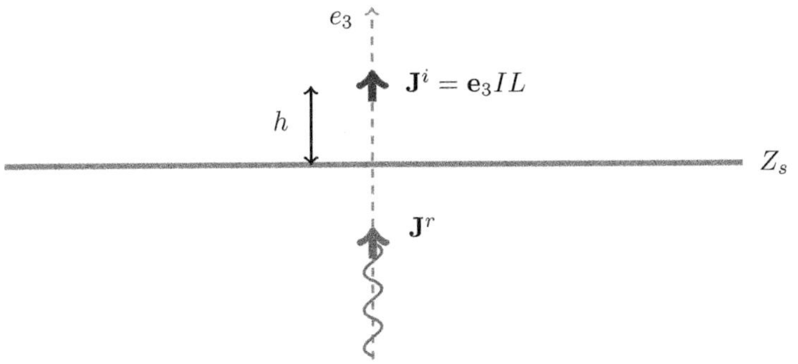

Figure 3.3: The image of an electric dipole above an IBC boundary consists of a point dipole at the mirror image point of the original dipole and a line source extending from the mirror image point to infinity.

which equals the original dipole source moved to the image point $x_3 = -h$. For the PEC boundary this is the only image since the second term, (3.119), vanishes with $Z_s = 0$.

The second term of the reflected wave (3.119), valid in the region $x_3 > 0$, can be expanded as

$$I_b^r(x_3) = \frac{2Z_s/\eta_o}{-j\beta - jk_o Z_s/\eta_o} \frac{ILK^2}{2j\beta} e^{-j\beta(x_3+h)}, \tag{3.122}$$

whence it satisfies

$$(\partial_{x_3} - jk_o Z_s/\eta_o) I_b^r(x_3) = \frac{2Z_s}{\eta_o} \frac{ILK^2}{2j\beta} e^{-j\beta(x_3+h)}. \tag{3.123}$$

The image source $J_{3b}^r(x_3)$ corresponding to $I_b^r(x_3)$ satisfies a similar condition,

$$\begin{aligned}(\partial_{x_3} - jk_o Z_s/\eta_o) J_{3b}^r(x_3) &= -j\frac{k_o}{K^2}(\partial_{x_3}^2 + \beta^2)(\partial_{x_3} - jk_o Z_s/\eta_o) I^r(x_3) \\ &= -j\frac{2k_o ILZ_s}{\eta_o}(\partial_{x_3}^2 + \beta^2)\frac{e^{-j\beta|x_3+h|}}{2j\beta} \\ &= j\frac{2k_o ILZ_s}{\eta_o}\delta(x_3 + h). \end{aligned} \tag{3.124}$$

Here, again, we have replaced $(x_3 + h)$ by $|x_3 + h|$ because the reflected wave is considered in the region $x_3 > 0$. This is a first-order differential equation for the image source $J_{3b}^r(x_3)$ under the constraint that it must vanish in the region $x_3 > 0$. The solution can be written as

$$J_{3b}^r(x_3) = -2ILk_o(Z_s/\eta_o)e^{-j(Z_s/\eta_o)k_o x_3}\Theta(-(x_3 + h)), \tag{3.125}$$

where $\Theta(x)$ is the Heaviside unit step function. It consists of a line current with exponential dependence extending from $x_3 = -h$ to $-\infty$. When computing fields from this image source for complex values of the impedance Z_s, one can extend the integration path from $x_3 = -h$ into complex space to obtain exponential convergence for the integral [35].

The complete image source for the vertical dipole above an impedance boundary is, thus,

$$\mathbf{J}^r(\mathbf{r}) = \mathbf{e}_3 IL\big(\delta(x_3+h) - 2k_o(Z_s/\eta_o)e^{-j(Z_s/\eta_o)k_o x_3}\Theta(-(x_3+h))\big)\delta(\boldsymbol{\rho}). \quad (3.126)$$

For the special case of the PEC boundary with $Z_s \to 0$ only the first term of (3.126) survives. For the PMC boundary with $Z_s \to \infty$, from (3.124) we obtain

$$J^r_{3b}(x_3) \to -2j(\partial^2_{x_3} + \beta^2)\frac{IL}{2j\beta}e^{-j\beta(x_3+h)} = -2IL\delta(x_3+h), \quad (3.127)$$

whence the image becomes

$$J^r_3(x_3) = J^r_{3a}(x_3) + J^r_{3b}(x_3) = -IL\delta(x_3+h). \quad (3.128)$$

This equals the image of the PEC boundary with opposite sign.

Image representations for the impedance boundary have been proposed in the past by Sommerfeld [103] and Felsen [12]. Corresponding image theory can be formed to the soft-and-hard (SH) boundary [36]. There exist also image theories for spherical boundaries and static sources, [34, 49].

3.6 General Isotropic Boundary

In the general case, an isotropic boundary contains both simple-isotropic and antisymmetric (PEMC) parts,

$$\overline{\overline{\mathsf{Z}}}_t = Z_s\overline{\overline{\mathsf{I}}}_t + Z_n\mathbf{n}\times\overline{\overline{\mathsf{I}}}, \quad (3.129)$$

whence the boundary conditions have the form

$$\mathbf{E}_t = Z_s\mathbf{n}\times\mathbf{H}_t - Z_n\mathbf{H}_t. \quad (3.130)$$

The relation between the impedance Z_n and the PEMC admittance M is

$$Z_n = \frac{1}{M}. \quad (3.131)$$

Expanding the normal component of the Poynting vector at the boundary yields

$$\begin{aligned}\mathbf{n}\cdot\mathbf{S} &= \frac{1}{2}\mathbf{n}\cdot\mathbf{E}_t\times\mathbf{H}^*_t \\ &= -\frac{1}{2}(Z_s\mathbf{n}\times\mathbf{H}_t - Z_n\mathbf{H}_t)\cdot(\mathbf{n}\times\mathbf{H}^*) \\ &= -\frac{1}{2}(Z_s + jZ_n\mathbf{n}\cdot\mathbf{p}(\mathbf{H}_t))|\mathbf{H}_t|^2. \end{aligned} \quad (3.132)$$

$\mathbf{p}(\mathbf{H}_t)$ is the polarization vector of the magnetic field (see Appendix B)

$$\mathbf{p}(\mathbf{H}_t) = \frac{\mathbf{H}_t \times \mathbf{H}_t^*}{j\mathbf{H}_t \cdot \mathbf{H}_t^*}, \tag{3.133}$$

which is parallel to $\mathbf{n}$ and has positive and negative values depending on the polarization of $\mathbf{H}_t$.

Power absorbed by the boundary is represented by

$$P = -\Re(\mathbf{n} \cdot \mathbf{S}) = \frac{1}{2}(R_s - X_n \mathbf{n} \cdot \mathbf{p}(\mathbf{H}_t))|\mathbf{H}_t|^2, \tag{3.134}$$

with $R_s = \Re(Z_s)$ and $X_n = \Im(Z_n)$. The boundary is lossless when $P = 0$ for any fields, which requires $R_s = 0$ and $X_n = 0$, i.e., $Z_s = jX_s$ and $Z_n = R_n$.

For the boundary to be lossy, we must have $P > 0$, which requires

$$R_s > X_n \mathbf{n} \cdot \mathbf{p}(\mathbf{H}_t), \tag{3.135}$$

for all fields $\mathbf{H}_t$. From this we may conclude:

- Because $\mathbf{n} \cdot \mathbf{p}(\mathbf{H}_t)$ has positive and negative values depending on the polarization of $\mathbf{H}_t$, a pure PEMC boundary cannot be lossy for all fields.
- Because $\mathbf{p}(\mathbf{a})$ is a real vector whose magnitude is less or equal than unity, the condition (3.135) requires $R_s > |X_s|$. This requires that the simple-isotropic component have enough resistivity to compensate the effect of the imaginary part of the PEMC component for any fields.

Plane-Wave Reflection

For the general isotropic boundary, the reflection dyadic (3.50) can be expanded as

$$\begin{aligned} \Delta &= \mathrm{det}_t(Z_s\overline{\overline{\mathsf{I}}}_t + Z_n\mathbf{n} \times \overline{\overline{\mathsf{I}}}) + \eta_o^2 - \eta_o \mathrm{tr}(Z_s\mathbf{n} \times \overline{\overline{\mathsf{J}}}_t - Z_n\overline{\overline{\mathsf{J}}}_t) \\ &= \frac{1}{k_o k_n}(k_o Z_s + k_n\eta_o)(k_n Z_s + k_o\eta_o) + Z_n^2, \end{aligned} \tag{3.136}$$

$$\begin{aligned} \Delta\overline{\overline{\mathsf{R}}}_t &= (\Delta - 2\eta_o^2)\overline{\overline{\mathsf{I}}}_t + 2\eta_o\overline{\overline{\mathsf{J}}}_t \cdot \overline{\overline{\mathsf{Z}}}_t^T \times \mathbf{n} \\ &= \frac{1}{k_o k_n k_t^2}((k_n Z_s - k_o\eta_o)(k_o Z_s + k_n\eta_o)(\mathbf{n} \times \mathbf{k})(\mathbf{n} \times \mathbf{k}) \\ &+ (k_o Z_s - k_n\eta_o)(k_n Z_s + k_o\eta_o)\mathbf{k}_t\mathbf{k}_t) + Z_n^2\overline{\overline{\mathsf{I}}}_t + 2\eta_o Z_n\overline{\overline{\mathsf{J}}}_t. \end{aligned} \tag{3.137}$$

Details of the derivation are left as an exercise. Let us check (3.137) for the two special cases.

- For the simple isotropic boundary with $Z_n = 0$ we have

$$\Delta = \frac{1}{k_o k_n}(k_o Z_s + k_n\eta_o)(k_n Z_s + k_o\eta_o), \tag{3.138}$$

and

$$\overline{\overline{\mathsf{R}}}_t = \frac{1}{k_t^2}\left(\frac{k_n Z_s - k_o\eta_o}{k_n Z_s + k_o\eta_o}(\mathbf{n}\times\mathbf{k})(\mathbf{n}\times\mathbf{k}) + \frac{k_o Z_s - k_n\eta_o}{k_o Z_s + k_n\eta_o}\mathbf{k}_t\mathbf{k}_t\right), \tag{3.139}$$

which coincides with (3.75).

- For the PEMC boundary with $Z_s = 0$ we have

$$\Delta = Z_n^2 + \eta_o^2, \tag{3.140}$$

$$\overline{\overline{\mathsf{R}}}_t = \frac{1}{Z_n^2+\eta_o^2}((Z_n^2 - \eta_o^2)\overline{\overline{\mathsf{I}}}_t + 2\eta_o Z_n \overline{\overline{\mathsf{J}}}_t)$$

$$= \frac{1}{Z_n^2+\eta_o^2}(Z_n\overline{\overline{\mathsf{I}}}_t + \eta_o\overline{\overline{\mathsf{J}}}_t)^2, \tag{3.141}$$

which coincides with (2.30) for $M = 1/Z_n$.

Eigenwaves

The eigenvectors of $\overline{\overline{\mathsf{R}}}_t$ in (3.50) are the same as for the eigenproblem

$$(\overline{\overline{\mathsf{J}}}_t\cdot\overline{\overline{\mathsf{Z}}}_t^T\times\mathbf{n})\cdot\mathbf{z}_t = \lambda\mathbf{z}_t. \tag{3.142}$$

Substituting $\mathbf{z}_t = A\mathbf{k}_t + B\mathbf{n}\times\mathbf{k}_t$, we can solve for the eigenvalues,

$$\lambda_\pm = -\frac{(k_o^2+k_n^2)Z_s}{2k_ok_n} \pm \sqrt{\left(\frac{(k_o^2-k_n^2)Z_s}{2k_ok_n}\right)^2 - Z_n^2}, \tag{3.143}$$

whence the eigenvectors become

$$\mathbf{z}_{t\pm} = (k_oZ_s + k_n\lambda_\pm)\mathbf{k}_t + k_oZ_n(\mathbf{n}\times\mathbf{k}). \tag{3.144}$$

The corresponding eigenvalues for the reflection dyadic $\overline{\overline{\mathsf{R}}}_t$ are

$$R_\pm = \frac{1}{\Delta}(\Delta - 2\eta_o^2 + 2\eta_o\lambda_\pm). \tag{3.145}$$

Details of the derivation are left as exercises.

Let us again check the two special cases. For the simple isotropic boundary with $Z_n = 0$, we obtain Δ from (3.138) and (3.143) yields

$$\lambda_+ = -Z_sk_n/k_o, \quad \lambda_- = -Z_sk_o/k_n. \tag{3.146}$$

Thus, the eigenvalues of $\overline{\overline{\mathsf{R}}}_t$, obtained from (3.54), take the form

$$R_+ = 1 + \frac{2\eta_o}{\Delta}(\lambda_+ - \eta_o) = \frac{k_oZ_s - k_n\eta_o}{k_oZ_s + k_n\eta_o}, \tag{3.147}$$

$$R_- = 1 + \frac{2\eta_o}{\Delta}(\lambda_- - \eta_o) = \frac{k_nZ_s - k_o\eta_o}{k_nZ_s + k_o\eta_o}. \tag{3.148}$$

These reproduce the coefficients of (3.83) and (3.84).

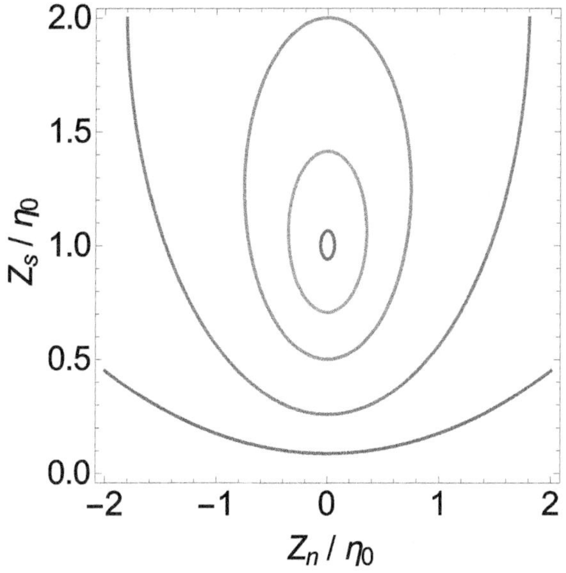

Figure 3.4: Dependence of $k_n = k_o \cos\vartheta$ on the impedances Z_s and Z_n for the dispersion equation (3.150) of the general isotropic boundary. Values of ϑ for curves extending from the center: $20^o, 45^o, 60^o, 75^o$ and 85°.

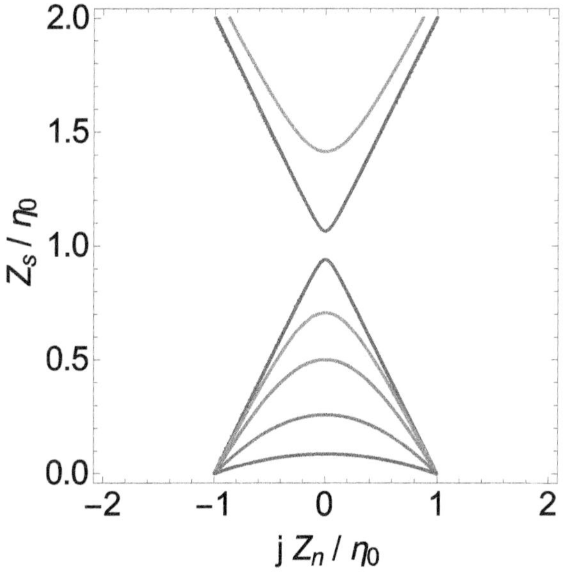

Figure 3.5: Same as Figure 3.4 but for imaginary Z_n values. The curves correspond to the same ϑ values and they share the same points for $Z_n = 0$.

Matched Waves

The dispersion equation for matched waves (3.72) is reduced for the general isotropic boundary to

$$k_o k_n(\eta_o^2 + Z_s^2 + Z_n^2) + \eta_o(Z_s k_t^2 + 2Z_s k_n^2) = 0, \tag{3.149}$$

which can be written as

$$(k_o Z_s + k_n \eta_o)(k_n Z_s + k_o \eta_o) + k_o k_n Z_n^2 = 0. \tag{3.150}$$

For the simple isotropic boundary, $Z_n = 0$, this coincides with (3.73). For the PEMC boundary, $Z_s = 0$, we obtain $k_n(Z_n^2 + \eta_o^2) = 0$. This yields the lateral plane wave $k_n = 0$ when excluding the cases $Z_n = \pm j\eta_o$.

In Figures 3.4 and 3.5 the dispersion equation (3.150) is visualized for a range of real Z_s values and real or imaginary Z_n values.

3.7 Perfectly Anisotropic Boundary

The class of perfectly anisotropic (PA) boundary surfaces is defined by requiring that the impedance dyadic $\overline{\overline{\mathsf{Z}}}_t = \overline{\overline{\mathsf{Z}}}_{ta}$ be symmetric and trace free,

$$\overline{\overline{\mathsf{Z}}}_{ta}^T = \overline{\overline{\mathsf{Z}}}_{ta}, \qquad \mathrm{tr}\overline{\overline{\mathsf{Z}}}_{ta} = 0. \tag{3.151}$$

An equivalent condition to define the class of PA boundary dyadics is

$$\overline{\overline{\mathsf{Z}}}_{ta} {}^{\times}_{\times} \mathbf{nn} = -\overline{\overline{\mathsf{Z}}}_{ta}. \tag{3.152}$$

Because of the symmetric impedance dyadic, any PA boundary is reciprocal. Expanding $\overline{\overline{\mathsf{Z}}}_{ta}$ in its real and imaginary parts,

$$\overline{\overline{\mathsf{Z}}}_{ta} = \overline{\overline{\mathsf{R}}}_{ta} + j\overline{\overline{\mathsf{X}}}_{ta}, \tag{3.153}$$

both of which are symmetric and trace-free dyadics, from (3.27) we conclude that a PA boundary is lossy when $\overline{\overline{\mathsf{R}}}_{ta}$ is positive definite and active when $\overline{\overline{\mathsf{R}}}_{ta}$ is negative definite. It is lossless when the impedance dyadic is pure imaginary, $\overline{\overline{\mathsf{R}}}_{ta} = 0$.

Applying the 2D Cayley-Hamilton equation for the dyadic $\overline{\overline{\mathsf{Z}}}_{ta}$,

$$\overline{\overline{\mathsf{Z}}}_{ta}^2 - \mathrm{tr}\overline{\overline{\mathsf{Z}}}_{ta}\,\overline{\overline{\mathsf{Z}}}_{ta} + \det{}_t\overline{\overline{\mathsf{Z}}}_{ta}\,\overline{\overline{\mathsf{I}}}_t = 0, \tag{3.154}$$

the following properties are seen to follow:

$$\overline{\overline{\mathsf{Z}}}_{ta}^2 = -\det{}_t\overline{\overline{\mathsf{Z}}}_{ta}\,\overline{\overline{\mathsf{I}}}_t, \tag{3.155}$$

$$\det\overline{\overline{\mathsf{Z}}}_{ta} = -\frac{1}{2}\mathrm{tr}\overline{\overline{\mathsf{Z}}}_{ta}^2, \tag{3.156}$$

$$\overline{\overline{\mathsf{Z}}}_{ta}^{-1} = -\frac{1}{\det{}_t\overline{\overline{\mathsf{Z}}}_{ta}}\overline{\overline{\mathsf{Z}}}_{ta}. \tag{3.157}$$

In terms of a basis of orthonormal vectors $\mathbf{e}_1, \mathbf{e}_2$ and $\mathbf{e}_3 = \mathbf{n}$, a basis of four 2D dyadics can be constructed as [52]

$$\overline{\overline{\mathsf{I}}}_t = \mathbf{e}_1\mathbf{e}_1 + \mathbf{e}_2\mathbf{e}_2, \tag{3.158}$$

$$\mathbf{n}\times\overline{\overline{\mathsf{I}}} = \mathbf{e}_2\mathbf{e}_1 - \mathbf{e}_1\mathbf{e}_2, \tag{3.159}$$

$$\overline{\overline{\mathsf{K}}} = \mathbf{e}_1\mathbf{e}_1 - \mathbf{e}_2\mathbf{e}_2, \tag{3.160}$$

$$\overline{\overline{\mathsf{L}}} = \mathbf{e}_1\mathbf{e}_2 + \mathbf{e}_2\mathbf{e}_1, \tag{3.161}$$

whence any PA boundary can be represented by the dyadic

$$\overline{\overline{\mathsf{Z}}}_{ta} = Z_K\overline{\overline{\mathsf{K}}} + Z_L\overline{\overline{\mathsf{L}}}. \tag{3.162}$$

While the dyadics $\overline{\overline{\mathsf{I}}}_t$ and $\mathbf{n}\times\overline{\overline{\mathsf{I}}}$ are independent of any 2D basis, the dyadics $\overline{\overline{\mathsf{K}}}$ and $\overline{\overline{\mathsf{L}}}$ are dependent on the choice of vectors $\mathbf{e}_1$ and $\mathbf{e}_2$. Aactually, one can show that, for a given PA boundary dyadic, it is possible to choose the orthonormal 2D basis of unit vectors so that either $Z_K = 0$ or $Z_L = 0$. Details are left as an exercise. For example, any PA dyadic can be expressed as

$$\overline{\overline{\mathsf{Z}}}_{ta} = Z_K\overline{\overline{\mathsf{K}}} = Z_K(\mathbf{e}_1\mathbf{e}_1 - \mathbf{e}_2\mathbf{e}_2), \tag{3.163}$$

where $\mathbf{e}_1$ and $\mathbf{e}_2$ are eigenvectors of the dyadic $\overline{\overline{\mathsf{Z}}}_{ta}$.

The dyadic $\overline{\overline{\mathsf{K}}}$ can be interpreted as a 2D reflection dyadic because it changes the sign of the $\mathbf{e}_2$ component of a vector. It satisfies the properties

$$\overline{\overline{\mathsf{K}}}^2 = \overline{\overline{\mathsf{I}}}_t, \quad \mathrm{tr}\overline{\overline{\mathsf{K}}} = 0, \quad \det{}_t\overline{\overline{\mathsf{K}}} = -1. \tag{3.164}$$

Reflection from PA Boundary

As a simple example, let us consider reflection of a normally incident plane wave from a PA boundary defined by (3.163). Substituting $\mathbf{k}^i = -k_o\mathbf{n}$ in (1.43), we have

$$\overline{\overline{\mathsf{J}}}_t = \mathbf{n}\times\overline{\overline{\mathsf{I}}}, \tag{3.165}$$

and the expression (3.50) of the reflection dyadic can be given the form

$$\overline{\overline{\mathsf{R}}}_t = \frac{Z_K - \eta_o}{Z_K + \eta_o}\mathbf{e}_1\mathbf{e}_1 + \frac{Z_K + \eta_o}{Z_K - \eta_o}\mathbf{e}_2\mathbf{e}_2, \tag{3.166}$$

$$\Delta = -(Z_K^2 - \eta_o^2). \tag{3.167}$$

Details of the derivation are left as an exercise.

Assuming a lossless PA boundary (3.163), defined by

$$Z_K = j\eta_o \tan\psi, \tag{3.168}$$

the reflection dyadic can be simplified to

$$\overline{\overline{\mathsf{R}}}_t = -(e^{-j2\psi}\mathbf{e}_1\mathbf{e}_1 + e^{j2\psi}\mathbf{e}_2\mathbf{e}_2) \tag{3.169}$$

$$= -\cos 2\psi\;\overline{\overline{\mathsf{I}}}_t + j\sin 2\psi\;\overline{\overline{\mathsf{K}}} = -\exp(-2j\psi\overline{\overline{\mathsf{K}}}). \tag{3.170}$$

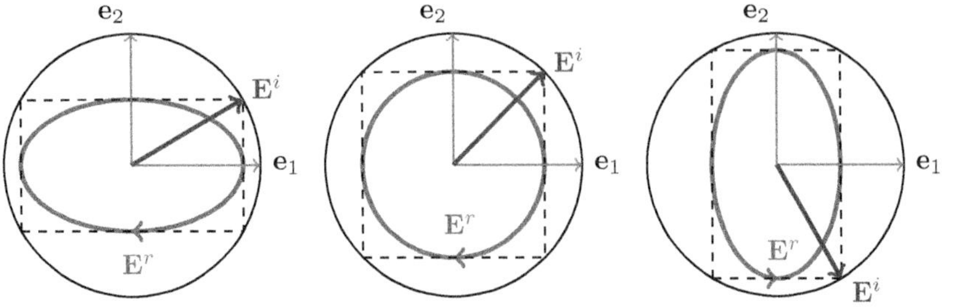

Figure 3.6: A linearly polarized field with normal incidence is reflected from a PA-boundary surface with impedance dyadic (3.163), as elliptically polarized. The handedness and axial ratio depend on the angle of the incident polarization with respect to the eigendirections of the impedance dyadic.

The polarizations $\mathbf{e}_1$ and $\mathbf{e}_2$ are eigenvectors of $\overline{\overline{\mathsf{R}}}_t$ and they are reflected co-polarized. More generally, the polarization is changed in reflection. For an incident linearly-polarized field

$$\mathbf{E}^i = E(\mathbf{e}_1 \cos\varphi + \mathbf{e}_2 \sin\varphi) \tag{3.171}$$

the reflected field is elliptically polarized as

$$\begin{aligned} \mathbf{E}^r &= -E(\mathbf{e}_1 e^{-j2\psi} \cos\varphi + \mathbf{e}_2 e^{j2\psi} \sin\varphi) \\ &= -E \cos 2\psi(\mathbf{e}_1 \cos\varphi + \mathbf{e}_2 \sin\varphi) \\ &\quad +jE \sin 2\psi(\mathbf{e}_1 \cos\varphi - \mathbf{e}_2 \sin\varphi). \end{aligned} \tag{3.172}$$

The polarization vector of the reflected field becomes

$$\mathbf{p}(\mathbf{E}^r) = \frac{\mathbf{E}^r \times \mathbf{E}^{r*}}{j\mathbf{E}^r \cdot \mathbf{E}^{r*}} = -\sin 4\psi \sin 2\varphi\ \mathbf{n}. \tag{3.173}$$

Choosing $\sin 4\psi = 1$, $\psi = \pi/8$, which corresponds to $Z_K/\eta_o = \tan(\pi/8) = \sqrt{2} - 1$, we have $\mathbf{p}(\mathbf{E}^r) = -\sin 2\varphi\ \mathbf{n}$, whence the reflected field can obtain any ellipticity for a linearly polarized incident field with a proper choice the angle φ. Circular polarizations are obtained for $\sin 2\varphi = \pm 1$, i.e., for $\varphi = \pm\pi/4$. This effect is depicted by Figure 3.6. Because the polarization of a wave can be continuously changed by rotating the reflecting PA boundary surface, such a PA boundary has the potential application as a polarization transformer.

Matched Waves at PA boundary

To find matched waves at a PA boundary, we can apply the dispersion equation (3.72) for $\overline{\overline{\mathsf{Z}}}_t = Z_K \overline{\overline{\mathsf{K}}}$. Substituting $\mathrm{det}_t \overline{\overline{\mathsf{K}}} = -1$ and $k_n = \sqrt{k_o^2 - \mathbf{k}_t \cdot \mathbf{k}_t}$ we obtain an equation for $\mathbf{k}_t$,

$$(\overline{\overline{\mathsf{K}}} : \mathbf{k}_t \mathbf{k}_t)^2 = k_o^2 (k_o^2 - \mathbf{k}_t \cdot \mathbf{k}_t) X^2, \quad X = \frac{\eta_o}{Z_K} - \frac{Z_K}{\eta_o}. \tag{3.174}$$

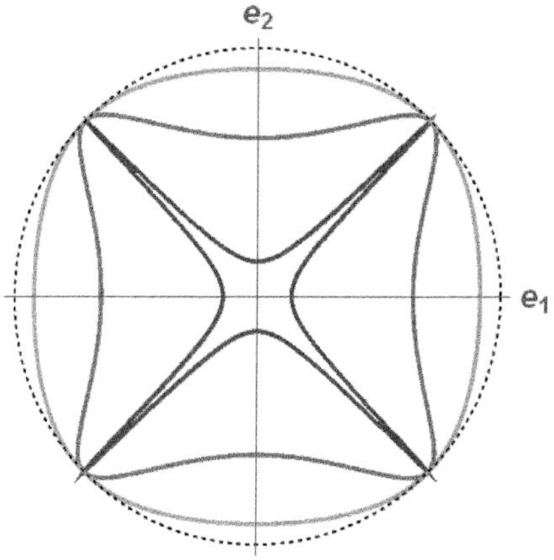

Figure 3.7: Dispersion diagrams depicting $\mathbf{k}_t/k_o$ for waves matched to a perfectly anisotropic boundary. In radially extending order, the impedance parameters values are $Z_K/\eta_o = 1.01, 1.3$ and 2.5. Points at the dotted unit circle correspond to lateral matched waves with $k_n = 0$.

For $\overline{\overline{\mathsf{K}}} = \mathbf{e}_1\mathbf{e}_1 - \mathbf{e}_2\mathbf{e}_2$ the dispersion equation can be written in the form

$$(k_1^2 - k_2^2)^2 = k_o^2 X^2 (k_o^2 - k_1^2 - k_2^2). \tag{3.175}$$

From this one can note the special cases:

- For $Z_K = 0$(PMC) or $1/Z_K = 0$ (PEC) we have $1/X = 0$, whence the dispersion diagram is a circle, $k_n = 0$, $k_t^2 = k_1^2 + k_2^2 = k_o^2$. Thus, any matched wave is a lateral wave.
- For $Z_K = \pm\eta_o$, we have $X = 0$, whence the dispersion diagram is made of two orthogonal lines, $k_1^2 = k_2^2$.

These properties can be observed from Figure 3.7.

3.8 Generalized Soft-and-Hard (GSH) Boundary

As another example of an impedance boundary, let us consider one defined by an impedance dyadic of the form

$$\overline{\overline{\mathsf{Z}}}_t = \eta_o(-\delta\mathbf{e}_1\mathbf{e}_1 + \frac{1}{\delta}\mathbf{e}_2\mathbf{e}_2), \tag{3.176}$$

where δ is a scalar parameter. Since the impedance dyadic is symmetric, there is no PEMC component. The conditions can be written in the form of (3.20) as

$$\mathbf{e}_1 \cdot \mathbf{E}_t - \delta\mathbf{e}_2 \cdot \eta_o\mathbf{H} = 0, \tag{3.177}$$

$$\delta\mathbf{e}_2 \cdot \mathbf{E} - \mathbf{e}_1 \cdot \eta_o\mathbf{H} = 0. \tag{3.178}$$

For $\delta \to 0$, these become conditions of the soft-and-hard (SH) boundary [25],

$$\mathbf{e}_1 \cdot \mathbf{E} = 0, \qquad \mathbf{e}_1 \cdot \mathbf{H} = 0. \tag{3.179}$$

The SH boundary is anisotropic, but not perfectly anisotropic, since $\mathrm{tr}\overline{\overline{\mathsf{Z}}}_t \neq 0$.

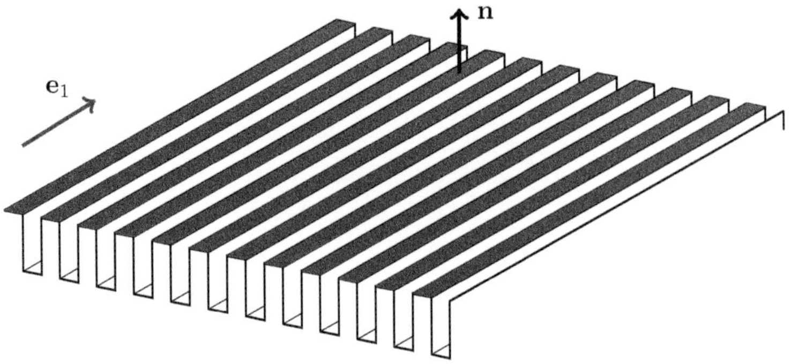

Figure 3.8: A soft-and-hard boundary can be approximated by a corrugated conducting surface where the troughs are subwavelength in width but have depth of a quarter-wavelength.

The concept of soft-and-hard boundary has grown from various engineering applications making use of its various realizations, starting from corrugated surfaces in the 1940's [26, 27, 38].

The conditions of the SH boundary (3.179) can be straightforwardly generalized to

$$\mathbf{a}_t \cdot \mathbf{E} = 0, \qquad \mathbf{b}_t \cdot \mathbf{H} = 0, \tag{3.180}$$

where $\mathbf{a}_t$ and $\mathbf{b}_t$ are two arbitrary tangential vectors. (3.180) can be called the conditions of the Generalized Soft-and-Hard (GSH) boundary [41, 43]. Possible realizations for the GSH boundary have been suggested in [17, 111].

Conditions (3.180) define a special case of impedance-boundary. They can be shown to arise from the impedance dyadic

$$\overline{\overline{\mathsf{Z}}}_t = \delta\eta_o \mathbf{b}_t \mathbf{a}_t + \frac{1}{\delta}\eta_o(\mathbf{n} \times \mathbf{a}_t)(\mathbf{n} \times \mathbf{b}_t). \tag{3.181}$$

In fact, inserted in (3.1), (3.181) yields the boundary conditions

$$\mathbf{a}_t \cdot \mathbf{E}_t = \delta(\mathbf{a}_t \cdot \mathbf{b}_t)(\mathbf{a}_t \times \mathbf{n} \cdot \eta_o \mathbf{H}_t), \tag{3.182}$$

$$\delta(\mathbf{n} \times \mathbf{b}_t) \cdot \mathbf{E}_t = (\mathbf{n} \times \mathbf{b}_t) \cdot (\mathbf{n} \times \mathbf{a}_t)(\mathbf{n} \cdot \eta_o \mathbf{H}_t). \tag{3.183}$$

Letting $\delta \to 0$, the GSH conditions (3.180) can be seen to emerge.

To find the condition of reciprocity for the GSH boundary, let us require

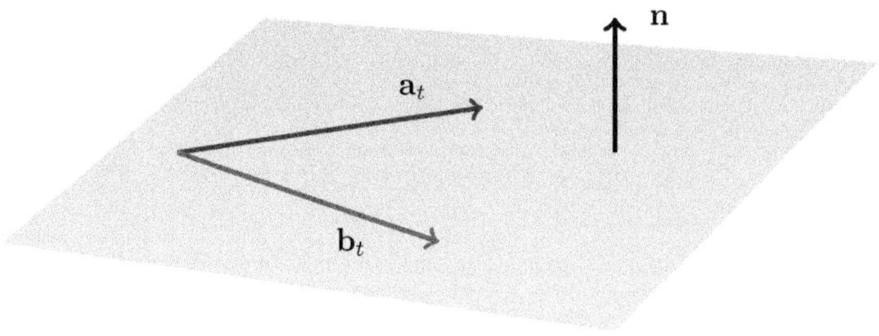

Figure 3.9: A generalized soft-and-hard (GSH) boundary defined by two tangential vectors $\mathbf{a}_t$ and $\mathbf{b}_t$. Their magnitudes have no importance.

vanishing of the antisymmetric part of the impedance dyadic (3.181) $\overline{\overline{\mathsf{Z}}}_s$:

$$\frac{1}{\eta_o}(\overline{\overline{\mathsf{Z}}}_t - \overline{\overline{\mathsf{Z}}}_t^T) = \delta(\mathbf{b}_t\mathbf{a}_t - \mathbf{a}_t\mathbf{b}_t) + \frac{1}{\delta}\mathbf{nn}{}^{\times}_{\times}(\mathbf{a}_t\mathbf{b}_t - \mathbf{b}_t\mathbf{a}_t) \tag{3.184}$$

$$= \delta(\mathbf{a}_t \times \mathbf{b}_t)\times\overline{\overline{\mathsf{I}}} + \frac{1}{\delta}\mathbf{nn}{}^{\times}_{\times}((\mathbf{b}_t \times \mathbf{a}_t)\times\overline{\overline{\mathsf{I}}}) \tag{3.185}$$

$$= (\mathbf{n}\cdot\mathbf{a}_t \times \mathbf{b}_t)\left(\delta(\mathbf{n}\times\overline{\overline{\mathsf{I}}}) - \frac{1}{\delta}\mathbf{nn}{}^{\times}_{\times}(\mathbf{n}\times\overline{\overline{\mathsf{I}}})\right) \tag{3.186}$$

$$= (\mathbf{n}\cdot\mathbf{a}_t \times \mathbf{b}_t)\left(\delta + \frac{1}{\delta}\right)(\mathbf{n}\times\overline{\overline{\mathsf{I}}}) = 0. \tag{3.187}$$

This implies $\mathbf{n}\cdot\mathbf{a}_t \times \mathbf{b}_t = 0$, whence $\mathbf{a}_t$ and $\mathbf{b}_t$ must be linearly dependent. For $\delta \to 0$ we conclude that a GSH boundary is reciprocal exactly when it actually is an SH boundary. This can be verified by finding the decomposition (3.41) for the impedance dyadic (3.181), which is left as an exercise.

TE/TM Decomposition

To find reflection of a plane wave from the GSH boundary, let us consider decomposition of a plane wave in two parts in terms of two triples of vectors $(\mathbf{a}_t, (\mathbf{b}_t \times \mathbf{k}), \mathbf{k})$ and $(\mathbf{b}_t, (\mathbf{a}_t \times \mathbf{k}), \mathbf{k})$. Each of them defines a base of vectors when the vectors satisfy

$$\begin{aligned}\Delta &= \mathbf{a}_t \times (\mathbf{b}_t \times \mathbf{k})\cdot\mathbf{k} = \mathbf{b}_t \times (\mathbf{a}_t \times \mathbf{k})\cdot\mathbf{k} = -(\mathbf{a}_t \times \mathbf{k})\cdot(\mathbf{b}_t \times \mathbf{k}) \\ &= (\mathbf{a}_t\cdot\mathbf{k}_t)(\mathbf{b}_t\cdot\mathbf{k}_t) - k_o^2(\mathbf{a}_t\cdot\mathbf{b}_t) \neq 0.\end{aligned} \tag{3.188}$$

The unit dyadic can be expressed in either of these two sets of basis vectors as (see Appendix C)

$$\overline{\overline{\mathsf{I}}} = \frac{1}{\Delta}(((\mathbf{b}_t \times \mathbf{k})\times\mathbf{k})\mathbf{a}_t + (\mathbf{k}\times\mathbf{a}_t)(\mathbf{b}_t \times \mathbf{k}) + (\mathbf{a}_t \times (\mathbf{b}_t \times \mathbf{k}))\mathbf{k}) \tag{3.189}$$

$$\overline{\overline{\mathsf{I}}} = \frac{1}{\Delta}(((\mathbf{a}_t \times \mathbf{k})\times\mathbf{k})\mathbf{b}_t + (\mathbf{k}\times\mathbf{b}_t)(\mathbf{a}_t \times \mathbf{k}) + (\mathbf{b}_t \times (\mathbf{a}_t \times \mathbf{k}))\mathbf{k}) \tag{3.190}$$

Applying $\mathbf{k}\cdot\mathbf{E}=0$ and $\mathbf{k}\cdot\mathbf{H}=0$, we can expand the field vectors as

$$\begin{aligned}\mathbf{E} &= \frac{1}{\Delta}(((\mathbf{b}_t\times\mathbf{k})\times\mathbf{k})(\mathbf{a}_t\cdot\mathbf{E}_t)+(\mathbf{k}\times\mathbf{a}_t)((\mathbf{b}_t\times\mathbf{k})\cdot\mathbf{E}))\\ &= \frac{1}{\Delta}(((\mathbf{b}_t\times\mathbf{k})\times\mathbf{k})(\mathbf{a}_t\cdot\mathbf{E}_t)+k_o(\mathbf{k}\times\mathbf{a}_t)(\mathbf{b}_t\cdot\eta_o\mathbf{H}_t)) \qquad (3.191)\end{aligned}$$

$$\begin{aligned}\eta_o\mathbf{H} &= \frac{1}{\Delta}(((\mathbf{a}_t\times\mathbf{k})\times\mathbf{k})(\mathbf{b}_t\cdot\eta_o\mathbf{H}_t)+(\mathbf{k}\times\mathbf{b}_t)((\mathbf{a}_t\times\mathbf{k})\cdot\eta_o\mathbf{H}))\\ &= \frac{1}{\Delta}(((\mathbf{a}_t\times\mathbf{k})\times\mathbf{k})(\mathbf{b}_t\cdot\eta_o\mathbf{H}_t)-k_o(\mathbf{k}\times\mathbf{b}_t)(\mathbf{a}_t\cdot\mathbf{E}_t)). \qquad (3.192)\end{aligned}$$

In view of these equations, any plane wave can be split in two parts, TE_a and TM_b waves, as

$$(\mathbf{E},\mathbf{H})=(\mathbf{E}_a,\mathbf{H}_a)+(\mathbf{E}_b,\mathbf{H}_b), \qquad (3.193)$$

satisfying

$$\mathbf{a}_t\cdot\mathbf{E}_a=0, \qquad \mathbf{b}_t\cdot\mathbf{H}_b=0. \qquad (3.194)$$

For a given plane-wave with fields $\mathbf{E},\mathbf{H}$, the decomposed plane-wave fields are obtained as

$$\mathbf{E}_a = \frac{k_o}{\Delta}(\mathbf{k}\times\mathbf{a}_t)(\mathbf{b}_t\cdot\eta_o\mathbf{H}_t), \qquad (3.195)$$

$$\eta_o\mathbf{H}_a = \frac{1}{\Delta}((\mathbf{a}_t\times\mathbf{k})\times\mathbf{k})(\mathbf{b}_t\cdot\eta_o\mathbf{H}_t), \qquad (3.196)$$

$$\mathbf{E}_b = \frac{1}{\Delta}((\mathbf{b}_t\times\mathbf{k})\times\mathbf{k})(\mathbf{a}_t\cdot\mathbf{E}_t), \qquad (3.197)$$

$$\eta_o\mathbf{H}_b = -\frac{k_o}{\Delta}((\mathbf{k}\times\mathbf{b}_t)(\mathbf{a}_t\cdot\mathbf{E}_t). \qquad (3.198)$$

One can show that the TE_a part and the TM_b part of a plane wave are two independent plane waves, each satisfying plane-wave equations of their own. The quantities $\mathbf{a}_t\cdot\mathbf{E}_t$ and $\mathbf{b}_t\cdot\eta_o\mathbf{H}_t$ can be conceived as potential quantities in terms of which the fields can be expressed. When $\mathbf{a}_t$ and $\mathbf{b}_t$ are linearly dependent, the GSH boundary reduces to the SH boundary, in which case the expressions (3.195) - (3.198) can be shown to coincide with the results of [38].

Reflection from GSH Boundary

Considering a TE_a plane wave incident to the GSH boundary, from the condition (3.180) we obtain

$$\mathbf{a}_t\cdot\mathbf{E}=\mathbf{a}_t\cdot(\mathbf{E}_a^i+\mathbf{E}^r)=\mathbf{a}_t\cdot\mathbf{E}^r=0. \qquad (3.199)$$

Similarly, for a TM_b wave, we obtain

$$\mathbf{b}_t\cdot\mathbf{H}=\mathbf{b}_t\cdot(\mathbf{H}_b^i+\mathbf{H}^r)=\mathbf{b}_t\cdot\mathbf{H}^r=0. \qquad (3.200)$$

Thus, a TE_a wave is reflected as a TE_a wave and a TM_b wave is reflected as a TM_b wave.

Because the TE_a and TM_b waves already satisfy one half of the boundary conditions (3.180), $\mathbf{a}_t \cdot (\mathbf{E}_a^i + \mathbf{E}_a^r) = 0$ and $\mathbf{b}_t \cdot (\mathbf{H}_b^i + \mathbf{H}_b^r) = 0$, the remaining boundary conditions serve as additional conditions for the fields. From (3.188) one finds that Δ is the same for $\mathbf{k} = \mathbf{k}^i$ and $\mathbf{k} = \mathbf{k}^r$,

$$\Delta = -(\mathbf{a}_t \times \mathbf{k}^r) \cdot (\mathbf{b}_t \times \mathbf{k}^r) = -(\mathbf{a}_t \times \mathbf{k}^i) \cdot (\mathbf{b}_t \times \mathbf{k}^i). \tag{3.201}$$

Invoking (3.196) for both incident and reflected TE_a waves, let us expand

$$\begin{aligned} \Delta\mathbf{n} \times (\mathbf{H}_a^i + \mathbf{H}_a^r) &= \mathbf{n} \times ((\mathbf{a}_t \times \mathbf{k}^i) \times \mathbf{k}^i)(\mathbf{b}_t \cdot \mathbf{H}_a^i) \\ &\quad + \mathbf{n} \times ((\mathbf{a}_t \times \mathbf{k}^r) \times \mathbf{k}^r)(\mathbf{b}_t \cdot \mathbf{H}_a^r), \qquad (3.202) \\ &= \mathbf{n} \times (\mathbf{k}_t(\mathbf{a}_t \cdot \mathbf{k}_t) - k_o^2 \mathbf{a}_t)\mathbf{b}_t \cdot (\mathbf{H}_a^i + \mathbf{H}_a^r) = 0 \qquad (3.203) \end{aligned}$$

At the last step we have applied the GSH boundary condition for the total field, $\mathbf{b}_t \cdot \mathbf{H}_a = 0$. Similarly, we obtain

$$\Delta\mathbf{n} \times (\mathbf{E}_b^i + \mathbf{E}_b^r) = \mathbf{n} \times (\mathbf{k}_t(\mathbf{b}_t \cdot \mathbf{k}_t) - k_o^2 \mathbf{b}_t)\mathbf{a}_t \cdot (\mathbf{E}_b^i + \mathbf{E}_b^r) = 0. \tag{3.204}$$

The normal components can be expanded as

$$\begin{aligned} \Delta\mathbf{n} \cdot (\mathbf{E}_a^i + \mathbf{E}_a^r) &= k_o(\mathbf{n} \cdot \mathbf{k}_t \times \mathbf{a}_t)\mathbf{b}_t \cdot \eta_o(\mathbf{H}_a^i + \mathbf{H}_a^r) = 0, \qquad (3.205) \\ \Delta\mathbf{n} \cdot \eta_o(\mathbf{H}_b^i + \mathbf{H}_b^r) &= -k_o(\mathbf{n} \cdot \mathbf{k}_t \times \mathbf{b}_t)\mathbf{a}_t \cdot (\mathbf{E}_b^i + \mathbf{E}_b^r) = 0. \qquad (3.206) \end{aligned}$$

From these conditions we can make two conclusions:

- For a TE_a wave, the GSH boundary appears as the PMC boundary.
- For a TM_b wave, the GSH boundary appears as the PEC boundary.

Similar conclusions remain valid for the special case of the SH boundary. One can state that these boundaries have the property of *PEC/PMC equivalence*. More general examples of boundaries with the same property will be presented in Chapter 5.

From (3.199) and (3.205) on one hand, and (3.200) and (3.206) on the other hand, the total TE_a and TM_b plane-wave fields at the GSH boundary must be polarized as

$$\mathbf{E}_a^i + \mathbf{E}_a^r = E_a \mathbf{n} \times \mathbf{a}_t, \quad \mathbf{H}_b^i + \mathbf{H}_b^r = H_b \mathbf{n} \times \mathbf{b}_t. \tag{3.207}$$

To find the relation between the reflected and incident fields, applying (3.195) - (3.198), let us expand

$$\begin{aligned} \Delta\mathbf{E}^r &= \Delta(\mathbf{E}_a^r + \mathbf{E}_b^r) \\ &= k_o(\mathbf{k}^r \times \mathbf{a}_t)(\mathbf{b}_t \cdot \eta_o \mathbf{H}_t^r) + ((\mathbf{b}_t \times \mathbf{k}^r) \times \mathbf{k}^r)(\mathbf{a}_t \cdot \mathbf{E}_t^r) \\ &= -k_o(\mathbf{k}^r \times \mathbf{a}_t)(\mathbf{b}_t \cdot \eta_o \mathbf{H}_t^i) - ((\mathbf{b}_t \times \mathbf{k}^r) \times \mathbf{k}^r)(\mathbf{a}_t \cdot \mathbf{E}_t^i) \\ &= -(\mathbf{k}^r \times \mathbf{a}_t)(\mathbf{b}_t \times \mathbf{k}^i) \cdot \mathbf{E}^i - ((\mathbf{b}_t \times \mathbf{k}^r) \times \mathbf{k}^r)(\mathbf{a}_t \cdot \mathbf{E}_t^i) \\ &= \Delta\overline{\overline{\mathsf{R}}} \cdot \mathbf{E}^i. \end{aligned} \tag{3.208}$$

The reflection dyadic can be identified as

$$\overline{\overline{\mathsf{R}}} = \frac{-1}{\Delta}((\mathbf{k}^r \times \mathbf{a}_t)(\mathbf{b}_t \times \mathbf{k}^i) + ((\mathbf{b}_t \times \mathbf{k}^r) \times \mathbf{k}^r)\mathbf{a}_t), \tag{3.209}$$

or,

$$\overline{\overline{\mathsf{R}}} = \frac{1}{k_o^2\Delta}(k_o^2(\mathbf{a}_t\mathbf{b}_t {}^\times_\times \mathbf{k}^r\mathbf{k}^i) + (\mathbf{b}_t\mathbf{a}_t {}^\times_\times \mathbf{k}^r\mathbf{k}^i) {}^\times_\times \mathbf{k}^r\mathbf{k}^i), \tag{3.210}$$

The latter more symmetric expression is equivalent to the former in the sense that it maps $\mathbf{E}^i$ to $\mathbf{E}^r$ in the same way.

Equation (3.210) generalizes the result for the SH boundary derived in [38]. It is left as an exercise to check that, substitutions $\mathbf{E}^i = \mathbf{E}^i_a$ and $\mathbf{E}^i = \mathbf{E}^i_b$ yield respectively $\mathbf{E}^r = \mathbf{E}^r_a$ and $\mathbf{E}^r = \mathbf{E}^r_b$.

As a simple example, let us consider normal incidence, $\mathbf{k}^i = -\mathbf{n}k_o$. In this case, the reflection dyadic (3.209) can be expanded as

$$\overline{\overline{\mathsf{R}}} = \frac{1}{\mathbf{a}_t \cdot \mathbf{b}_t}((\mathbf{n} \times \mathbf{a}_t)(\mathbf{n} \times \mathbf{b}_t) - \mathbf{b}_t\mathbf{a}_t). \tag{3.211}$$

Assuming an incident field polarized as

$$\mathbf{E}^i = E(\cos\alpha\mathbf{b}_t + \sin\alpha(\mathbf{n} \times \mathbf{a}_t)), \tag{3.212}$$

the reflected field becomes

$$\begin{aligned} \mathbf{E}^r &= \frac{E}{\mathbf{a}_t \cdot \mathbf{b}_t}(\sin\alpha(\mathbf{n} \times \mathbf{a}_t)(\mathbf{n} \times \mathbf{b}_t) \cdot (\mathbf{n} \times \mathbf{a}_t) - \cos\alpha\mathbf{b}_t(\mathbf{a}_t \cdot \mathbf{b}_t)) \\ &= -E(\cos\alpha\mathbf{b}_t - \sin\alpha(\mathbf{n} \times \mathbf{a}_t)). \end{aligned} \tag{3.213}$$

Assuming $\mathbf{a}_t$ and $\mathbf{b}_t$ are unit vectors, from the product

$$\mathbf{E}^i \cdot \mathbf{E}^r = -E^2(\cos^2\alpha - \sin^2\alpha) = -E^2\cos 2\alpha, \tag{3.214}$$

we note that, choosing $\cos 2\alpha = 0$, the reflected field appears cross polarized. The same is valid for the SH boundary, $\mathbf{b}_t = \mathbf{a}_t$.

Matched Waves at GSH Boundary

The expression (3.210) fails when $\Delta(\mathbf{k}) = 0$. From (3.201) we obtain the corresponding condition for the wave vector of a possible matched wave,

$$(\mathbf{a}_t \times \mathbf{k}) \cdot (\mathbf{b}_t \times \mathbf{k}) = (\mathbf{a}_t \cdot \mathbf{b}_t)k_o^2 - (\mathbf{a}_t \cdot \mathbf{k}_t)(\mathbf{b}_t \cdot \mathbf{k}_t) = 0. \tag{3.215}$$

The same dispersion equation can be obtained by starting from (3.72) and substituting the expression (3.181) with $\delta \to 0$.

Assuming $\mathbf{a}_t \cdot \mathbf{a}_t \neq 0$ and $\mathbf{b}_t \cdot \mathbf{b}_t \neq 0$, without losing generality, we can set

$$\mathbf{a}_t = \mathbf{e}_1\cos\alpha + \mathbf{e}_2\sin\alpha, \tag{3.216}$$

$$\mathbf{b}_t = \mathbf{e}_1\cos\alpha - \mathbf{e}_2\sin\alpha, \tag{3.217}$$

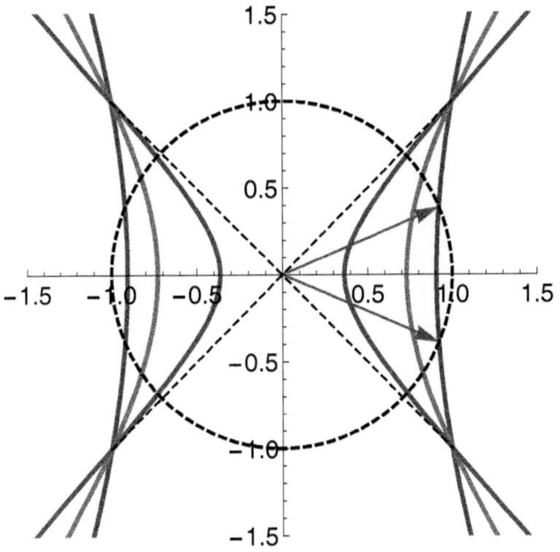

Figure 3.10: Dispersion diagrams depicting $\mathbf{k}_t/k_o$ for waves matched to a GSH boundary for three different pairs of unit vectors $\mathbf{a}_t$ and $\mathbf{b}_t$ corresponding to values of $\alpha = 0.75, 0.6$ and 0.5. Real vectors $\mathbf{a}_t$ and $\mathbf{b}_t$ of unit magnitude and corresponding to $\alpha = 0.5$ are shown in the diagram. The curves have common points at $\varphi = \pm\pi/4 + n\pi/2$, $k_t = k_o\sqrt{2}$. Points outside the unit circle are associated with imaginary values of the normal component $k_n = \sqrt{k_o^2 - k_t^2}$.

in terms of an angle α. Expressing also

$$\mathbf{k}_t = k_t(\mathbf{e}_1 \cos\varphi + \mathbf{e}_2 \sin\varphi), \tag{3.218}$$

the dispersion equation (3.215) can be given the form

$$k_t^2(\varphi) = k_o^2 \frac{\cos 2\alpha}{\cos(\varphi + \alpha)\cos(\varphi - \alpha)}, \tag{3.219}$$

details of which are left as an exercise.

Polarizations of the tangential field components for the matched waves are obtained from the GSH conditions (3.180) as

$$\mathbf{E}_t \sim \mathbf{n} \times \mathbf{a}_t, \qquad \mathbf{H}_t \sim \mathbf{n} \times \mathbf{b}_t. \tag{3.220}$$

Let us consider some special cases.

- For the SH boundary with $\mathbf{a}_t = \mathbf{b}_t = \mathbf{e}_1$, the dispersion equation (3.215) is reduced to $(\mathbf{k} \cdot \mathbf{e}_1)^2 = k_o^2$. For a real solution we must have $\mathbf{k} = \pm\mathbf{e}_1 k_o$. Thus, such a matched wave is a lateral wave propagating along the vector $\mathbf{a}_t$.

- For $\alpha = \pi/4$ we have $\mathbf{a}_t \cdot \mathbf{b}_t = 0$. In this case, (3.215) reduces to $(\mathbf{a}_t \cdot \mathbf{k})(\mathbf{b}_t \cdot \mathbf{k}) = 0$ which corresponds to the two dashed lines in the dispersion diagram, Fig. 3.10. Thus, $\mathbf{k}_t$ of the matched wave may have any magnitude.

- The matched wave is a lateral wave with $k_t = k_o$ for any α when $\varphi = \pm\alpha$, i.e., when $\mathbf{k} = k_t$ is along $\pm\mathbf{a}_t$ or $\pm\mathbf{b}_t$.

- From (3.219) one may notice that, choosing $\varphi = \pi/4 + n\pi$, we have $k_t^2 = 2k_o^2$ for any α, i.e., for any vectors $\mathbf{a}_t$ and $\mathbf{b}_t$. In Figure 3.10 this is evidenced by the common crossing points of the dispersion curves. Because the special points appear outside the unit circle, the corresponding matched waves have imaginary k_n values.

3.9 Duality Transformation of Impedance Boundaries

Applying the duality transformation (1.12), with transformation parameters satisfying $AD - BC = 1$, to the fields in the impedance boundary condition (3.1), yields

$$(D\eta_o\overline{\overline{\mathsf{I}}}_t + C\overline{\overline{\mathsf{Z}}}_t \times \mathbf{n}) \cdot \mathbf{E}_{td} = (B\eta_o\overline{\overline{\mathsf{I}}}_t + A\overline{\overline{\mathsf{Z}}}_t \times \mathbf{n}) \cdot \eta_o\mathbf{H}_{td}. \tag{3.221}$$

This can be written as

$$\mathbf{E}_{td} = \overline{\overline{\mathsf{Z}}}_{td} \cdot (\mathbf{n} \times \mathbf{H}_{td}), \tag{3.222}$$

with the dual of the impedance- dyadic $\overline{\overline{\mathsf{Z}}}_t$ represented by

$$\overline{\overline{\mathsf{Z}}}_{td} = \eta_o(D\eta_o\overline{\overline{\mathsf{I}}}_t + C\overline{\overline{\mathsf{Z}}}_t \times \mathbf{n})^{-1} \cdot (-B\eta_o\overline{\overline{\mathsf{I}}}_t \times \mathbf{n} + A\overline{\overline{\mathsf{Z}}}_t). \tag{3.223}$$

Applying the rule (3.9) for the 2D inverse, we can expand

$$\begin{aligned}
\overline{\overline{\mathsf{Z}}}_{td} &= \frac{\eta_o}{\Delta}((D\eta_o\overline{\overline{\mathsf{I}}}_t + C\overline{\overline{\mathsf{Z}}}_t \times \mathbf{n})^T {}^{\times}_{\times}\mathbf{nn}) \cdot (-B\eta_o\overline{\overline{\mathsf{I}}}_t \times \mathbf{n} + A\overline{\overline{\mathsf{Z}}}_t) \\
&= -\frac{\eta_o}{\Delta}(BD\eta_o^2\overline{\overline{\mathsf{I}}} \times \mathbf{n} + BC\overline{\overline{\mathsf{Z}}}_t^T - AD\overline{\overline{\mathsf{Z}}}_t + AC\overline{\overline{\mathsf{Z}}}_t^T \cdot (\mathbf{n} \times \overline{\overline{\mathsf{Z}}}_t)) \\
&= -\frac{\eta_o}{\Delta}((BD\eta_o^2 + AC\mathrm{det}\overline{\overline{\mathsf{Z}}}_t)\mathbf{n} \times \overline{\overline{\mathsf{I}}} + BC\overline{\overline{\mathsf{Z}}}_t^T - AD\overline{\overline{\mathsf{Z}}}_t),
\end{aligned} \tag{3.224}$$

where we have made use of the identity (see Appendix C)

$$\overline{\overline{\mathsf{A}}}_t^T \cdot (\mathbf{n} \times \overline{\overline{\mathsf{A}}}_t) = \mathrm{det}_t\overline{\overline{\mathsf{A}}}_t(\mathbf{n} \times \overline{\overline{\mathsf{I}}}), \tag{3.225}$$

for $\overline{\overline{\mathsf{A}}}_t = \overline{\overline{\mathsf{Z}}}_t$. In the expression (3.224), we must substitute

$$\begin{aligned}
\Delta &= \mathrm{det}_t(D\eta_o\overline{\overline{\mathsf{I}}}_t + C\overline{\overline{\mathsf{Z}}}_t \times \mathbf{n}) \\
&= D^2\eta_o^2 + CD\eta_o\mathrm{tr}(\overline{\overline{\mathsf{Z}}}_t \times \mathbf{n}) + C^2\mathrm{det}_t\overline{\overline{\mathsf{Z}}}_t.
\end{aligned} \tag{3.226}$$

The dual-impedance dyadic (3.224) can be further expanded by introducing the decomposition (3.41),

$$\overline{\overline{\mathsf{Z}}}_t = Z_s \overline{\overline{\mathsf{I}}}_t + Z_n \mathbf{n} \times \overline{\overline{\mathsf{I}}} + \overline{\overline{\mathsf{Z}}}_{ta}, \tag{3.227}$$

and applying the properties (proofs left as exercises)

$$\mathrm{tr}(\overline{\overline{\mathsf{Z}}}_t \times \mathbf{n}) = -2Z_n, \quad \mathrm{det}_t \overline{\overline{\mathsf{Z}}}_t = Z_s^2 + Z_n^2 + \mathrm{det}_t \overline{\overline{\mathsf{Z}}}_{ta}, \tag{3.228}$$

$$\Delta = (D\eta_o - CZ_n)^2 + C^2(Z_s^2 + \mathrm{det}_t \overline{\overline{\mathsf{Z}}}_{ta}). \tag{3.229}$$

Combining (3.224) and (3.229) we arrive at the duality transformation of the dyadic $\overline{\overline{\mathsf{Z}}}_t$, which can be decomposed as

$$\overline{\overline{\mathsf{Z}}}_{dt} = Z_{ds} \overline{\overline{\mathsf{I}}}_t + Z_{dn} \mathbf{n} \times \overline{\overline{\mathsf{I}}}_t + \overline{\overline{\mathsf{Z}}}_{dta}, \tag{3.230}$$

with

$$Z_{ds} = \frac{\eta_o^2}{\Delta} Z_s, \tag{3.231}$$

$$Z_{dn} = -\frac{\eta_o}{\Delta}((AZ_n - B\eta_o)(CZ_n - D\eta_o) + AC(Z_s^2 + \mathrm{det}_t \overline{\overline{\mathsf{Z}}}_{ta})), \tag{3.232}$$

$$\overline{\overline{\mathsf{Z}}}_{dta} = \frac{\eta_o^2}{\Delta} \overline{\overline{\mathsf{Z}}}_{ta}. \tag{3.233}$$

A few conclusions can be drawn from the above results,

- A PEMC boundary $\overline{\overline{\mathsf{Z}}}_t = Z_n \mathbf{n} \times \overline{\overline{\mathsf{I}}}$ is transformed to a PEMC boundary $\overline{\overline{\mathsf{Z}}}_{dt} = Z_{dn} \mathbf{n} \times \overline{\overline{\mathsf{I}}}$ with

$$Z_{dn} = -\eta_o \frac{AZ_n - B\eta_o}{CZ_n - B\eta_o}. \tag{3.234}$$

  Writing $Z_n = 1/M$ and $Z_{dn} = 1/M_d$, (3.234) can be seen to be equivalent to (1.16).

- An isotropic boundary with $\overline{\overline{\mathsf{Z}}}_{ta} = 0$ is transformed to an isotropic boundary $\overline{\overline{\mathsf{Z}}}_{dt} = Z_{ds} \overline{\overline{\mathsf{I}}}_t + Z_{dn} \mathbf{n} \times \overline{\overline{\mathsf{I}}}$ with

$$Z_{ds} = \frac{\eta_o}{\Delta} \eta_o Z_s, \tag{3.235}$$

$$Z_{dn} = -\frac{\eta_o}{\Delta}(ACZ_s^2 + (AZ_n - B\eta_o)(CZ_n - D\eta_o)), \tag{3.236}$$

$$\Delta = C^2 Z_s^2 + (CZ_n - D\eta_o)^2. \tag{3.237}$$

  Thus, the class of isotropic boundaries is closed in the duality transformation.

- A perfectly anisotropic boundary ($Z_s = Z_n = 0$) is transformed to

$$\overline{\overline{\mathsf{Z}}}_{dta} = \frac{\eta_o}{C^2 \mathrm{det}_t \overline{\overline{\mathsf{Z}}}_{ta} + D^2 \eta_o^2}(\eta_o \overline{\overline{\mathsf{Z}}}_{ta} - (BC\eta_o^2 + AC \mathrm{det}_t \overline{\overline{\mathsf{Z}}}_{ta}) \mathbf{n} \times \overline{\overline{\mathsf{I}}}), \tag{3.238}$$

which, when the transformation parameters satisfy $BD\eta_o^2 + AC\mathrm{det}_t\overline{\overline{\mathsf{Z}}}_{ta} = 0$, is transformed to a similar perfectly anisotropic boundary with different magnitude.

- By choosing the transformation parameters properly, it is possible to transform any impedance dyadic to a symmetric impedance dyadic. As an example, it is possible to transform the PEMC boundary to a simple-isotropic boundary.

Duality transformation is useful when a problem can be transformed to a problem which can be solved with less effort. In the case of transforming boundary conditions, this requires that the medium is not transformed to a more complicated one. It is known from (1.19) that a simple isotropic medium (ϵ_o, μ_o) is invariant in the transformation defined by a parameter φ as $A = D = \cos\varphi$ and $B = -C = \sin\varphi$. It is left as an exercise to show that, in terms of this transformation, a general isotropic boundary with impedance dyadic $\overline{\overline{\mathsf{Z}}}_t = Z_s\overline{\overline{\mathsf{I}}}_t + Z_n\mathbf{n}\times\overline{\overline{\mathsf{I}}}$ can be transformed to a simple isotropic boundary with impedance dyadic $\overline{\overline{\mathsf{Z}}}_{td} = Z_{sd}\overline{\overline{\mathsf{I}}}_t$.

3.10 Realization of Impedance Boundaries

It is known that any impedance boundary can be realized by a slab of certain medium, called wave-guiding medium [51]. Such a medium acts as a continuous distribution of parallel waveguides along which a wave can travel without interaction in the transverse direction. How to realize such a medium is a problem not discussed here. There exist other possible realizations using metamaterial structures for some special cases of impedance boundaries, for example, in [76].

The wave-guiding medium assumes uniaxial anisotropy with axis defined by the unit vector $\mathbf{e}_3$, defined by dyadic medium parameters of the form

$$\overline{\overline{\epsilon}} = \epsilon_o\overline{\overline{\mathsf{A}}}_t + \epsilon_3\mathbf{e}_3\mathbf{e}_3, \tag{3.239}$$

$$\overline{\overline{\mu}} = \mu_o\overline{\overline{\mathsf{A}}}_t^T + \mu_3\mathbf{e}_3\mathbf{e}_3, \tag{3.240}$$

where $\overline{\overline{\mathsf{A}}}_t$ is a 2D dyadic. Such a medium has been called affinely uniaxial in the past [35] since the dyadic parameters satisfy a relation of the form[1].

$$\overline{\overline{\epsilon}}^{-1}\cdot\overline{\overline{\mu}}^T = A\overline{\overline{\mathsf{I}}}_t + B\mathbf{e}_3\mathbf{e}_3, \tag{3.241}$$

the right side of which is a uniaxial dyadic. We assume that the dyadic $\overline{\overline{\mathsf{A}}}_t$ satisfies $\mathrm{det}_t\overline{\overline{\mathsf{A}}}_t \neq 0$, whence it has the 2D inverse (3.9).

Choosing $\epsilon_3 = \epsilon_o/\delta$ and $\mu_3 = \mu_o/\delta$ and assuming $\delta \to 0$, both ϵ_3 and μ_3 become infinite in magnitude. In this case, $E_3 = \delta D_3/\epsilon_o$ and $H_3 = \delta B_3/\mu_o$ will decrease in magnitude and, for $\delta = 0$, eventually become zero. Thus, the fields

[1] Note that the transpose sign is missing in eq. (5.60) of [35]

$\mathbf{E} = \mathbf{E}_t$ and $\mathbf{H} = \mathbf{H}_t$ have no axial components in such a medium, while D_3 and B_3 may have finite magnitudes.

Considering propagating waves of the form

$$\mathbf{E}_t(\mathbf{r}) = \mathbf{E}_t f(\mathbf{r}_t) e^{-j\beta x_3}, \tag{3.242}$$

$$\mathbf{H}_t(\mathbf{r}) = \mathbf{H}_t f(\mathbf{r}_t) e^{-j\beta x_3}, \tag{3.243}$$

where $\mathbf{r} = \mathbf{r}_t + \mathbf{e}_3 x_3$ and $f(\mathbf{r}_t)$ is any function defining the variation of fields in the transverse dimension, the axial and transverse components of the Maxwell equations can be expressed as

$$\mathbf{e}_3 \cdot (\nabla f(\mathbf{r}_t) \times \mathbf{E}_t) = -j\omega B_3(\mathbf{r}_t), \tag{3.244}$$

$$\mathbf{e}_3 \cdot (\nabla f(\mathbf{r}_t) \times \mathbf{H}_t) = j\omega D_3(\mathbf{r}_t), \tag{3.245}$$

$$\beta \mathbf{e}_3 \times \mathbf{E}_t = k_o \overline{\overline{\mathsf{A}}}_t^T \cdot \eta_o \mathbf{H}_t, \tag{3.246}$$

$$\beta \mathbf{e}_3 \times \eta_o \mathbf{H}_t = -k_o \overline{\overline{\mathsf{A}}}_t \cdot \mathbf{E}_t. \tag{3.247}$$

Applying (3.246) and (3.247), we can expand

$$\begin{aligned} \beta^2 \mathbf{E}_t &= -\beta^2 \mathbf{e}_3 \times (\mathbf{e}_3 \times \mathbf{E}_t) \\ &= -\beta k_o \mathbf{e}_3 \times \overline{\overline{\mathsf{A}}}_t^T \cdot \eta_o \mathbf{H}_t \\ &= \beta k_o (\mathbf{e}_3 \times \overline{\overline{\mathsf{A}}}_t^T \times \mathbf{e}_3) \cdot \mathbf{e}_3 \times \eta_o \mathbf{H}_t \\ &= k_o^2 (\overline{\overline{\mathsf{A}}}_t^T {}^\times_\times \mathbf{e}_3 \mathbf{e}_3) \cdot \overline{\overline{\mathsf{A}}}_t \cdot \mathbf{E}_t \\ &= k_o^2 (\mathrm{det}_t \overline{\overline{\mathsf{A}}}_t) \mathbf{E}_t. \end{aligned} \tag{3.248}$$

At the last step we have made use of (3.9).

In conclusion, the equation for β becomes

$$\beta^2 - k_o^2 \mathrm{det}_t \overline{\overline{\mathsf{A}}}_t = 0, \tag{3.249}$$

without any restriction to $\mathbf{E}_t$. Thus, in a wave-guiding medium defined by (3.239) and (3.240), for any transverse vector $\mathbf{E}_t$ and function $f(\mathbf{r}_t)$ there are two waves propagating with

$$\beta_\pm = \pm\beta, \quad \beta = k_o \sqrt{\mathrm{det}_t \overline{\overline{\mathsf{A}}}_t}. \tag{3.250}$$

Because the transverse distribution of the fields in the wave are unchanged, one can imagine the wave-guiding medium as consisting of tiny parallel waveguides with no tranverse coupling.

For given $\mathbf{E}_{t\pm}$, and $\beta_\pm$ from (3.250), the fields $\mathbf{H}_{t\pm}$ are obtained from (3.247) as

$$\begin{aligned} \eta_o \mathbf{H}_{t\pm} &= \frac{k_o}{\beta_\pm} \mathbf{e}_3 \times \overline{\overline{\mathsf{A}}}_t \cdot \mathbf{E}_{t\pm} \\ &= \pm \frac{1}{\sqrt{\mathrm{det}_t \overline{\overline{\mathsf{A}}}_t}} (\overline{\overline{\mathsf{A}}}_t {}^\times_\times \mathbf{e}_3 \mathbf{e}_3) \cdot (\mathbf{e}_3 \times \mathbf{E}_{t\pm}) \\ &= \pm \frac{\beta}{k_o} \overline{\overline{\mathsf{A}}}_t^{-1T} \cdot (\mathbf{e}_3 \times \mathbf{E}_{t\pm}), \end{aligned} \tag{3.251}$$

while the components $B_{3\pm}$ and $D_{3\pm}$ of the two waves can be found from (3.244) and (3.245).

Omitting the function $f(\mathbf{r}_t)$, the combined fields are

$$\mathbf{E}_t(x_3) = \mathbf{E}_{t+}e^{-j\beta x_3} + \mathbf{E}_{t-}e^{j\beta x_3}, \tag{3.252}$$

$$\eta_o\mathbf{H}_t(x_3) = \sqrt{\mathrm{det}_t\overline{\overline{\mathsf{A}}}_t}(\overline{\overline{\mathsf{A}}}_t^{-1T} \times \mathbf{e}_3)\cdot(\mathbf{E}_{t+}e^{-j\beta x_3} - \mathbf{E}_{t-}e^{j\beta x_3}). \tag{3.253}$$

Let us consider a slab of wave-guiding medium $0 > x_3 > -d$, with PEC termination at $x_3 = -d$. Requiring $\mathbf{E}_t(-d) = 0$, yields the relation

$$\mathbf{E}_{t+} = -\mathbf{E}_{t-}e^{-2j\beta d}, \tag{3.254}$$

whence the fields at $x_3 = 0$ can be expressed as

$$\mathbf{E}_t(0) = (1 - e^{-2j\beta d})\mathbf{E}_{t-} = 2je^{-j\beta d}\sin\beta d\,\mathbf{E}_{t-}, \tag{3.255}$$

$$\begin{aligned}\eta_o\mathbf{H}_t(0) &= -\frac{\beta}{k_o}(1+e^{-2j\beta d})(\overline{\overline{\mathsf{A}}}_t^{-1T}\times\mathbf{e}_3)\cdot\mathbf{E}_{t-}\\ &= j\frac{\beta}{k_o}\cot\beta d\,\overline{\overline{\mathsf{A}}}_t^{-1T}\cdot(\mathbf{e}_3\times\mathbf{E}_t(0)).\end{aligned} \tag{3.256}$$

For $\mathbf{e}_3 = \mathbf{n}$, the relation (3.256) is of the form (3.5),

$$\mathbf{H}_t = -\overline{\overline{\mathsf{Y}}}_t\cdot(\mathbf{n}\times\mathbf{E}_t), \tag{3.257}$$

where $\overline{\overline{\mathsf{Y}}}_t$ can be recognized as the admittance dyadic $\overline{\overline{\mathsf{Y}}}_t$ of an impedance boundary. It is related to the parameters of the slab of the wave-guiding medium as

$$\eta_o\overline{\overline{\mathsf{Y}}}_t = j\frac{\beta}{k_o}\cot\beta d\,\overline{\overline{\mathsf{A}}}_t^{-1T}, \quad \beta = k_o\sqrt{\mathrm{det}_t\overline{\overline{\mathsf{A}}}_t}. \tag{3.258}$$

In conclusion, the interface of a slab of a wave-guiding medium with PEC termination may serve as an impedance boundary of admittance dyadic $\overline{\overline{\mathsf{Y}}}_t$ defined by (3.258). This analysis suggests that any given impedance boundary can be realized by a layer of suitable wave-guiding medium, because the dyadic $\overline{\overline{\mathsf{A}}}_t$ defining the medium can be chosen from (3.258) to match a required boundary admittance dyadic $\overline{\overline{\mathsf{Y}}}_t$.

Let us consider some special cases of the condition (3.258).

- For $d \to 0$, or $\beta d = n\pi$, the boundary becomes that of PEC with infinite magnitude of $\overline{\overline{\mathsf{Y}}}_t$.
- For $\beta d = n\pi + \pi/2$, the interface becomes a PMC boundary with $\overline{\overline{\mathsf{Y}}}_t = 0$.
- The PEMC boundary condition involving an antisymmetric surface admittance dyadic $\overline{\overline{\mathsf{Y}}}_t = M\mathbf{e}_3 \times \overline{\overline{\mathsf{I}}}$ can be realized by a slab of anisotropic wave-guiding medium defined by medium dyadics

$$\overline{\overline{\epsilon}} = \epsilon_t\mathbf{n}\times\overline{\overline{\mathsf{I}}} + \epsilon_3\mathbf{e}_3\mathbf{e}_3, \tag{3.259}$$

$$\overline{\overline{\mu}} = -\mu_t\mathbf{n}\times\overline{\overline{\mathsf{I}}} + \mu_3\mathbf{e}_3\mathbf{e}_3, \tag{3.260}$$

with $\epsilon_3 \to \infty$ and $\mu_3 \to \infty$. The relation between ϵ_t, μ_t and the antisymmetric dyadic $\overline{\overline{\mathsf{A}}}_t = A\mathbf{n} \times \overline{\overline{\mathsf{I}}}$ can be found from (3.258). Such a realization requires a wave-guiding medium with gyrotropy in both permittivity and permeability [47]. However, it is also possible to make a realization with gyrotropic permeability and permittivity with simple transverse isotropy. Such a realization in terms of cylinders of high μ and conductivity (complex ϵ) in a base medium with gyrotropic $\overline{\overline{\mu}}$ (requiring a static magnetic field parallel to the cylinder axes) is depicted in Figure 3.11. Diameters of the cylinders and their distances from one another are much smaller than the wavelength.

Practical realizations of the PEMC boundary in terms of metasurface have been discussed in [5, 11, 90].

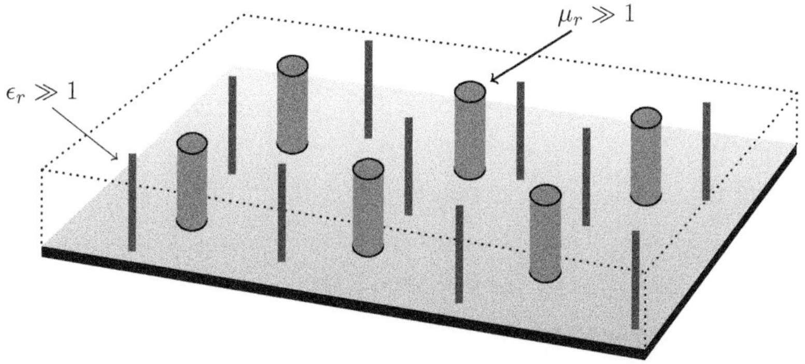

Figure 3.11: A subwavelength view of a possible realization of the PEMC boundary with high-permeability and high-permittivity cylinders in a gyrotropic substrate backed by a PEC plane.

3.11 Problems

3.1 Expand $\mathrm{det}_t\overline{\overline{\mathsf{Z}}}_t$ for $\overline{\overline{\mathsf{Z}}}_t = Z_s\overline{\overline{\mathsf{I}}}_s + Z_n\mathbf{n} \times \overline{\overline{\mathsf{I}}}_t + \overline{\overline{\mathsf{Z}}}_{ta}$.

3.2 Show that, for the PEMC boundary, the reflection dyadic (3.48) reduces to the form (2.30).

3.3 Show that, for the simple-isotropic boundary dyadic (3.74), the reflection dyadic (3.75) can be obtained from the expression (3.50).

3.4 Show that, substituting $\mathbf{E}^i = \mathbf{E}^i_a$ and $\mathbf{E}^i = \mathbf{E}^i_b$ in (3.210), the fields satisfy the GSH boundary conditions in both cases.

3.5 Find conditions for which the boundary admittance dyadic $\overline{\overline{\mathsf{Y}}}_t$ and the impedance dyadic $\overline{\overline{\mathsf{Z}}}_t$ are 2D inverses of each other.

3.6 For the 2D eigenproblem

$$\overline{\overline{\mathsf{A}}}_t \cdot \mathbf{x}_t = \lambda \overline{\overline{\mathsf{C}}}_t \cdot \mathbf{x}_t,$$

show that the eigenvalue satisfies the equation

$$(\overline{\overline{\mathsf{A}}}_t - \lambda \overline{\overline{\mathsf{C}}}_t)^{(2)} = 0.$$

Assuming λ known, show that the corresponding eigenvector is obtained from

$$\mathbf{x}_t = \mathbf{a}_t \cdot (\overline{\overline{\mathsf{A}}}_t - \lambda \overline{\overline{\mathsf{C}}}_t) \times \mathbf{n}$$

for any $\mathbf{a}_t$ yielding $\mathbf{x}_t \neq 0$.

3.7 Show that the TE$_a$ and TM$_b$ decomposition of a plane wave (3.195) - (3.198) yields two independent plane waves, each satisfying plane-wave equations.

3.8 Show that it is possible to choose an orthonormal 2D vector basis $\mathbf{e}'_1, \mathbf{e}'_2$ so that a given perfectly anisotropic (PA) boundary impedance dyadic $\overline{\overline{\mathsf{Z}}}_{ta}$ (3.162) can expressed in the form $\overline{\overline{\mathsf{Z}}}_{ta} = Z'_K \overline{\overline{\mathsf{K}}}'$, with $\overline{\overline{\mathsf{K}}}' = \mathbf{e}'_1\mathbf{e}'_1 - \mathbf{e}'_2\mathbf{e}'_2$.

3.9 Derive the expressions (3.167) and (3.166).

3.10 Derive conditions for the incident wave matched to a boundary defined by an impedance dyadic $\overline{\overline{\mathsf{Z}}}_t$ starting from the boundary condition (3.1).

3.11 Derive the expression (3.137) for the reflection dyadic corresponding to the general isotropic boundary.

3.12 Show that the general isotropic boundary dyadic $\overline{\overline{\mathsf{Z}}}_t = Z_s \overline{\overline{\mathsf{I}}}_t + Z_n \mathbf{n} \times \overline{\overline{\mathsf{I}}}$ can be transformed to a simple isotropic boundary dyadic $\overline{\overline{\mathsf{Z}}}_{td} = Z_{sd} \overline{\overline{\mathsf{I}}}_t$ through the duality transformation defined by parameters $A = D = \cos\varphi$, $B = -C = \sin\varphi$.

3.13 Find the solution for the eigenproblem (3.142).

3.14 Derive the dispersion equation (3.219) for the matched wave of the GSH boundary. Find the condition for a lateral matched wave, $k_t = k_o$.

3.15 Show that if $\overline{\overline{\mathsf{Z}}}_t = \overline{\overline{\mathsf{Z}}}_{ta}$ is a perfectly anisotropic impedance dyadic, $\mathbf{n} \times \overline{\overline{\mathsf{Z}}}_t$ and $\overline{\overline{\mathsf{Z}}}_t \times \mathbf{n}$ are also perfectly anisotropic impedance dyadics.

3.16 Derive the properties (3.228).

3.17 Derive the expressions (3.50) and (3.51) defining the reflection dyadic for an impedance boundary.

3.18 Derive the dispersion equation (3.72) for matched waves associated with the impedance boundary defined by the impedance dyadic $\overline{\overline{\mathsf{Z}}}_t$.

3.19 Show that, in the representation (3.17), (3.18), the condition of reciprocity for the impedance boundary can be expressed as

$$\mathbf{a}_{1t} \times \mathbf{a}_{2t} = \mathbf{b}_{1t} \times \mathbf{b}_{2t}.$$

3.20 Writing $\mathbf{E}_t^r = R\mathbf{E}_t^i$, find the possible reflection coefficients R for the GSH boundary starting from the boundary conditions (3.180).

3.21 Expand the impedance dyadic of the GSH boundary (3.181) in its simple-isotropic, antisymmetric and anisotropic parts as (3.41), by first deriving the expansion

$$\overline{\overline{\mathsf{I}}}_t = \frac{1}{\mathbf{a}_t \cdot \mathbf{b}_t}(\mathbf{b}_t\mathbf{a}_t + (\mathbf{n} \times \mathbf{a}_t)(\mathbf{n} \times \mathbf{b}_t)).$$

3.22 Show that, in the representation (3.17), (3.18), the impedance dyadic $\overline{\overline{\mathsf{Z}}}_t$ takes the decomposed form (3.41) with (3.43), (3.44) and (3.45) inserted.

3.23 Assuming real $\mathbf{k}^i$, show by expanding $\mathbf{E}^i = \mathbf{E}^i_{TE} + \mathbf{E}^r_{TM}$ that, for imaginary Z_s, the fields reflected from a simple-impedance boundary with imaginary Z_s, satisfy

$$|\mathbf{E}_t^r|^2 = |\mathbf{E}_t^i|^2,$$

whence the boundary is lossless.

3.24 Show that the dispersion equation (3.215) can be obtained from (3.72) when the GSH boundary is defined in terms of the impedance dyadic (3.181) with $\delta \to 0$.

Chapter 4

DB Boundary

4.1 Boundary Conditions Involving Normal Field Components

A class of electromagnetic boundaries not belonging to that of impedance boundaries can be defined in terms of normal field components of fields as

$$\mathbf{n} \cdot \mathbf{D} = 0, \qquad \mathbf{n} \cdot \mathbf{B} = 0. \tag{4.1}$$

Such a class was originally introduced by Victor Rumsey in 1959 [87]. Conditions of the form (4.1) did not raise engineering attention before the 21th century when they found application in cloaking problems [27, 114, 116, 117, 122]. Conditions (4.1) were simultaneously introduced as forming an essential part of basic electromagnetic theory [54, 55, 58, 59], and were labeled as DB conditions. Over the intervening years, mathematicians have also paid attention to existence and uniqueness conditions in boundary conditions of the form (4.1) [118, 85, 86, 16].

The DB boundary is self dual, because the conditions (4.1) are invariant in any duality transformation. In fact, the two conditions (4.1) can be replaced by any linear combinations as

$$\begin{pmatrix} A & B \\ C & D \end{pmatrix} \begin{pmatrix} \mathbf{n} \cdot \mathbf{D} \\ \mathbf{n} \cdot \mathbf{B} \end{pmatrix} = \begin{pmatrix} \mathbf{n} \cdot (A\mathbf{D} + B\mathbf{B}) \\ \mathbf{n} \cdot (C\mathbf{D} + D\mathbf{B}) \end{pmatrix} = \begin{pmatrix} 0 \\ 0 \end{pmatrix}, \tag{4.2}$$

when $AD - BC \neq 0$. The self-dual property has the interesting consequence that any object with DB boundary and certain symmetry appears invisible for monostatic radar [62].

The form (4.1) is not alone in representing boundary conditions in terms of normal field components. For a planar boundary with constant unit normal vector $\mathbf{n}$, we can define

$$\partial_n D_n = (\mathbf{n} \cdot \nabla)(\mathbf{n} \cdot \mathbf{D}) = 0, \qquad \partial_n B_n = (\mathbf{n} \cdot \nabla)(\mathbf{n} \cdot \mathbf{B}) = 0. \tag{4.3}$$

which have been called conditions of the D'B' boundary [60, 61]. It turns out that, for a boundary surface not necessarily planar, the D'B' conditions must actually be written in the form [60]

$$\nabla\cdot(\mathbf{nn}\cdot\mathbf{D}) = 0, \qquad \nabla\cdot(\mathbf{nn}\cdot\mathbf{B}) = 0, \tag{4.4}$$

where the unit vector $\mathbf{n}(\mathbf{r})$ is associated to a boundary surface $f(\mathbf{r}) = 0$ as $\mathbf{n}(\mathbf{r}) = \nabla f(\mathbf{r})/\sqrt{\nabla f(\mathbf{r})\cdot\nabla f(\mathbf{r})}$. For example, for a spherical boundary, the D'B' boundary conditions (4.4) take the form

$$\partial_r(r^2 D_r) = 0, \qquad \partial_r(r^2 B_r) = 0, \tag{4.5}$$

with $D_r = \mathbf{e}_r\cdot\mathbf{D}$, $B_r = \mathbf{e}_r\cdot\mathbf{B}$. Unlike the conditions of the DB boundary, those of the D'B' boundary are not local, since they involve a differential operator. However, the D'B' conditions are also self dual.

The previous DB and D'B' conditions can be used to define two other sets of boundary conditions involving normal field components as

$$\mathbf{n}\cdot\mathbf{D} = 0, \qquad \nabla\cdot(\mathbf{nn}\cdot\mathbf{B}) = 0, \tag{4.6}$$

and

$$\nabla\cdot(\mathbf{nn}\cdot\mathbf{D}) = 0, \qquad \mathbf{n}\cdot\mathbf{B} = 0, \tag{4.7}$$

which have been respectively labeled as DB' and D'B boundary conditions [60]. They are swapped in duality transformation. It is left as an exercise to show that DB' and D'B boundaries are equivalent to the respective PMC and PEC boundaries.

4.2 Reflection from DB Boundary

Assuming a sum of incident and reflected plane-wave fields, the DB boundary conditions become

$$\begin{aligned} -\omega\mathbf{n}\cdot\mathbf{D} &= \mathbf{n}\cdot(\mathbf{k}^i\times\mathbf{H}^i + \mathbf{k}^r\times\mathbf{H}^r) \\ &= (\mathbf{n}\times\mathbf{k}_t)\cdot(\mathbf{H}_t^i + \mathbf{H}_t^r) = 0, \qquad (4.8) \\ \omega\mathbf{n}\cdot\mathbf{B} &= \mathbf{n}\cdot(\mathbf{k}^i\times\mathbf{E}^i + \mathbf{k}^r\times\mathbf{E}^r) \\ &= (\mathbf{n}\times\mathbf{k}_t)\cdot(\mathbf{E}_t^i + \mathbf{E}_t^r) = 0. \qquad (4.9) \end{aligned}$$

For the simple-isotropic medium we can apply the field relations (1.47), (1.48) and (1.50), whence (4.8) yields

$$\begin{aligned} \eta_o(\mathbf{n}\times\mathbf{k}_t)\cdot(\mathbf{H}_t^i + \mathbf{H}_t^r) &= (\mathbf{n}\times\mathbf{k}_t)\cdot\overline{\overline{\mathsf{J}}}_t\cdot(\mathbf{E}_t^r - \mathbf{E}_t^i) \\ &= \frac{k_o}{k_n}\mathbf{k}_t\cdot(\mathbf{E}_t^r - \mathbf{E}_t^i) = 0. \end{aligned} \tag{4.10}$$

Conditions (4.9) and (4.10) represent two relations between the tangential electric-field components:

$$\mathbf{k}_t\cdot\mathbf{E}_t^r = \mathbf{k}_t\cdot\mathbf{E}_t^i \tag{4.11}$$

$$(\mathbf{n}\times\mathbf{k}_t)\cdot\mathbf{E}_t^r = -(\mathbf{n}\times\mathbf{k}_t)\cdot\mathbf{E}_t^i, \tag{4.12}$$

Excluding normal incidence, $\mathbf{k}_t \cdot \mathbf{k}_t = k_t^2 \neq 0$, the tangential component of the reflected electric field can be expressed in terms of the unit dyadic (1.41) as

$$\begin{aligned} \mathbf{E}_t^r &= \frac{1}{k_t^2}(\mathbf{k}_t\mathbf{k}_t + (\mathbf{n}\times\mathbf{k}_t)(\mathbf{n}\times\mathbf{k}_t))\cdot\mathbf{E}_t^r \\ &= \frac{1}{k_t^2}(\mathbf{k}_t\mathbf{k}_t - (\mathbf{n}\times\mathbf{k}_t)(\mathbf{n}\times\mathbf{k}_t)))\cdot\mathbf{E}_t^i) \\ &= \overline{\overline{\mathsf{R}}}_t\cdot\mathbf{E}_t^i. \end{aligned} \tag{4.13}$$

The reflection dyadic of the DB boundary

$$\overline{\overline{\mathsf{R}}}_t = \frac{1}{k_t^2}(\mathbf{k}_t\mathbf{k}_t - (\mathbf{n}\times\mathbf{k}_t)(\mathbf{n}\times\mathbf{k}_t)), \tag{4.14}$$

differs from the expression of the unit dyadic by the minus sign of the latter dyad. The expression (4.14) has a similarity with the reflection dyadic of the SH boundary, (3.210) with $\mathbf{b}_t = \mathbf{a}_t$, when $\mathbf{a}_t$ and $\mathbf{k}_t$ are parallel vectors. This is why the DB boundary has been sometime called the isotropic soft-and-hard boundary [27].

The eigenvalues of the reflection dyadic can be found by inspection as

$$\overline{\overline{\mathsf{R}}}_t\cdot\mathbf{E}_{t\pm}^i = R_\pm\mathbf{E}_\pm^i, \quad R_\pm = \pm 1, \tag{4.15}$$

with the corresponding tangential eigenfields defined by

$$\mathbf{E}_{t+}^i = A_+\mathbf{k}_t, \qquad \mathbf{E}_{t+}^r = A_+\mathbf{k}_t, \tag{4.16}$$

$$\mathbf{E}_{t-}^i = A_-\mathbf{n}\times\mathbf{k}_t, \qquad \mathbf{E}_{t-}^r = -A_-\mathbf{n}\times\mathbf{k}_t. \tag{4.17}$$

From

$$\mathbf{k}^i\cdot\mathbf{E}_-^i = (\mathbf{n}\cdot\mathbf{k}^i)(\mathbf{n}\cdot\mathbf{E}_-^i) = 0 \quad\Rightarrow\quad \mathbf{n}\cdot\mathbf{E}_-^i = 0, \tag{4.18}$$

it follows that the eigenfield $\mathbf{E}_-^i, \mathbf{H}_-^i$ is a TE field for $\mathbf{n}\cdot\mathbf{k}^i \neq 0$.

The corresponding magnetic eigenfields are obtained from (1.47) and (1.48) as

$$\eta_o\mathbf{H}_{t+}^i = -\frac{k_o}{k_n}A_+\mathbf{n}\times\mathbf{k}_t, \qquad \eta_o\mathbf{H}_{t+}^r = \frac{k_o}{k_n}A_+\mathbf{n}\times\mathbf{k}_t, \tag{4.19}$$

$$\eta_o\mathbf{H}_{t-}^i = \frac{k_n}{k_o}A_-\mathbf{k}_t, \qquad \eta_o\mathbf{H}_{t-}^r = \frac{k_n}{k_o}A_-\mathbf{k}_t. \tag{4.20}$$

From

$$\mathbf{k}^i\cdot\mathbf{H}_+^i = (\mathbf{n}\cdot\mathbf{k}^i)(\mathbf{n}\cdot\mathbf{H}_+^i) = 0, \quad\Rightarrow\quad \mathbf{n}\cdot\mathbf{H}_+^i = 0, \tag{4.21}$$

it follows that the eigenfield $\mathbf{E}_+^i, \mathbf{H}_+^i$ is a TM field for $\mathbf{n}\cdot\mathbf{k}^i \neq 0$.

The total tangential eigenfields at the DB boundary become

$$\mathbf{E}_{t+}^i + \mathbf{E}_{t+}^r = 2A_+\mathbf{k}_t, \qquad \mathbf{H}_{t+}^i + \mathbf{H}_{t+}^r = 0 \tag{4.22}$$

$$\mathbf{E}_{t-}^i + \mathbf{E}_{t-}^r = 0, \qquad \mathbf{H}_{t-}^i + \mathbf{H}_{t-}^r = 2\frac{k_n}{k_o}A_-\mathbf{k}_t. \tag{4.23}$$

From the above analysis we can make the following conclusions.

- From (4.22) and (4.23) we see that any TE wave is an eigenwave which is reflected from the DB boundary as from the PEC boundary satisfying $(\mathbf{E}_-^i + \mathbf{E}_-^r)_t = 0$. Since this is a linear property valid for any $\mathbf{k}$ vector, it is valid for any sum of TE waves and, ultimately, for any TE field which can be expressed as a sum or integral of plane waves. Thus, the DB boundary is equivalent to a PEC boundary for sources which produce a TE-polarized field. For example, for a vertical magnetic current source, which radiates a field with no vertical component of electric field, a horizontal DB plane acts as a PEC plane and can be replaced by a PEC image source.

- Similarly, any TM wave is reflected from the DB boundary as from the PMC boundary satisfying $(\mathbf{H}_+^i + \mathbf{E}_-^r)_t = 0$. For the TM wave the DB boundary appears a the PMC boundary and, for a vertical electric current source, it can be replaced by the PMC image source.

- In conclusion, just like the GSH boundary in Section 3.8, the DB boundary has the property of PEC/PMC equivalence.

Reflection from a DB plane for both eigenpolarizations is visualized by Figures 4.1 and 4.2.

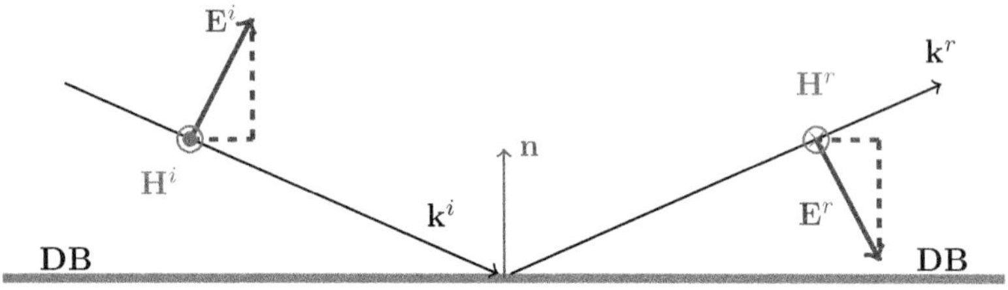

Figure 4.1: The TM-polarized plane-wave reflection from a DB boundary is identical to reflection from a PMC boundary. The normal component of the total electric field (sum of the incident and reflected fields) vanishes. For the magnetic field, the DB condition is automatically satisfied for a TM wave. The magnetic field changes sign in reflection, meaning that the reflection is like that from a PMC boundary in which the magnetic field is short-circuited.

In the preceding analysis we have assumed $\mathbf{k}_t \neq 0$, because otherwise the dyadic $\overline{\overline{\mathsf{R}}}_t$ does not exist. Because a matched wave above a DB boundary in simple-isotropic medium satisfies from (4.1)

$$\mathbf{n} \cdot \mathbf{E} = \mathbf{n} \cdot \mathbf{H} = 0, \tag{4.24}$$

it is actually a TEM wave with $\mathbf{k} = \mathbf{n}k_o$. Since for a normally incident plane wave there is no reflection, a DB boundary cannot be lossless. In fact, assuming

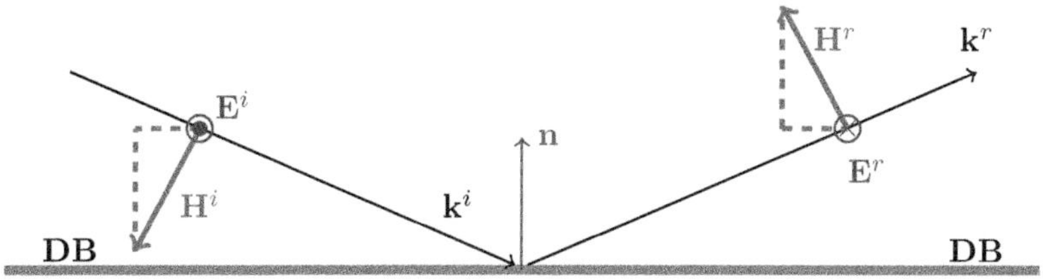

Figure 4.2: The same as in Figure 4.1, for the TE polarized wave (no normal component for the electric field, and the DB condition satisfied by the electric field). The normal component of the reflected magnetic field cancels that of the incident field, meaning that the total tangential electric field vanishes at the boundary (the PEC condition).

$\mathbf{k}^i = -\mathbf{n}k_o$, $\mathbf{E}^i = \mathbf{E}^i_t$ and $\mathbf{H}^i = -\mathbf{n}\times\mathbf{E}^i/\eta_o$, at the DB boundary we obtain

$$-\mathbf{n}\cdot\mathbf{E}^i\times\mathbf{H}^{i*} = \frac{1}{\eta_o}|\mathbf{E}^i|^2. \tag{4.25}$$

whence the DB boundary acts as a perfect absorber for a normally incident plane wave.

4.3 Realization of DB Boundary

The most obvious way to realize a planar surface on which the fields satisfy the DB boundary conditions (4.1) is by an interface of an anisotropic medium with dyadics satisfying $\mathbf{n}\cdot\overline{\overline{\epsilon}} = 0$ and $\mathbf{n}\cdot\overline{\overline{\mu}} = 0$. Such a medium has been called zero axial parameter (ZAP) medium in the past [56]. Because in such a medium the fields satisfy $\mathbf{n}\cdot\mathbf{D} = 0$ and $\mathbf{n}\cdot\mathbf{B} = 0$ everywhere, the DB conditions are satisfied at the interface. Actually, it is sufficient to have just a thin layer of ZAP medium to produce a DB boundary.

Uniaxial Anisotropic Medium

To study the effect of a layer of uniaxial medium which approaches the ZAP medium, let us assume a medium defined by the permittivity and permeability dyadics of the form

$$\overline{\overline{\epsilon}} = \epsilon_t\overline{\overline{\mathsf{I}}}_t + \epsilon_3\mathbf{e}_3\mathbf{e}_3 = \epsilon_t(\overline{\overline{\mathsf{I}}}_t + e\mathbf{e}_3\mathbf{e}_3), \tag{4.26}$$

$$\overline{\overline{\mu}} = \mu_t\overline{\overline{\mathsf{I}}}_t + \mu_3\mathbf{e}_3\mathbf{e}_3 = \mu_t(\overline{\overline{\mathsf{I}}}_t + m\mathbf{e}_3\mathbf{e}_3),, \tag{4.27}$$

where $\mathbf{e}_3$ is a unit vector defining the axis of the medium. m and e denote the relative axial parameters

$$m = \mu_3/\mu_t, \qquad e = \epsilon_3/\epsilon_t. \tag{4.28}$$

All of the four parameters are assumed to have real non-negative values.

Since the fields of a plane wave in any medium satisfy $\mathbf{E}\cdot\mathbf{B} = 0$ and $\mathbf{H}\cdot\mathbf{D} = 0$, for the uniaxial medium we have

$$\epsilon_t \mathbf{E}\cdot\mathbf{B} - \mu_t \mathbf{H}\cdot\mathbf{D} = \epsilon_t\mu_t(m-e)E_3H_3 = 0. \tag{4.29}$$

Excluding the special case $m = e$, or,

$$\mu_t\epsilon_3 - \epsilon_t\mu_3 = 0, \tag{4.30}$$

from (4.29) it follows that any plane wave in the uniaxially anisotropic medium must satisfy either $E_3 = 0$ or $H_3 = 0$. This means that we can consider waves polarized TE and TM to $\mathbf{e}_3$ separately in the medium. Because the components D_3 and B_3 are continuous at an interface normal to $\mathbf{e}_3$, TE and TM polarizations are preserved through any interface orthogonal to $\mathbf{e}_3$. Thus, the two polarizations can be handled separately for layered problems.

Let us assume that a plane wave of either polarization is incident to an interface of uniaxial medium with wave vector, see Figure 4.3,

$$\mathbf{k}^i = -\mathbf{e}_3 k_o \cos\theta^i + \mathbf{k}_t, \quad \mathbf{k}_t\cdot\mathbf{k}_t = k_o^2\sin^2\theta^i. \tag{4.31}$$

The wave vector of a plane wave in the general anisotropic medium is known to satisfy the dispersion equation [35]

$$\det(\omega^2\overline{\overline{\epsilon}} - \mathbf{k}\mathbf{k}{}^{\times}_{\times}\overline{\overline{\mu}}^{-1}) = 0. \tag{4.32}$$

Writing $\mathbf{k} = \mathbf{k}_t + \mathbf{e}_3\beta$ and substituting (4.26) and (4.27), (4.32) can be solved for the axial propagation factor β corresponding to the TE and TM polarizations as

$$\beta_{TE}^2 = \omega^2\mu_t\epsilon_t - \frac{1}{m}\mathbf{k}_t\cdot\mathbf{k}_t = \omega^2\mu_t\epsilon_t - \frac{k_o^2}{m}\sin\theta^i, \tag{4.33}$$

$$\beta_{TM}^2 = \omega^2\mu_t\epsilon_t - \frac{1}{e}\mathbf{k}_t\cdot\mathbf{k}_t = \omega^2\mu_t\epsilon_t - \frac{k_o^2}{e}\sin\theta^i. \tag{4.34}$$

Details of the derivation are left as an exercise.

The dispersion surface for $\mathbf{k}(\theta)$ is a spheroid in both cases when m and e have finite positive values. For the special case $m = e$ the two surfaces coincide. For $m \to 0$ and $e \to 0$ the two spheroids become prolate needles and the medium approaches a ZAP medium. In this case, from (4.33) and (4.34) it follows that real β is possible only for small magnitudes of $\mathbf{k}_t$, which corresponds to waves propagating in almost axial direction, $\theta \approx 0$.

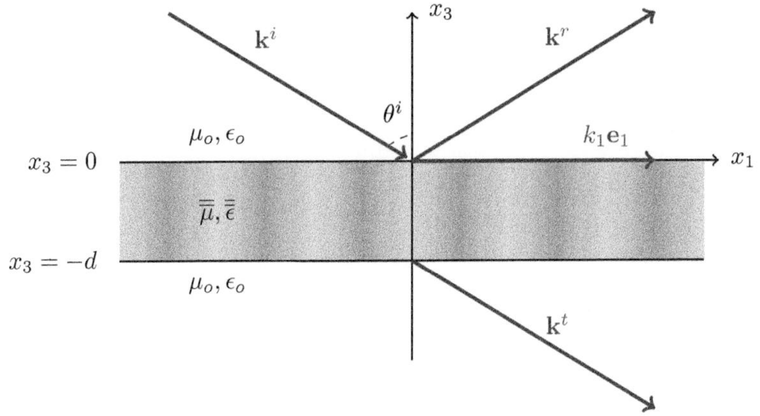

Figure 4.3: Incident, reflected, and transmitted plane waves associated to a layer of uniaxial anisotropic medium with thickness d. The axial components of the $\mathbf{k}$ vectors of the TE and TM waves in the layer are denoted by β_{TE} and β_{TM}.

Layer of Uniaxial Anisotropic Medium

To visualize a realization of the DB boundary, let us consider a layer of uniaxial medium $0 > x_3 > -d$ defined by (4.26) and (4.27) above a simple-isotropic half space $x_3 < -d$, see Figure 4.3. For convenience we denote

$$\mathbf{k}^i = k_1\mathbf{e}_1 - k_3\mathbf{e}_3 = \mathbf{k}^t, \quad \mathbf{k}^r = k_1\mathbf{e}_1 + k_3\mathbf{e}_3, \tag{4.35}$$

whence the incident and reflected TE fields in the upper half space satisfy

$$\mathbf{E}^i = \mathbf{e}_2 E^i, \quad k_o\eta_o\mathbf{H}^i = (k_1\mathbf{e}_3 + k_3\mathbf{e}_1)E^i, \tag{4.36}$$

$$\mathbf{E}^r = \mathbf{e}_2 E^r, \quad k_o\eta_o\mathbf{H}^r = (k_1\mathbf{e}_3 - k_3\mathbf{e}_1)E^r, \tag{4.37}$$

while the transmitted fields in the lower half space are

$$\mathbf{E}^t = \mathbf{e}_2 E^t, \quad k_o\eta_o\mathbf{H}^t = (k_1\mathbf{e}_3 + k_3\mathbf{e}_1)E^t. \tag{4.38}$$

Considering the TE case, the two waves propagating in opposite directions in the layer have the $\mathbf{k}$ vectors

$$\mathbf{k}^+ = k_1\mathbf{e}_1 - \beta_{TE}\mathbf{e}_3, \quad \mathbf{k}^- = k_1\mathbf{e}_1 + \beta_{TE}\mathbf{e}_3, \tag{4.39}$$

where β_{TE} is obtained from (4.33). The fields of the two waves satisfy

$$\mathbf{E}^+ = \mathbf{e}_2 E^+, \quad \omega\mu_t(\mathbf{e}_1 H_1^+ + m\mathbf{e}_3 H_3^+) = (\beta_{TE}\mathbf{e}_1 + k_1\mathbf{e}_3)E^+ \tag{4.40}$$

$$\mathbf{E}^- = \mathbf{e}_2 E^-, \quad \omega\mu_t(\mathbf{e}_1 H_1^- + m\mathbf{e}_3 H_3^-) = (\beta_{TE}\mathbf{e}_1 - k_1\mathbf{e}_3)E^-. \tag{4.41}$$

Requiring continuity of the tangential components of the total fields at both interfaces, we can find the reflected and transmitted fields of the TE wave as

$$E^r = R_{TE}E^i, \quad E^t = T_{TE}E^i, \tag{4.42}$$

where the two coefficients can be shown to have the form

$$R_{TE} = \frac{j(k_3^2\mu_t^2 - \beta_{TE}^2\mu_o^2)\sin\beta_{TE}d}{2k_3\beta_{TE}\mu_t\mu_o\cos\beta_{TE}d + j(k_3^2\mu_t^2 + \beta_{TE}^2\mu_o^2)\sin\beta_{TE}d}, \tag{4.43}$$

$$T_{TE} = \frac{2k_3\beta_{TE}\mu_t\mu_o e^{-jk_od}}{2k_3\beta_{TE}\mu_t\mu_o\cos\beta_{TE}d + j(k_3^2\mu_t^2 + \beta_{TE}^2\mu_o^2)\sin\beta_{TE}d}. \tag{4.44}$$

After similar steps, the corresponding coefficients for the TM wave can be constructed as

$$R_{TM} = \frac{j(k_3^2\epsilon_t^2 - \beta_{TM}^2\epsilon_o^2)\sin\beta_{TM}d}{2k_3\beta_{TM}\epsilon_t\epsilon_o\cos\beta_{TM}d + j(k_3^2\epsilon_t^2 + \beta_{TM}^2\epsilon_o^2)\sin\beta_{TM}d}, \tag{4.45}$$

$$T_{TM} = \frac{2k_3\beta_{TM}\epsilon_t\epsilon_o e^{-jk_od}}{2k_3\beta_{TM}\epsilon_t\epsilon_o\cos\beta_{TM}d + j(k_3^2\epsilon_t^2 + \beta_{TM}^2\epsilon_o^2)\sin\beta_{TM}d}. \tag{4.46}$$

Let us consider some special cases of the expressions (4.43) - (4.46).

- For $d \to 0$ we obtain $R_{TE} \to R_{TM} \to 0$ and $T_{TE} \to T_{TM} \to 1$, which corresponds to no layer at all.

- For $m \to 0$, $e \to 0$ and $\mathbf{k}_t \cdot \mathbf{k}_t \neq 0$, (4.33) and (4.34) yield imaginary β_{TE} and β_{TM} with magnitudes tending to infinity. In this case, we have $|\sin\beta d| \to |\cos\beta d|$ for both β_{TE} and β_{TM}, whence the reflection coefficients become $R_{TE} \to R_{TM} \to -1$, corresponding to those of PEC and PMC boundaries for the respective TE and TM waves and DB conditions for any combination of TE and TM fields. Also, we obtain $T_{TE} \to T_{TM} \to 0$ with no transmission for oblique incidence.

- It is possible to choose the distance d so that $\sin\beta d = 0$ for either of the TE and TM waves. In such a case there is no reflected TE or TM wave incident at a certain angle θ^i, while the corresponding magnitude of T_{TE} or T_{TM} is unity.

The total TE and TM fields at the interface $x_3 = 0$ are

$$\mathbf{E} = \mathbf{e}_2(1 + R_{TE})E^i \tag{4.47}$$

$$k_o\eta_o\mathbf{H} = (\mathbf{e}_3 k_1(1 + R_{TE}) + \mathbf{e}_1 k_3(1 - R_{TE}))E^i, \tag{4.48}$$

$$\mathbf{H} = \mathbf{e}_2(1 + R_{TM})H^i \tag{4.49}$$

$$k_o\mathbf{E} = -(\mathbf{e}_3 k_1(1 + R_{TM}) + \mathbf{e}_1 k_3(1 - R_{TM}))\eta_o H^i. \tag{4.50}$$

The DB conditions $\mathbf{e}_3 \cdot \mathbf{E} = 0$ and $\mathbf{e}_3 \cdot \mathbf{H} = 0$ are satisfied at the interface for $R_{TE} = R_{TM} = -1$.

Figures 4.4 and 4.5 represent reflection from and transmission through a layer of uniaxial medium for the TE polarization for different values of the medium parameters. One can notice that for small values of the parameter $m = \mu_3/\mu_t = \mu_3/\mu_o$, the layer reflects the TE wave like a PEC plane except when the angle of incidence is close to normal. Waves with normal incidence pass through with no reflection. Actually, a layer of uniaxial medium with small parameter m acts as a spatial filter for TE waves. Similar figures apply for TM waves when the parameter $e = \epsilon_3/\epsilon_t$ is small.

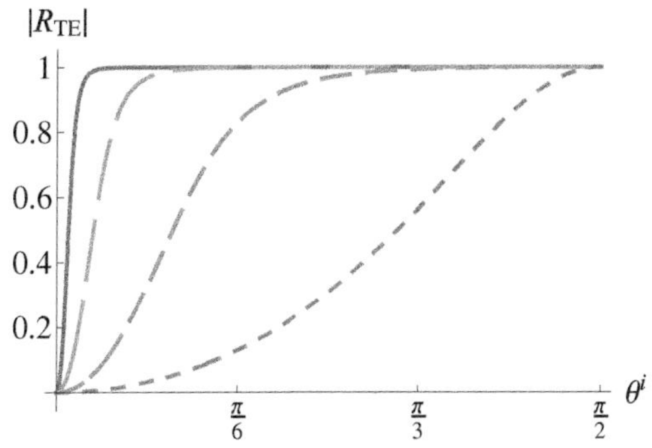

Figure 4.4: Magnitude of the reflection coefficient R_{TE} for a layer of uniaxial medium for angles of incidence $\theta^i = 0, ..\pi/2$. The parameters satisfy $\mu_t = \mu_o$, $\epsilon_t = \epsilon_o$ and $k_o d = 0.1$. The relative axial parameter is varied as $m = 0.1, 0.01, 0.001$ and 0.0001. For $m \to 0$ the slab acts as a PEC plane for TE waves except for almost normal incidence $\theta^i \approx 0$.

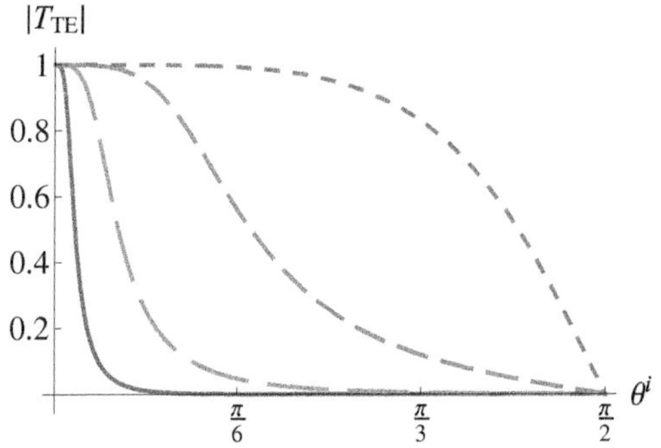

Figure 4.5: Magnitude of the transmission coefficient T_{TE} for a layer of uniaxial medium for the same set of parameter values as in the previous Figure. The slab acts as a narrow-beam spatial filter for TE waves when $m \to 0$.

In conclusion, a layer of uniaxial medium can serve as a realization for the DB boundary when the axial parameters have small values compared to those of the transverse parameters.

Realization by Uniaxial Bianisotropic Medium

The DB boundary can be realized by the interface of a bianisotropic medium defined by conditions of the form [59]

$$\begin{pmatrix} \mathbf{D} \\ \mathbf{H} \end{pmatrix} = \begin{pmatrix} \overline{\overline{\alpha}} & \overline{\overline{\epsilon}}' \\ \overline{\overline{\mu}}^{-1} & \overline{\overline{\beta}} \end{pmatrix} \cdot \begin{pmatrix} \mathbf{B} \\ \mathbf{E} \end{pmatrix}. \tag{4.51}$$

In this case, the four medium dyadics are assumed to be of the uniaxial form

$$\overline{\overline{\alpha}} = a\overline{\overline{\mathsf{I}}}_t + b\mathbf{nn}, \tag{4.52}$$

$$\overline{\overline{\epsilon}}' = e\mathbf{n}\times\overline{\overline{\mathsf{I}}}, \tag{4.53}$$

$$\overline{\overline{\mu}}^{-1} = m\mathbf{n}\times\overline{\overline{\mathsf{I}}}, \tag{4.54}$$

$$\overline{\overline{\beta}} = c\overline{\overline{\mathsf{I}}}_t + d\mathbf{nn}, \tag{4.55}$$

where a, b, c, d, e, m are six scalar parameters. Because $\overline{\overline{\epsilon}}'$ and $\overline{\overline{\mu}}^{-1}$ are antisymmetric 3D dyadics, they do not have inverses, whence it is not possible to express the medium conditions in the form (1.3). The conditions of the above medium can be expressed for tangential and normal components as

$$\mathbf{D}_t = a\mathbf{B}_t + e\mathbf{n}\times\mathbf{E}_t, \tag{4.56}$$

$$D_n = bB_n, \tag{4.57}$$

$$\mathbf{H}_t = m\mathbf{n}\times\mathbf{B}_t + c\mathbf{E}_t, \tag{4.58}$$

$$H_n = dE_n. \tag{4.59}$$

If the parameters satisfy the condition $b + c = a + d$, the medium has a particularly natural and simple definition in terms of four-dimensional formalism [72] as it is a special case of skewon medium [20].

Fields of a plane wave in the medium satisfy the conditions

$$\omega\mathbf{B}_t = k_n\mathbf{n}\times\mathbf{E}_t + \mathbf{k}_t\times\mathbf{n}E_n \tag{4.60}$$

$$\omega B_n = \mathbf{n}\cdot\mathbf{k}_t\times\mathbf{E}_t, \tag{4.61}$$

$$\omega\mathbf{D}_t = -k_n\mathbf{n}\times\mathbf{H}_t - \mathbf{k}_t\times\mathbf{n}H_n, \tag{4.62}$$

$$\omega D_n = -\mathbf{n}\cdot\mathbf{k}_t\times\mathbf{H}_t. \tag{4.63}$$

Applying (4.56) - (4.59), equations (4.62) and (4.63) can be written as

$$(\omega a - k_n m)\mathbf{B}_t = -(\omega e + k_n c)\mathbf{n}\times\mathbf{E}_t + d(\mathbf{n}\times\mathbf{k}_t)E_n, \tag{4.64}$$

$$\omega bB_n = -m\mathbf{k}_t\cdot\mathbf{B}_t - c(\mathbf{n}\times\mathbf{k}_t)\cdot\mathbf{E}_t. \tag{4.65}$$

Eliminating $\mathbf{B}_t$, $\mathbf{n}\times\mathbf{E}_t$ and E_n from the four equations (4.60), (4.61), (4.64) and (4.65), leaves us the equation

$$(\omega b + \omega c - mk_n)B_n = 0. \tag{4.66}$$

Omitting the special case when the bracketed factor vanishes, which man happen for some special medium and direction of propagation, it turns out from (4.66) and (4.57) that a plane wave propagating in the medium satisfies the conditions

$$B_n = 0, \qquad D_n = 0. \tag{4.67}$$

Because of continuity, the fields at the interface of the medium satisfy the conditions (4.1), whence the interface of such a medium acts as the DB boundary.

As yet another possibility, the DB boundary can be shown to be realizable by the interface of a special bianisotropic medium defined by medium conditions of the form ('simple-skewon medium', [66])

$$\mathbf{D} = N\mathbf{B}, \qquad \mathbf{H} = N\mathbf{E}, \tag{4.68}$$

where N is a scalar parameter. Proof of this is left as an exercise.

In the previous cases there remains the problem of realizing the media with suggested parameters. More practical realizations of the DB boundary have been proposed in terms of suitably tailored metasurfaces in [120, 121].

4.4 Spherical DB Resonator

As an example of a structure defined by the DB boundary, let us consider a spherical resonator of radius a [57]. The medium parameters of the resonator are ϵ_o and μ_o.

It is known that fields in a homogeneous and isotropic medium can be decomposed in TE$_r$ and TM$_r$ parts with respect to the radial direction $\mathbf{e}_r = \mathbf{r}/r$, in terms of the respective vector potentials $\mathbf{F}(\mathbf{r}) = \mathbf{e}_r F(\mathbf{r})$ and $\mathbf{A}(\mathbf{r}) = \mathbf{e}_r A(\mathbf{r})$. In fact, we can write [2]

$$\mathbf{E}(\mathbf{r}) = -\frac{1}{\epsilon_o}\nabla\times(\mathbf{F}(\mathbf{r})) + \frac{1}{j\omega\mu_o\epsilon_o}\nabla\times(\nabla\times(\mathbf{A}(\mathbf{r}))), \tag{4.69}$$

$$\mathbf{H}(\mathbf{r}) = \frac{1}{\mu_o}\nabla\times(\mathbf{A}(\mathbf{r})) + \frac{1}{j\omega\mu_o\epsilon_o}\nabla\times(\nabla\times(\mathbf{F}(\mathbf{r}))). \tag{4.70}$$

More explicitly, the dependence of the spherical components of the respective TE$_r$ and TM$_r$ fields can be expressed as

$$E_r = 0, \tag{4.71}$$

$$E_\theta = -\frac{1}{\epsilon_o r\sin\theta}\partial_\varphi F, \tag{4.72}$$

$$E_\varphi = \frac{1}{\epsilon_o r}\partial_\theta F, \tag{4.73}$$

$$H_r = \frac{1}{j\omega\mu_o\epsilon_o}(\partial_r^2 + k_o^2)F, \tag{4.74}$$

$$H_\theta = \frac{1}{j\omega\mu_o\epsilon_o r}\partial_r\partial_\theta F, \tag{4.75}$$

$$H_\varphi = \frac{1}{j\omega\mu_o\epsilon_o r\sin\theta}\partial_r\partial_\varphi F, \tag{4.76}$$

and

$$H_r = 0, \tag{4.77}$$

$$H_\theta = \frac{1}{\mu_o r\sin\theta}\partial_\varphi A, \tag{4.78}$$

$$H_\varphi = -\frac{1}{\mu_o r}\partial_\theta A, \tag{4.79}$$

$$E_r = \frac{1}{j\omega\mu_o\epsilon_o}(\partial_r^2 + k_o^2)A, \tag{4.80}$$

$$E_\theta = \frac{1}{j\omega\mu_o\epsilon_o r}\partial_r\partial_\theta A, \tag{4.81}$$

$$E_\varphi = \frac{1}{j\omega\mu_o\epsilon_o r\sin\theta}\partial_r\partial_\varphi A. \tag{4.82}$$

The radial vector potentials satisfy the Helmholtz equations in the form [2]

$$(\nabla^2 + k_o^2)\frac{F}{r} = 0, \quad (\nabla^2 + k_o^2)\frac{A}{r} = 0. \tag{4.83}$$

TE_r and TM_r modes

The resonator modes satisfying (4.83) are obtained from potentials of the functional form

$$F_{mnp}(r,\theta,\varphi) = rj_n(k_o r)P_n^m(\cos\theta)\begin{pmatrix}\cos m\varphi\\ \sin m\varphi\end{pmatrix}, \tag{4.84}$$

$$A_{mnp}(r,\theta,\varphi) = rj_n(k_o r)P_n^m(\cos\theta)\begin{pmatrix}\cos m\varphi\\ \sin m\varphi\end{pmatrix}, \tag{4.85}$$

with $0 \leq m \leq n$. The amplitude factors are normalized to unity. The index $p = 1,2,3...$ refers to the corresponding resonance wavenumber $k_o = k_{mnp}$. For $n = 0$, the potentials F and A become multiples of $rj_0(k_o r)$. From (4.71) – (4.129) we see that, in this case, all fields vanish and we can ignore that possibility.

The spherical Bessel function satisfies the differential equation

$$\partial_r(r^2\partial_r j_n(k_o r)) + [(k_o r)^2 - n(n+1)]j_n(k_o r) = 0, \tag{4.86}$$

which can also be written as

$$(r^2\partial_r^2 + 2r\partial_r + k_o^2 r^2 - n(n+1))j_n(k_o r) = 0. \tag{4.87}$$

Expanding

$$(\partial_r^2 + k_o^2)(rj_n(k_o r)) = (r\partial_r^2 + 2\partial_r + k_o^2 r)j_n(k_o r), \tag{4.88}$$

yields the relation

$$(\partial_r^2 + k_o^2)(r j_n(k_o r)) = \frac{n(n+1)}{r} j_n(k_o r). \tag{4.89}$$

Thus, the expressions (4.74) and (4.80) for the TE_r and TM_r radial field components can be rewritten as

$$H_r = \frac{n(n+1)}{j\omega\mu_o\epsilon_o r} F, \tag{4.90}$$

$$E_r = \frac{n(n+1)}{j\omega\mu_o\epsilon_o r} A, \tag{4.91}$$

for the *mnp* mode in question.

The DB boundary conditions now require that the potentials satisfy

$$H_r(a,\theta,\varphi) = 0 \quad \Rightarrow \quad F(a,\theta,\varphi) = 0, \tag{4.92}$$

$$E_r(a,\theta,\varphi) = 0 \quad \Rightarrow \quad A(a,\theta,\varphi) = 0. \tag{4.93}$$

For the TE_r modes (4.93) is automatically satisfied. From (4.92), (4.72), (4.73) and

$$F(a,\theta,\varphi) = 0, \quad \Rightarrow \quad \partial_\theta F(a,\theta,\varphi) = 0, \quad \partial_\varphi F(a,\theta,\varphi) = 0 \tag{4.94}$$

we obtain

$$E_\theta(a,\theta,\varphi) = 0, \qquad E_\varphi(a,\theta,\varphi) = 0. \tag{4.95}$$

Because of the operator ∂_r in (4.75) and (4.76), the fields H_θ and H_φ do not vanish at the boundary. The condition (4.95) equals the PEC condition $\mathbf{n} \times \mathbf{E} = 0$. This leads to the conclusion that the TE_r modes in a spherical DB resonator equal those of the corresponding PEC resonator.

From the symmetry of equations we can conclude the dual case: TM_r modes in a spherical resonator equal those of the corresponding PMC resonator. Thus, the spherical DB resonator can be conceived as a kind of combination of PEC and PMC resonators.

Because the PEC or PMC boundary does not couple TE_r and TM_r modes, from the previous it follows that the DB boundary does not couple TE and TM modes. In contrast, the general impedance condition is known to couple TE and TM fields at the boundary surface, whence they cannot exist as independent modes.

Dominant modes

The dominant TM_{011} mode of the spherical PEC resonator corresponding to the lowest resonance wavenumber ($ka = 2.744$) [18] does not exist in the DB resonator. The lowest TE_r mode in the PEC resonator with resonance wavenumber

$$j_1(k_o a) = 0, \quad \Rightarrow \quad k_o a = 4.493, \tag{4.96}$$

corresponds to the lowest TE_r mode in the DB resonator. From (4.84) we find three linearly independent potential functions (omitting amplitude factors)

$$F_{011}(r,\theta,\varphi) = rj_1(4.493r/a)P_1^0(\cos\theta), \tag{4.97}$$

$$F_{111}(r,\theta,\varphi) = rj_1(4.493r/a)P_1^1(\cos\theta)\left(\begin{array}{c}\cos\varphi \\ \sin\varphi\end{array}\right), \tag{4.98}$$

with

$$P_1^0(\cos\theta) = \cos\theta, \quad P_1^1(\cos\theta) = \sin\theta. \tag{4.99}$$

The TM_r modes in PMC resonators have the same wavenumbers and the potential functions (4.85) are

$$A_{011}(r,\theta,\varphi) = rj_1(4.493r/a)P_1^0(\cos\theta), \tag{4.100}$$

$$A_{111}(r,\theta,\varphi) = rj_1(4.493r/a)P_1^1(\cos\theta)\left(\begin{array}{c}\cos\varphi \\ \sin\varphi\end{array}\right), \tag{4.101}$$

which means that there is a six-fold degeneracy at the dominant resonance frequency.

The general resonance field is any sum of partial fields arising from these potential functions. Choosing the amplitude factors for the six potentials allows a lot of freedom to set conditions to the resonance field which may have application in the study of electromagnetic properties of materials. For example, with suitable excitations two resonance fields can be formed to satisfy

$$\mathbf{E}_+ = j\eta_o\mathbf{H}_+, \quad \mathbf{E}_- = -j\eta_o\mathbf{H}_-. \tag{4.102}$$

Because such fields have the form of wavefields [33], the eigenfields of a chiral medium, they can be applied in the measurement of chirality parameters of a material sample. In fact, it is known that a material sample is polarized differently in these two fields when the chirality parameter is not zero, whence the shift of the resonance frequency is different for the two eigenfields.

4.5 Circular DB Waveguide

As another example, let us consider a waveguide with circular DB boundary of radius $\rho = a$ and isotropic medium of parameters ϵ_o, μ_o, with axis of the waveguide along the coordinate x_3 [63]. Because of the geometry, the dependence of the modal fields on polar coordinates ρ, φ, x_3 can be represented as

$$\mathbf{E}(\mathbf{r}) = \mathbf{E}(\rho)e^{jn\varphi}e^{-j\beta x_3}, \qquad \mathbf{H}(\mathbf{r}) = \mathbf{H}(\rho)e^{jn\varphi}e^{-j\beta x_3}, \tag{4.103}$$

where n is a positive or negative integer. For $n = 0$ the fields are rotationally symmetric. For $n > 0$ and $n < 0$ the field vectors $\mathbf{E}$ and $\mathbf{H}$ have respective left- and right-hand polarizations when looking into the direction $\mathbf{e}_3$.

Field equations

While all resonance modes of the spherical DB-resonator can be decomposed in two sets which coincide with the TE_r and TM_r modes of the respective PEC and PMC resonators, in the case of the circular waveguide, modes polarized TE_ρ and TM_ρ with respect to the radial direction $\boldsymbol{\rho}$ do not form a complete set. The basic reason for this lies in the fact that the general fields cannot be derived from the vector potential components F_ρ and A_ρ. They can, however, be expressed in terms of the axial components E_3, H_3 of the fields.

Starting from fields (4.103) and eliminating radial and circumferential field components from the Maxwell equations, we can obtain equations for the axial field components as

$$\frac{1}{\rho}\partial_\rho(\rho\partial_\rho E_3(\rho)) + (k_c^2 - \frac{n^2}{\rho^2})E_3(\rho) = 0, \tag{4.104}$$

$$\frac{1}{\rho}\partial_\rho(\rho\partial_\rho H_3(\rho)) + (k_c^2 - \frac{n^2}{\rho^2})H_3(\rho) = 0. \tag{4.105}$$

Since these have the form of Bessel differential equation, the solutions finite at $\rho = 0$ can be expressed as

$$E_3(\rho) = EJ_n(k_c\rho), \qquad H_3(\rho) = HJ_n(k_c\rho) \tag{4.106}$$

with

$$\beta = \sqrt{k_o^2 - k_c^2}. \tag{4.107}$$

Substituting (4.106) in the Maxwell equations and eliminating the φ components of the fields, the radial components of the fields become

$$E_\rho(\rho) = \frac{nk_o}{k_c^2\rho}\eta_o HJ_n(k_c\rho) - \frac{j\beta}{k_c}EJ_n'(k_c\rho), \tag{4.108}$$

$$\eta H_\rho(\rho) = -\frac{nk_o}{k_c^2\rho}EJ_n(k_c\rho) - \frac{j\beta}{k_c}\eta_o HJ_n'(k_c\rho), \tag{4.109}$$

where the prime denotes differentiation with respect to the argument $k_c\rho$.

Dispersion equation

Substituting the field expressions (4.108) and (4.109) in the conditions of the DB boundary, $E_\rho(a) = 0$ and $H_\rho(a) = 0$, we obtain relations between the amplitude coefficiets E and H as

$$\begin{pmatrix} j\beta k_c aJ_n'(k_c a) & -nk_o J_n(k_c a) \\ nkJ_n(k_c a) & j\beta k_c aJ_n'(k_c a) \end{pmatrix} \begin{pmatrix} E \\ \eta_o H \end{pmatrix} = 0. \tag{4.110}$$

Requiring vanishing of the determinant of the matrix leads to the dispersion equation. It can be split in two equations as

$$\beta k_c aJ_n'(k_c a) \pm nk_o J_n(k_c a) = 0, \tag{4.111}$$

defining two sets of modal solutions. Assuming positive values for k_c and applying properties of the Bessel functions,

$$J_{-n}(x) = (-1)^n J_n(x), \qquad J'_{-n}(x) = (-1)^n J'_n(x) \tag{4.112}$$

the set of $'+'$ solutions for n are seen to coincide with the set of $'-'$ solutions for $-n$. Also, a $'+'$ solution for β corresponds to a $'-'$ solution for $-\beta$.

Inserting (4.107) in (4.111), the latter becomes an equation for $k_c a$. Let us separate two cases.

- For the rotationally symmetric fields with $n = 0$, there is no dependence of $k_c a$ on $k_o a$, because (4.111) reduces to

  $$J'_0(k_c a) = -J_1(k_c a) = 0. \tag{4.113}$$

  In this case, the solutions are the roots of $J_1(x)$,

  $$k_c a = x_{1m}, \qquad m = 1, 2, 3, \cdots, \tag{4.114}$$

  with

  $$x_{11} = 3.832, \qquad x_{12} = 7.016, \qquad x_{13} = 10.173, \cdots \tag{4.115}$$

  The corresponding propagation coefficients are obtained from (4.107). In the rotationally symmetric case, k_c can be interpreted as the cutoff wavenumber since the wave decays exponentially for $k_o < k_c$.

- For $n \neq 0$ the k_c values obtained from (4.111) are not fixed numbers but functions of k_o, i.e., the frequency and parameters ϵ_o, μ_o of the medium inside the guide. In this case, k_c cannot be called the cutoff wavenumber, because propagation in the waveguide is possible for $k_o < k_c$.

  Applying rules for the Bessel functions,

  $$2nJ_n(x) = x(J_{n-1}(x) + J_{n+1}(x)), \tag{4.116}$$

  $$2J'_n(x) = J_{n-1}(x) - J_{n+1}(x), \tag{4.117}$$

  the dispersion condition (4.111) can be rewritten in the form

  $$(k_o \pm \beta)J_{n-1}(k_c a) + (k_o \mp \beta)J_{n+1}(k_c a) = 0, \tag{4.118}$$

  whence (4.118) can be further expressed as the inverted relation $k_o a(k_c a)$,

  $$k_o a = \frac{k_c a|J_{n-1}(k_c a) - J_{n+1}(k_c a)|}{\sqrt{-4J_{n-1}(k_c a)J_{n+1}(k_c a)}}. \tag{4.119}$$

  Here, the square root is assumed positive when it is real. k_c as a function of k_o is independent of the sign of n while the sign of β depends on the sign of n as was noted above.

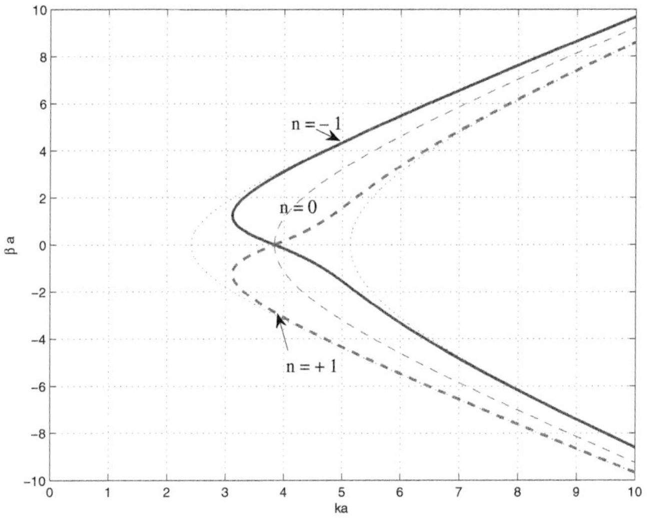

Figure 4.6: Dispersion diagrams $\beta a(k_o a)$ for the three basic modes in a circular DB waveguide of radius a corresponding to $n = 0$ and $n = \pm 1$ with the asymptotic curves of the latter (the dotted lines). The graphs denoted by $n = 1$ and $n = -1$ correspond to the self-dual mode $\mathbf{E}_+, \mathbf{H}_+$. For the other solution $\mathbf{E}_-, \mathbf{H}_-$ the $n = 1$ and $n = -1$ graphs must be interchanged. Backward-wave regions can be spotted for $n = 1$ and $n = -1$ in the region $3.112 < k_o a < 3.832$. From [63], with permission from the IEEE.

Propagation of Basic Modes

To find the propagation coefficient (4.107) for a particular mode, we first have to solve the dispersion equation (4.111), or invert (4.119), to find $k_c a$ for a given value of $k_o a$. Considering the modes corresponding to $n = 1$ and $n = -1$, the lowest range of $k_c a$ values for which the square root in (4.119) is real lies between the first zero of J_0, ($k_c a = x_{01} = 2.405$), and the first zero of J_2, ($k_c a = x_{21} = 5.136$).

From (4.111) we can see that, at $k_o a = k_c a = x_{11} = 3.832$, we have $\beta = 0$ whence the $n = \pm 1$ modes do not propagate. This coincides with the cutoff of the lowest rotationally symmetric $n = 0$ mode, Figure 4.6. One can show that when $k_c a$ is close to the value x_{11}, we have $\beta a \approx n(k_c a - x_{11})$, whence $\beta a(k_o a)$ has a positive slope for $n = 1$ and negative slope for $n = -1$. This helps us to distinguish the two dispersion curves $n = 1$ and $n = -1$ from one another in Figure 4.6. In fact, for $n = 1$ the upper curve corresponds to $'-'$ and the lower curve to $'+'$, while for $n = -1$ the opposite is the case.

Interestingly, in the region $3.1124 < k_o a < 3.832$ we have modes with opposite phase and group velocities. Close to $k_o a = 3.832$ for $n = 1$ and $\beta a > 0$ the group velocity has the value

$$v_g = \frac{d\omega}{d\beta} = \frac{1}{\sqrt{\mu_o \epsilon_o}} \frac{dk_o a}{d\beta a} \approx -\frac{\omega}{k_o}. \qquad (4.120)$$

Waves with phase and group velocities in opposite directions have been called backward waves in the past [23] and they have found applications in the design of microwave oscillators and amplifiers [22]. While backward waves are usually associated with dispersive media or periodic structures, such restrictions do not appear in the DB waveguide.

Modal Fields

Modes $n = 0$

For the rotationally symmetric modes corresponding to the dispersion equation (4.113) the field amplitudes E and H can be chosen independently. The basic linearly independent solutions are obtained by choosing $E = 0$ and $H = 0$. The fields in the former case have the form ($\mathrm{TE}_{\rho 0m}$ or $\mathrm{TE}_{x_3 0m}$ modes)

$$\mathbf{E}(\rho) = -\mathbf{e}_\varphi \frac{jk_o}{k_c} \eta H J_1(k_c \rho), \qquad k_c a = x_{1m}, \qquad (4.121)$$

$$\mathbf{H}(\rho) = \mathbf{e}_\rho \frac{j\beta}{k_c} H J_1(k_c \rho) + \mathbf{e}_3 H J_0(k_c \rho), \qquad (4.122)$$

while in the latter case they are of the form ($\mathrm{TM}_{\rho 0m}$ or $\mathrm{TM}_{x_3 0m}$ modes)

$$\mathbf{H}(\rho) = \mathbf{e}_\varphi \frac{jk}{k_c \eta} E J_1(k_c \rho), \qquad k_c a = x_{1m}, \qquad (4.123)$$

$$\mathbf{E}(\rho) = \mathbf{e}_\rho \frac{j\beta}{k_c} E J_1(k_c \rho) + \mathbf{e}_3 E J_0(k_c \rho). \qquad (4.124)$$

For the same index m the TE and TM modes have the same propagation factor, whence any linear combinations of them are modes as well.

One can note that the $\mathrm{TE}_{\rho 0m}$ or $\mathrm{TE}_{x_3 0m}$ modes satisfy the PEC conditions $\mathbf{e}_\rho \times \mathbf{E} = 0$ at $\rho = a$ whence the rotationally symmetric $\mathrm{TE}_{x_3 0m}$ modes of a circular PEC waveguide are also modes of the DB waveguide. Similarly, the $\mathrm{TM}_{x_3 0m}$ modes of a PMC waveguide are also modes of the DB waveguide.

Modes $n \neq 0$

From (4.110) we can write the relation

$$\frac{E}{\eta H} = -j \frac{\beta k_c a J_n'(k_c a)}{n k_o J_n(k_c a)} = \pm j. \qquad (4.125)$$

The last form is due to the dispersion equations (4.111). Thus, there are two possible relations between the field amplitudes which can be denoted as

$$E_+ = j\eta_o H_+, \qquad E_- = -j\eta_o H_-, \qquad (4.126)$$

defining the '+' and '−' sets of modes in the DB waveguide. Let us consider them simultaneously with the double sign $\mathbf{E}_\pm, \mathbf{H}_\pm$.

From (4.106) the axial field components become

$$E_{3\pm}(\rho) = \pm j\eta_o H_{3\pm}(\rho) = E_\pm J_n(k_c\rho). \tag{4.127}$$

The radial and angular field components can be constructed from the expressions (4.108), (4.109) and applying (4.116), (4.117) as

$$\begin{aligned} E_{\rho\pm}(\rho) &= \pm j\eta_o H_{\rho\pm}(\rho) \\ &= \pm\frac{j}{2k_c}[(k_o \mp \beta)J_{n-1}(k_c\rho) + (k_o \pm \beta)J_{n+1}(k_c\rho)]E_\pm, \quad (4.128) \\ E_{\varphi\pm}(\rho) &= \pm j\eta_o H_{\varphi\pm}(\rho) \\ &= \mp\frac{1}{2k_c}[(k_o \mp \beta)J_{n-1}(k_c\rho) - (k_o \pm \beta)J_{n+1}(k_c\rho)]E_\pm. \quad (4.129) \end{aligned}$$

From (4.128) and the dispersion condition (4.118) we can verify that the modal fields really satisfy the DB-boundary conditions $E_\rho(a) = 0$, $H_\rho(a) = 0$.

From the previous expressions we conclude that the total modal fields satisfy the simple relations

$$\mathbf{E}_\pm(\mathbf{r}) = \pm j\eta_o \mathbf{H}_\pm(\mathbf{r}) \tag{4.130}$$

for each $n \neq 0$. Fields satisfying conditions of the form (4.130) can be called self dual because there exist two linear transformations

$$\begin{pmatrix} \mathbf{E} \\ \mathbf{H} \end{pmatrix} \to \begin{pmatrix} \mathbf{E}_{d1} \\ \mathbf{H}_{d1} \end{pmatrix} = \begin{pmatrix} j\eta_o\mathbf{H} \\ \mathbf{E}/j\eta_o \end{pmatrix}, \tag{4.131}$$

$$\begin{pmatrix} \mathbf{E} \\ \mathbf{H} \end{pmatrix} \to \begin{pmatrix} \mathbf{E}_{d2} \\ \mathbf{H}_{d2} \end{pmatrix} = -\begin{pmatrix} j\eta_o\mathbf{H} \\ \mathbf{E}/j\eta_o \end{pmatrix}, \tag{4.132}$$

in terms of which one of the two modal fields is invariant (self dual) and the other one is anti-invariant (anti-self dual) [35]. The DB boundary and the isotropic medium are invariant (self dual) in both transformations.

The modal electric fields transverse to the waveguide axis can be expressed in the alternative forms

$$\begin{aligned} \mathbf{E}_{t\pm}(\boldsymbol{\rho}) &= \pm\frac{E_\pm}{2k_c}\big[-(k_o \mp \beta)(\mathbf{e}_\varphi - j\mathbf{e}_\rho)J_{n-1}(k_c\rho) \\ &\quad +(k_o \pm \beta)(\mathbf{e}_\varphi + j\mathbf{e}_\rho)J_{n+1}(k_c\rho)\big]e^{jn\varphi}, \quad (4.133) \\ &= \pm\frac{jE_\pm}{2k_c}\big[(k_o \mp \beta)(\mathbf{e}_1 + j\mathbf{e}_2)J_{n-1}(k_c\rho)e^{j(n-1)\varphi} \\ &\quad +(k_o \pm \beta)(\mathbf{e}_1 - j\mathbf{e}_2)J_{n+1}(k_c\rho)e^{j(n+1)\varphi}\big]. \quad (4.134) \end{aligned}$$

Because the β solutions are real, the waveguide is lossless. This fact can be checked by forming the expression of the Poynting vector at the boundary of the guide. It will turn out that it has only a normal component with an imaginary value which means that there is no power loss at the DB boundary for the waveguide modes.

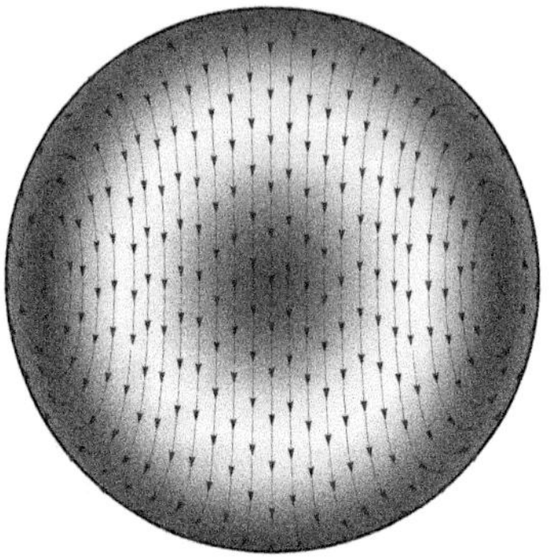

Figure 4.7: Field pattern of $\mathrm{Re}(\mathbf{E}_+)$ corresponding to the parameter values $n = 1$, $k_o a = 3.5$, $k_c a = 2.583$ and $\beta a = -2.362$ in a circular waveguide with DB boundary. The same pattern is obtained for $\mathrm{Im}(\mathbf{H}_+)$. From [63], with permission from the IEEE.

Modes $n = 1$ and $n = -1$

Let us finally consider in more detail the modes corresponding to $n = 1$ and $n = -1$. The transverse field expressions can be obtained from (4.134) as

$$\begin{aligned} \mathbf{E}_{t\pm}(\boldsymbol{\rho}) &= \pm\frac{jE_\pm}{2k_c}\big[(k_o \mp \beta)(\mathbf{e}_1 + j\mathbf{e}_2)J_0(k_c\rho) \\ &\quad +(k_o \pm \beta)(\mathbf{e}_1 - j\mathbf{e}_2)J_2(k_c\rho)e^{j2\varphi}\big], \quad \mathrm{n} = +1 \qquad (4.135) \\ \mathbf{E}_{t\pm}(\boldsymbol{\rho}) &= \pm\frac{jE_\pm}{2k_c}\big[(k_o \mp \beta)(\mathbf{e}_1 + j\mathbf{e}_2)J_2(k_c\rho)e^{-2j\varphi} \\ &\quad +(k_o \pm \beta)(\mathbf{e}_1 - j\mathbf{e}_2)J_0(k_c\rho)\big], \quad n = -1. \qquad (4.136) \end{aligned}$$

(4.135) and (4.136) represent the two self-dual modal fields as a sum of two circularly polarized components of opposite handedness. In particular, at the axis $\rho = 0$ where the axial component vanishes, the total fields are circularly polarized as

$$\begin{aligned} \mathbf{E}_\pm(0) &= \pm\frac{j(k_o \mp \beta)}{2k_c}(\mathbf{e}_1 + j\mathbf{e}_2)E_\pm, \quad n = 1, \qquad (4.137) \\ \mathbf{E}_\pm(0) &= \pm\frac{j(k_o \pm \beta)}{2k_c}(\mathbf{e}_1 - j\mathbf{e}_2)E_\pm, \quad n = -1. \qquad (4.138) \end{aligned}$$

To have an idea of the field patterns in a DB waveguide, let us choose $k_o a = 3.5$ which corresponds to four possible values of the propagation coefficient, $\beta a = -0.346$ and $\beta a = -2.362$ for $n = 1$ and $\beta a = 0.346$ and $\beta a = 2.362$ for $n = -1$, see Figure 4.6.

Figures 4.7 and 4.8 depict the real parts of the transverse electric field $\mathbf{E}_+$ for the two possible modes corresponding to $k_o a = 3.5$ and $n = 1$. The imaginary

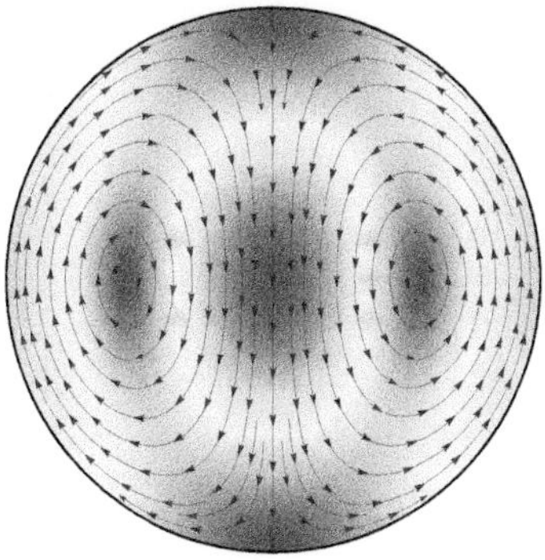

Figure 4.8: Field pattern of $\mathrm{Re}(\mathbf{E}_+)$ corresponding to the parameter values $n = 1$, $k_o a = 3.5$ $k_c a = 3.483$ and $\beta a = -0.346$ in a circular waveguide with DB boundary. The same pattern is obtained for $\mathrm{Im}(\mathbf{H}_+)$. From [63], with permission from the IEEE.

parts have somewhat similar form but they are rotated by $\pi/2$. For the magnetic field the patterns are exactly similar to those of the electric field so that the real part of the magnetic field has the same pattern as the imaginary part of the electric field and conversely.

There are four special points in each of the field patterns at which the real part of the transverse electric field vanishes. These points can be found from the expression (4.135) and the same points are found also from (4.136) for the case $n = -1$. Two of the points occur on the x_1 axis (horizontal in Figures 4.7 and 4.8) for ρ satisfying

$$(k_o \mp \beta)J_0(k_c\rho) - (k_o \pm \beta)J_2(k_c\rho) = 0, \tag{4.139}$$

and two on the x_2 axis for ρ satisfying

$$(k_o \mp \beta)J_0(k_c\rho) + (k_o \pm \beta)J_2(k_c\rho) = 0. \tag{4.140}$$

Comparing with (4.119) we can see that the latter points occur exactly at the boundary $\rho = a$. The solutions $\rho = \rho_o$ to (4.139) yield $\rho_o = 0.874a$ and $\rho_o = 0.550a$ on the x axis for the respective cases of Figures 4.7 and 4.8. Same results are obtained for the imaginary parts of the transverse fields but the positions of the zeros are rotated by $\pi/2$. It is easy to check that at these points of vanishing real and imaginary parts of the transverse fields the corresponding real and imaginary parts of the axial fields also vanish.

Comparing (4.139) and (4.129) one can see that $E_{\varphi\pm}(k_3\rho_o) = 0$ at the whole cylindrical surface of radius $\rho = \rho_o$. Thus, the effective boundary conditions

$$E_\varphi(k_c\rho_o) = 0, \qquad H_\varphi(k_c\rho_o) = 0, \tag{4.141}$$

are valid for the self-dual modes at the surface $\rho = \rho_o$. Actually, the form of (4.141) equals that of soft-and-hard waveguide [26] and the field patterns shown in Figures 4.7 and 4.8 inside the cylinder $\rho < \rho_o$ are those of the HE_{11} mode in a corrugated waveguide [7].

4.6 D'B' Boundary

The D'B' boundary is defined by the nonlocal conditions (4.4), which for a planar surface can be expressed as (4.3),

$$\partial_n D_n = 0, \qquad \partial_n B_n = 0. \tag{4.142}$$

For a combination of incident and reflected plane waves, the conditions become

$$k_n \mathbf{n} \cdot (\mathbf{D}^i - \mathbf{D}^r) = 0, \tag{4.143}$$
$$k_n \mathbf{n} \cdot (\mathbf{B}^r - \mathbf{B}^r) = 0. \tag{4.144}$$

Because these are satisfied for $k_n = 0$, any lateral incident wave is matched to the D'B' boundary. Assuming $k_n \neq 0$, we can write conditions similar to (4.8) and (4.9),

$$(\mathbf{n} \times \mathbf{k}_t) \cdot (\mathbf{H}_t^i - \mathbf{H}_t^r) = 0, \tag{4.145}$$
$$(\mathbf{n} \times \mathbf{k}_t) \cdot (\mathbf{E}_t^i - \mathbf{E}_t^r) = 0. \tag{4.146}$$

Expanding (4.145) as

$$(\mathbf{n} \times \mathbf{k}_t) \cdot \overline{\overline{\mathsf{J}}}_t \cdot (\mathbf{E}_t^i + \mathbf{E}_t^r) = \frac{k_o}{k_n} \mathbf{k}_t \cdot (\mathbf{E}_t^i + \mathbf{E}_t^r) = 0, \tag{4.147}$$

and combining with (4.146), we obtain

$$\mathbf{E}_t^r = \overline{\overline{\mathsf{R}}}_t \cdot \mathbf{E}_t^i, \tag{4.148}$$

with

$$\overline{\overline{\mathsf{R}}}_t = -\frac{1}{k_t^2}(\mathbf{k}_t\mathbf{k}_t - (\mathbf{n} \times \mathbf{k})(\mathbf{n} \times \mathbf{k})). \tag{4.149}$$

The reflection dyadic equals that of the DB boundary, (4.14), except for the minus sign in front. The eigenvalues are ± 1 and the electric eigenfields are

$$\mathbf{E}_{t+}^i = \mathbf{E}_{t+}^r = A_+ \mathbf{n} \times \mathbf{k}, \tag{4.150}$$

$$\mathbf{E}_{t-}^i = -\mathbf{E}_{t-}^r = A_- \mathbf{k}_t. \tag{4.151}$$

The corresponding magnetic eigenfields are

$$\eta_o \mathbf{H}_{t+}^i = -\eta_o \mathbf{H}_{t+}^r = -\overline{\overline{\mathsf{J}}}_t \cdot A_+(\mathbf{n} \times \mathbf{k}) = A_+ \frac{k_n}{k_o} \mathbf{k}_t, \tag{4.152}$$

$$\eta_o \mathbf{H}_{t-}^i = \eta_o \mathbf{H}_{t-}^r = -\overline{\overline{\mathsf{J}}}_t \cdot A_- \mathbf{k}_t = -A_- \frac{k_o}{k_n}(\mathbf{n} \times \mathbf{k}). \tag{4.153}$$

These expressions allow us to make the following conclusions:

- From (4.150) we have $\mathbf{n}\cdot\mathbf{E}^i_+ = \mathbf{n}\cdot\mathbf{E}^r_+ = 0$, whence the + wave is TE to $\mathbf{n}$. From (4.152) it follows that $\mathbf{H}^i_{t+} + \mathbf{H}^r_{t+} = 0$, whence the TE wave is reflected from the D'B' boundary as from the PMC boundary.

- From (4.153) we have $\mathbf{n}\cdot\mathbf{H}^i_- = \mathbf{n}\cdot\mathbf{H}^r_- = 0$, whence the - wave is TM to $\mathbf{n}$. from (4.150) it follows that $\mathbf{E}^i_{t-} + \mathbf{E}^r_{t-} = 0$, whence the TM wave is reflected as from the PEC boundary.

- Because of the preceding properties, the D'B' boundary belongs to the class of boundaries with PEC/PMC equivalence.

	TE	TM
DB	PEC	PMC
D'B'	PMC	PEC

Table 4.1: Both DB and D'B' boundary conditions can be replaced by effective PEC and PMC conditions for fields with TE and TM polarizations.

Realization of the D'B' Boundary

While the DB boundary can be realized by the interface of a uniaxial medium, there does not seem to exist a corresponding medium realizing the D'B' boundary. However, it turns out that one can transform a DB boundary to a D'B' boundary using a layer of uniaxial wave-guiding medium defined by (3.239) and (3.240) as [64, 65]

$$\overline{\overline{\epsilon}} = \epsilon_t \overline{\overline{\mathsf{I}}}_t + \epsilon_3 \mathbf{e}_3\mathbf{e}_3, \quad \overline{\overline{\mu}} = \mu_t \overline{\overline{\mathsf{I}}}_t + \mu_3 \mathbf{e}_3\mathbf{e}_3. \tag{4.154}$$

Assuming

$$\epsilon_3/\epsilon_t \to \infty, \quad \mu_3/\mu_t \to \infty, \tag{4.155}$$

for D_3 and B_3 to have finite magnitudes, we must have $E_3 \to 0$ and $H_3 \to 0$, whence $\mathbf{E} = \mathbf{E}_t$ and $\mathbf{H} = \mathbf{H}_t$. For plane waves traveling in opposite directions along the x_3 axis,

$$\mathbf{E}_{t\pm}(\mathbf{r}) = \mathbf{E}_{t\pm} e^{-j\mathbf{k}_t\cdot\mathbf{r}} e^{\mp j\beta x_3}, \tag{4.156}$$

$$\mathbf{H}_{t\pm}(\mathbf{r}) = \mathbf{H}_{t\pm} e^{-j\mathbf{k}_t\cdot\mathbf{r}} e^{\mp j\beta x_3}, \tag{4.157}$$

the Maxwell equations can be decomposed in axial and transverse parts as

$$\mathbf{e}_3\cdot\mathbf{k}_t\times\mathbf{E}_{t\pm} = \omega B_{3\pm}, \tag{4.158}$$

$$\pm\beta\mathbf{e}_3\times\mathbf{E}_{t\pm} = \omega\mu_t\mathbf{H}_{t\pm}, \tag{4.159}$$

$$\mathbf{e}_3\cdot\mathbf{k}_t\times\mathbf{H}_{t\pm} = -\omega D_{3\pm}, \tag{4.160}$$

$$\pm\beta\mathbf{e}_3\times\mathbf{H}_{t\pm} = -\omega\epsilon_t\mathbf{E}_{t\pm}. \tag{4.161}$$

For (4.159) and (4.161) to represent the same condition between $\mathbf{E}_{t\pm}$ and $\mathbf{H}_{t\pm}$, requires

$$\beta = \omega\sqrt{\mu_t \epsilon_t}. \tag{4.162}$$

The transverse fields can be represented in terms of the axial fields

$$D_3(x_3) = D_{3+}e^{-j\beta x_3} + D_{3-}e^{j\beta x_3}, \tag{4.163}$$
$$B_3(x_3) = B_{3+}e^{-j\beta x_3} + B_{3-}e^{j\beta x_3}, \tag{4.164}$$

by applying the orthogonality conditions

$$(\mathbf{k}_t \pm \beta \mathbf{e}_3)\cdot\mathbf{D}_\pm = \mathbf{k}_t\cdot\epsilon_t\mathbf{E}_{t\pm} \pm \beta D_{3\pm} = 0, \tag{4.165}$$
$$(\mathbf{k}_t \pm \beta \mathbf{e}_3)\cdot\mathbf{B}_\pm = \mathbf{k}_t\cdot\mu_t\mathbf{H}_{t\pm} \pm \beta B_{3\pm} = 0, \tag{4.166}$$

and (4.158), (4.160). In fact, we can expand

$$\mathbf{E}_{t\pm}(\mathbf{r}) = \frac{1}{k_t^2}(\mathbf{k}_t\mathbf{k}_t + (\mathbf{e}_3\times\mathbf{k}_t)(\mathbf{e}_3\times\mathbf{k}_t))\cdot\mathbf{E}_{t\pm}(\mathbf{r}) \tag{4.167}$$
$$= \frac{1}{\epsilon_t k_t^2}(\mp\beta\mathbf{k}_t D_{3\pm}(x_3) + \omega\epsilon_t(\mathbf{e}_3\times\mathbf{k}_t)B_{3\pm}(x_3))e^{-j\mathbf{k}_t\cdot\mathbf{r}}, \tag{4.168}$$

and, similarly,

$$\mathbf{H}_{t\pm}(\mathbf{r}) = \frac{1}{\mu_t k_t^2}(\mp\beta\mathbf{k}_t B_{3\pm}(x_3) - \omega\mu_t(\mathbf{e}_3\times\mathbf{k}_t)D_{3\pm}(x_3))e^{-j\mathbf{k}_t\cdot\mathbf{r}}. \tag{4.169}$$

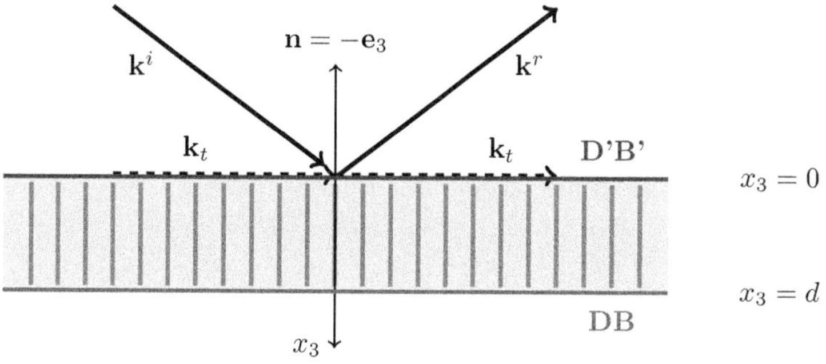

Figure 4.9: Quarter-wavelength layer of uniaxial wave-guiding medium can be used to transform the DB boundary to the D'B' boundary.

Requiring now that the fields satisfy DB conditions $D_3 = B_3 = 0$ at $x_3 = d$, Figure 4.9, we must have

$$D_{3-} = -D_{3+}e^{-2j\beta d}, \quad B_{3-} = -B_{3+}e^{-j2\beta d}, \tag{4.170}$$

whence the total transverse fields become

$$\mathbf{E}_t(\mathbf{r}) = \frac{2}{\epsilon_t k_t^2}\big[-\mathbf{k}_t\beta\cos\beta(d-x_3)\ D_{3+} + (\mathbf{e}_3\times\mathbf{k}_t)j\omega\mu_t\sin\beta(d-x_3)\ B_{3+}\big]e^{-j\beta d}e^{-j\mathbf{k}_t\cdot\mathbf{r}}, \quad (4.171)$$

$$\mathbf{H}_t(\mathbf{r}) = \frac{2}{\mu_t k_t^2}\big[-\mathbf{k}_t\beta\cos\beta(d-x_3)\ B_{3+} - (\mathbf{e}_3\times\mathbf{k}_t)j\omega\mu_t\sin\beta(d-x_3)\ D_{3+}\big]e^{-j\beta d}e^{-j\mathbf{k}_t\cdot\mathbf{r}}. \quad (4.172)$$

At the interface $x_3 = 0$ the tangential fields $\mathbf{E}_t, \mathbf{H}_t$ are continuous:

$$\mathbf{E}_t = \frac{2}{\epsilon_t k_t^2}\big[-\mathbf{k}_t\beta\cos\beta d\ D_{3+} + (\mathbf{e}_3\times\mathbf{k}_t)j\omega\mu_t\sin\beta d\ B_{3+}\big]e^{-j\beta d}e^{-j\mathbf{k}_t\cdot\mathbf{r}}, \quad (4.173)$$

$$\mathbf{H}_t = \frac{2}{\mu_t k_t^2}\big[-\mathbf{k}_t\beta\cos\beta d\ B_{3+} - (\mathbf{e}_3\times\mathbf{k}_t)j\omega\mu_t\sin\beta d\ D_{3+}\big]e^{-j\beta d}e^{-j\mathbf{k}_t\cdot\mathbf{r}}. \quad (4.174)$$

Because the fields have no sources at the boundary, they satisfy

$$\nabla\cdot\mathbf{D} = \partial_{x_3}D_3 - j\mathbf{k}_t\cdot\epsilon_o\mathbf{E}_t = 0 \quad (4.175)$$

$$\nabla\cdot\mathbf{B} = \partial_{x_3}B_3 - j\mathbf{k}_t\cdot\mu_o\mathbf{H}_t = 0, \quad (4.176)$$

whence we can write

$$\partial_{x_3}D_3 = -\frac{2j\beta\epsilon_o}{\epsilon_t}\cos\beta d\ D_{3+}e^{-j\beta d}e^{-j\mathbf{k}_t\cdot\mathbf{r}}, \quad (4.177)$$

$$\partial_{x_3}B_3 = -\frac{2j\beta\mu_o}{\mu_t}\cos\beta d\ B_{3+}e^{-j\beta d}e^{-j\mathbf{k}_t\cdot\mathbf{r}}. \quad (4.178)$$

For

$$\cos\beta d = 0, \quad \Rightarrow \quad d = \frac{\pi}{2\beta} = \frac{\pi}{2\omega\sqrt{\mu_t\epsilon_t}}, \quad (4.179)$$

the D'B' conditions are seen to be valid at the interface $x_3 = 0$.

In conclusion, the boundary with DB conditions at $x_3 = d$ is transformed to a boundary with D'B' conditions at $x_3 = 0$. Defining the wavelength λ in the wave-guiding medium by $\beta\lambda = 2\pi$, whence $d = \lambda/4$, the structure can be called wave-guiding quarter-wave transformer.

Since the transverse parameters of the wave-guiding medium can be freely chosen, in principle, the transformer length d can be made as small as one wishes by letting $\mu_t\epsilon_t$ grow large enough. Recalling that the realization of the DB boundary by a medium with vanishing axial parameters may also lead to a thin sheet of material, the final realization of the D'B' boundary may theoretically be achieved by a thin double sheet. Because the boundary conditions are local, the same thin-sheet implementation apparently remains valid for smoothly curved boundary surfaces as well.

A realization of the DB boundary by uniaxial material with zero axial components can also be exploited to fabricate another boundary condition. Cutting the surface of the uniaxial material in an oblique angle, the structure forces a combination of the normal and tangential components of **D** and **B** to vanish, leading to boundary conditions discussed in [106, 107].

4.7 Mixed-Impedance (DB/D'B') Boundary

Since any plane-wave in an isotropic medium can be split in TE and TM parts with fields defined by (1.55) - (1.58), we can define a mixed-impedance planar boundary by requiring that each of the two partial waves satisfy an impedance condition of their own, in terms of the respective parameters Z_{TE} and $Z_{TM} = 1/Y_{TM}$. The conditions defining such a boundary are, thus,

$$\mathbf{E}_{TE} = Z_{TE}\mathbf{n} \times \mathbf{H}_{TE}, \tag{4.180}$$

$$\mathbf{H}_{TM} = -Y_{TM}\mathbf{n} \times \mathbf{E}_{TM}. \tag{4.181}$$

For $Z_{TE} = 1/Y_{TM} = Z_s$ these reduce to the conditions of the simple-isotropic impedance boundary. In the general case, (4.180) and (4.181) define what can be called conditions of the mixed-impedance boundary [112]. For the two choices, $Z_{TE} = 1/Y_{TM} = 0$ and $Y_{TM} = 1/Z_{TE} = 0$, the mixed-impedance boundary reduces to the respective PEC and PMC boundaries.

Operating (4.180) by $\mathbf{n} \cdot \nabla \times ()$ with constant unit vector $\mathbf{n}$, we can expand

$$\begin{aligned} 0 &= \mathbf{n} \cdot (\nabla \times \mathbf{E}_{TE}) - Z_{TE}\mathbf{n} \cdot (\nabla \times (\mathbf{n} \times \mathbf{H}_{TE})) \\ &= -jk_o\eta_o\mathbf{n} \cdot \mathbf{H}_{TE} + Z_{TE}\nabla \cdot (\mathbf{n} \times (\mathbf{n} \times \mathbf{H}_{TE})). \end{aligned}$$

Applying $\nabla \cdot \mathbf{H}_{TE} = 0$ we obtain

$$(jk_o\eta_o - Z_{TE}\mathbf{n} \cdot \nabla)(\mathbf{n} \cdot \mathbf{H}_{TE}) = 0. \tag{4.182}$$

Because the steps can be reversed, (4.182) is actually equal to (4.180). Similarly, (4.181) is equal to

$$(jk_o - \eta_o Y_{TM}\mathbf{n} \cdot \nabla)(\mathbf{n} \cdot \mathbf{E}_{TM}) = 0. \tag{4.183}$$

Since $\mathbf{n} \cdot \mathbf{H}_{TM} = 0$ and $\mathbf{n} \cdot \mathbf{E}_{TE} = 0$, these conditions are actually valid for any plane waves, or linear combinations of plane waves, as

$$(jk_o\eta_o - Z_{TE}\partial_n)(\mathbf{n} \cdot \mathbf{H}) = 0, \tag{4.184}$$

$$(jk_o - Y_{TM}\eta_o\partial_n)(\mathbf{n} \cdot \mathbf{E}) = 0. \tag{4.185}$$

In this form, the mixed-impedance conditions can be recognized as a generalization of the DB and D'B' boundary conditions and can also be labeled as DB/D'B' conditions. In fact, the DB conditions (4.1) are obtained for the choice $Z_{TE} = Y_{TM} = 0$, i.e., when the boundary equals PEC for the TE wave and PMC for the TM wave. Similarly, the D'B' conditions (4.3) are obtained for $1/Z_{TE} = 1/Y_{TM} = 0$, in which case the boundary equals PMC for the TE wave and PEC for the TM wave.

Reflection Dyadic

For the sum of incident and reflected plane waves, the DB/D'B'-boundary conditions (4.184) and (4.185) become

$$k_o\eta_o\mathbf{n}\cdot(\mathbf{H}^i+\mathbf{H}^r)-k_nZ_{TE}\mathbf{n}\cdot(\mathbf{H}^i-\mathbf{H}^r) = 0, \quad (4.186)$$
$$k_o\mathbf{n}\cdot(\mathbf{E}^i+\mathbf{E}^r)-k_nY_{TM}\eta_o\mathbf{n}\cdot(\mathbf{E}^i-\mathbf{E}^r) = 0. \quad (4.187)$$

Substituting

$$\mathbf{n}\cdot\mathbf{E}^i=\frac{1}{k_n}\mathbf{k}_t\cdot\mathbf{E}_t^i, \qquad \mathbf{n}\cdot\mathbf{E}^r=-\frac{1}{k_n}\mathbf{k}_t\cdot\mathbf{E}_t^r, \quad (4.188)$$

and

$$\mathbf{n}\cdot\mathbf{H}^i=\frac{1}{k_o\eta_o}(\mathbf{n}\times\mathbf{k})\cdot\mathbf{E}_t^i, \qquad \mathbf{n}\cdot\mathbf{H}^r=\frac{1}{k_o\eta_o}(\mathbf{n}\times\mathbf{k})\cdot\mathbf{E}_t^r, \quad (4.189)$$

the above conditions become

$$(k_o\eta_o+k_nZ_{TE})(\mathbf{n}\times\mathbf{k})\cdot\mathbf{E}_t^r = -(k_o\eta_o-k_nZ_{TE})(\mathbf{n}\times\mathbf{k})\cdot\mathbf{E}_t^i \quad (4.190)$$
$$(k_o+k_nY_{TM}\eta_o)\mathbf{k}_t\cdot\mathbf{E}_t^r = (k_o-k_nY_{TM}\eta_o)\mathbf{k}_t\cdot\mathbf{E}_t^i \quad (4.191)$$

Applying the expansion of the 2D unit dyadic $\overline{\overline{\mathsf{I}}}_t$ in the vector basis $\mathbf{k}_t, \mathbf{n}\times\mathbf{k}$, we can construct the reflection dyadic as

$$\overline{\overline{\mathsf{R}}}_t=\frac{1}{k_t^2}\left(\frac{k_o-k_nY_{TM}\eta_o}{k_o+k_nY_{TM}\eta_o}\mathbf{k}_t\mathbf{k}_t-\frac{k_o\eta_o-k_nZ_{TE}}{k_o\eta_o+k_nZ_{TE}}(\mathbf{n}\times\mathbf{k})(\mathbf{n}\times\mathbf{k})\right). \quad (4.192)$$

The eigenvalues of the reflection dyadic can be identified as

$$R_{TE}=-\frac{k_o\eta_o-k_nZ_{TE}}{k_o\eta_o+k_nZ_{TE}}, \qquad R_{TM}=\frac{k_o-k_nY_{TM}\eta_o}{k_o+k_nY_{TM}\eta_o}, \quad (4.193)$$

and the corresponding eigenvectors are

$$\mathbf{E}_{TE}=A_{TE}\mathbf{n}\times\mathbf{k}, \qquad \mathbf{E}_{TMt}=A_{TM}\mathbf{k}_t. \quad (4.194)$$

It is left as an exercise to show that the reflection dyadic of the mixed-impedance boundary reduces to those of the simple-impedance boundary, the DB and D'B' boundaries, and the PEC and PMC boundaries.

The DB/D'B' boundary is self dual when the parameters satisfy $Z_{TE}/Y_{TM}=\eta_o^2$. In such a case the reflection dyadic has the form

$$\overline{\overline{\mathsf{R}}}_t=\left(\frac{k_o\eta_o-k_nZ_{TE}}{k_o\eta_o+k_nZ_{TE}}\right)\frac{1}{k_t^2}(\mathbf{k}_t\mathbf{k}_t-(\mathbf{n}\times\mathbf{k})(\mathbf{n}\times\mathbf{k})), \quad (4.195)$$

i.e., $R_{TE}=-R_{TM}$.

Matched Waves

For the DB/D'B' boundary, the conditions of the matched TE and TM polarized plane waves are obtained by requiring zero and infinite eigenvalues. Both conditions can be expressed for each eigenwave as

$$(k_o\eta_o)^2 = (k_n Z_{TE})^2, \qquad \Rightarrow \qquad \mathbf{n}\cdot\mathbf{k}_{TE} = \pm\frac{k_o\eta_o}{Z_{TE}}, \tag{4.196}$$

$$k_o^2 = (k_n Y_{TM}\eta_o)^2, \qquad \Rightarrow \qquad \mathbf{n}\cdot\mathbf{k}_{TM} = \pm\frac{k_o}{Y_{TM}\eta_o}. \tag{4.197}$$

There are no restrictions for the corresponding field polarizations beyond $\mathbf{n}\cdot\mathbf{E}_{TE} = 0$ and $\mathbf{n}\cdot\mathbf{H}_{TM} = 0$.

For the special case of the simple-isotropic boundary, $Z_{TE} = 1/Y_{TM} = Z_s$, the matched-wave dispersion conditions become

$$\mathbf{n}\cdot\mathbf{k}_{TE} = \pm\frac{k_o\eta_o}{Z_s}, \tag{4.198}$$

$$\mathbf{n}\cdot\mathbf{k}_{TM} = \pm\frac{k_o Z_s}{\eta_o}, \tag{4.199}$$

which coincide with the conditions (3.88) and (3.89).

4.8 Problems

4.1 Show that (4.32) reduces to (4.33) and (4.34) for plane waves respectively polarized TE and TM to the axial direction $\mathbf{e}_3$.

4.2 Derive the expressions of the reflection and transmission coefficients (4.43) and (4.44) corresponding to the zero axial parameter (ZAP) layer.

4.3 Show that the expressions (4.45) and (4.46) can be obtained from (4.43) and (4.44) through duality transformation (1.10) defined by (1.19) with parameter $\varphi = \pi/2$.

4.4 Show that the DB boundary can be realized by an interface of a bianisotropic medium defined by medium conditions of the form (4.68).

4.5 Assuming real $\mathbf{k}^i$ vector, show that the power in the plane wave reflected from the DB boundary equals that of the incident plane wave.

4.6 What happens to the polarization and handedness of a circularly or elliptically polarized plane wave in reflection from a planar DB boundary? Compute the reflected field for a right-handed circularly-polarized (RCP) incident field with incidence angle $\theta = 45°$.

4.7 Compare the expression (4.14) of the reflection dyadic for the DB boundary to to that of the SH boundary, (3.210) with $\mathbf{b}_t = \mathbf{a}_t$, in the case when $\mathbf{a}_t$ and $\mathbf{k}_t$ are parallel vectors.

4.8 Study the possibility of expressing the electromagnetic fields in terms of two vector potentials with only radial components orthogonal to $\mathbf{e}_3$ as $\mathbf{F} = \mathbf{e}_\varrho F$ and $\mathbf{A} = \mathbf{e}_\varrho A$, in the form (4.69) and (4.70).

4.9 Show that when $k_c a$ is close to the value x_{11} in the dispersion equation (4.111), $\beta a(k_o a)$ has a positive slope for $n = 1$ and negative slope for $n = -1$, which aids in distinguishing the two dispersion curves $n = 1$ and $n = -1$ from one another in Figure 4.6.

4.10 Show that the reflection dyadic (4.192) of the mixed-impedance boundary reduces to those of the special cases of simple-impedance boundary, PEC and PMC boundaries.

4.11 Show that the reflection dyadic (4.192) of the mixed-impedance boundary reduces to those of the special cases of DB and D'B' boundaries.

4.12 Show that the DB' and D'B boundaries, defined by the conditions (4.6) and (4.7), are equivalent to the respective PMC and PEC boundaries by considering plane-wave reflection from planar boundaries.

Chapter 5

General Boundary Conditions

5.1 Electromagnetic Sheet as Boundary Surface

Boundary surface can be pictured as a two-dimensional structure (sheet) in which the electromagnetic sources, induced by the external field, interact by some intrinsic mechanism. As sources we adopt electric and magnetic sheet currents, $\mathbf{J}_{es}, \mathbf{J}_{ms}$, and electric and magnetic sheet charges, $\varrho_{es}, \varrho_{ms}$. The interaction between these 2D sources in the sheet is assumed to obey the most general linear and local conditions of the form,

$$\alpha_1 c\varrho_{ms} + \beta_1 c\eta_o\varrho_{es} + \mathbf{a}_{1t}\cdot(\mathbf{n}\times\mathbf{J}_{ms}) - \mathbf{b}_{1t}\cdot\eta_o(\mathbf{n}\times\mathbf{J}_{es}) = 0, \quad (5.1)$$

$$\alpha_2 c\varrho_{ms} + \beta_2 c\eta_o\varrho_{es} + \mathbf{a}_{2t}\cdot(\mathbf{n}\times\mathbf{J}_{ms}) - \mathbf{b}_{2t}\cdot\eta_o(\mathbf{n}\times\mathbf{J}_{es}) = 0, \quad (5.2)$$

where the four scalars $\alpha_1 - \beta_2$ and the four 2D vectors $\mathbf{a}_{1t} - \mathbf{b}_{2t}$ are dimensionless quantities. The surface of the sheet is defined by $\mathbf{n}\cdot\mathbf{r} = 0$, and $c = 1/\sqrt{\mu_o\epsilon_o}$.

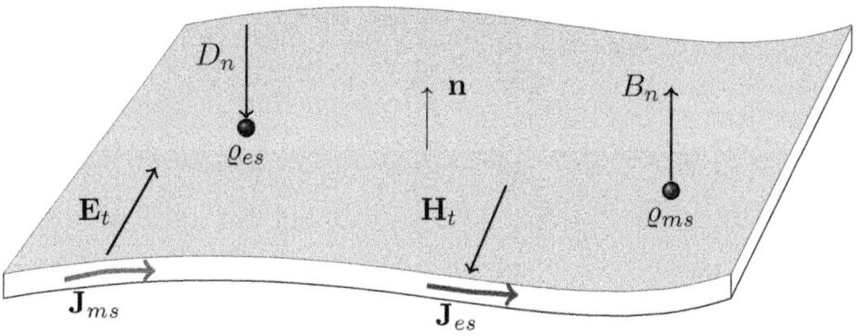

Figure 5.1: Boundary surface represented by a two-dimensional sheet. Interactions between two-dimensional electromagnetic sources give rise to boundary conditions for the exterior fields.

5.2 General Boundary Conditions (GBC)

The fields outside the sheet are related to the 2D sources by the conditions [29]

$$\mathbf{n}\times\mathbf{E} = -\mathbf{J}_{ms}, \qquad \mathbf{n}\times\mathbf{H} = \mathbf{J}_{es}, \tag{5.3}$$

$$\mathbf{n}\cdot\mathbf{D} = \varrho_{es}, \qquad \mathbf{n}\cdot\mathbf{B} = \varrho_{ms}. \tag{5.4}$$

For simplicity, we assume a planar boundary and constant unit vectors $\mathbf{e}_1, \mathbf{e}_2$ and $\mathbf{e}_3 = \mathbf{n}$ making an orthonormal basis. The sources obey the continuity conditions

$$\nabla\cdot\mathbf{J}_{es} = -j\omega\varrho_{es}, \qquad \nabla\cdot\mathbf{J}_{ms} = -j\omega\varrho_{ms}, \tag{5.5}$$

following from the Maxwell equations (1.1), (1.2), and (5.3), (5.4).

Applying the relations (5.3) and (5.4) to the source conditions (5.1) and (5.2) on the sheet, the following set of linear and local conditions for the fields at the sheet are obtained:

$$\alpha_1\mathbf{n}\cdot c\mathbf{B} + \beta_1\mathbf{n}\cdot c\eta_o\mathbf{D} + \mathbf{a}_{1t}\cdot\mathbf{E} + \mathbf{b}_{1t}\cdot\eta_o\mathbf{H} = 0, \tag{5.6}$$

$$\alpha_2\mathbf{n}\cdot c\mathbf{B} + \beta_2\mathbf{n}\cdot c\eta_o\mathbf{D} + \mathbf{a}_{2t}\cdot\mathbf{E} + \mathbf{b}_{2t}\cdot\eta_o\mathbf{H} = 0. \tag{5.7}$$

Boundary conditions of the form (5.6) and (5.7) generalize the impedance-boundary conditions (3.17), (3.18), and the DB-boundary conditions (4.1), to what have been called General Boundary Conditions (GBC) [75]. However, one must bear in mind that they are restricted by assumptions of linearity and locality. As examples of boundary conditions which are linear but not local, conditions of the D'B' boundary, (4.142), and the mixed-impedance (DB/D'B') boundary, (4.184), (4.185), can be mentioned.

Considering the basic problem of a field incident to the boundary, due to the Huygens principle, [29, p.670], the reflected field is uniquely determined when its two scalar components tangential to the surface are known [24, 104, 119]. Thus, the general boundary conditions must be of the form of two scalar conditions relating the normal components of **D** and **B** vectors and tangential components of **E** and **H** vectors at the boundary surface.

Assuming the simply isotropic medium with parameters μ_o, ϵ_o, the GBC boundary conditions (5.6), (5.7) can be written in the simpler form

$$\mathbf{a}_1\cdot\mathbf{E} + \mathbf{b}_1\cdot\eta_o\mathbf{H} = 0, \tag{5.8}$$

$$\mathbf{a}_2\cdot\mathbf{E} + \mathbf{b}_2\cdot\eta_o\mathbf{H} = 0, \tag{5.9}$$

with $\mathbf{a}_i = \beta_i\mathbf{n} + \mathbf{a}_{it} = a_{in}\mathbf{n} + \mathbf{a}_{it}$ and $\mathbf{b}_i = \alpha_i\mathbf{n} + \mathbf{b}_{it} = b_{in}\mathbf{n} + \mathbf{b}_{it}$ for $i = 1, 2$. The GBC conditions are similar to those of the impedance boundary, (3.17), (3.18), except that the vectors $\mathbf{a}_1\cdots\mathbf{b}_2$ are no longer restricted to be tangential to the boundary surface.

Another form of GBC conditions, equivalent to those of (5.6) and (5.7), or (5.8) and (5.9), can be obtained by successively eliminating the normal component of **E** and **H** from the two equations, whence they can be represented in the form

$$a_{1n}\mathbf{n}\cdot\mathbf{E}+\mathbf{a}_{1t}\cdot\mathbf{E}_t+\mathbf{b}_{1t}\cdot\eta_o\mathbf{H}_t = 0, \tag{5.10}$$

$$\mathbf{a}_{2t}\cdot\mathbf{E}_t+b_{2n}\mathbf{n}\cdot\eta_o\mathbf{H}+\mathbf{b}_{2t}\cdot\eta_o\mathbf{H}_t = 0. \tag{5.11}$$

This form shows us that, in the general case, the conditions are defined by the parameters $\mathbf{a}_{1t}/a_{1n}, \mathbf{b}_{1t}/a_{1n}, \mathbf{a}_{2t}/b_{2n}$ and $\mathbf{b}_{2t}/b_{2n}$ amounting to $4\times 2 = 8$, instead of 10, as suggested by (5.6) and (5.7).

Four-Dimensional Representation

The GBC conditions (5.6) and (5.7) can be given a natural form in terms of the four-dimensional coordinate-free formalism of G.A. Deschamps [10, 44, 72]. For the electromagnetic two-forms (see Appendix A)

$$\mathbf{\Phi} = c\mathbf{B}+\mathbf{E}\wedge\boldsymbol{\varepsilon}_4, \tag{5.12}$$

$$\mathbf{\Psi} = c\mathbf{D}-\mathbf{H}\wedge\boldsymbol{\varepsilon}_4, \tag{5.13}$$

they can be expressed as

$$\boldsymbol{\varepsilon}_3\wedge(\boldsymbol{\alpha}_1\wedge\mathbf{\Phi}+\boldsymbol{\beta}_1\wedge\eta_o\mathbf{\Psi}) = 0 \tag{5.14}$$

$$\boldsymbol{\varepsilon}_3\wedge(\boldsymbol{\alpha}_2\wedge\mathbf{\Phi}+\boldsymbol{\beta}_2\wedge\eta_o\mathbf{\Psi}) = 0, \tag{5.15}$$

where $\boldsymbol{\alpha}_1\cdots\boldsymbol{\beta}_2$ are four dimensionless one-forms defining the boundary. Here, $\boldsymbol{\varepsilon}_3$ is a spatial one-form which corresponds to the vector $\mathbf{n}$ (or $\mathbf{e}_3$) in defining the boundary surface by $\boldsymbol{\varepsilon}_3|\mathbf{x}=0$. The two four-form equations (5.14) and (5.15) are equivalent to scalar equations and they represent the simplest possible linear boundary conditions between the two electromagnetic two-forms $\mathbf{\Phi}$ and $\mathbf{\Psi}$.

As an obvious special case of the conditions (5.14) and (5.15) we can write

$$\boldsymbol{\varepsilon}_3\wedge\boldsymbol{\alpha}\wedge\mathbf{\Phi} = 0, \tag{5.16}$$

$$\boldsymbol{\varepsilon}_3\wedge\boldsymbol{\beta}\wedge\mathbf{\Psi} = 0, \tag{5.17}$$

and, as a further special case,

$$\boldsymbol{\varepsilon}_3\wedge\boldsymbol{\alpha}\wedge\mathbf{\Phi} = 0, \tag{5.18}$$

$$\boldsymbol{\varepsilon}_3\wedge\boldsymbol{\alpha}\wedge\mathbf{\Psi} = 0. \tag{5.19}$$

It will turn out that the conditions (5.18) and (5.19) generalize the conditions of the soft-and-hard boundary and the DB boundary, whence they can be called those of the soft-and-hard/DB (SHDB) boundary [69]. The conditions (5.16) and (5.16) in turn generalize the conditions (5.18) and (5.19), whence they have been called those of the generalized soft-and-hard/DB (GSHDB) boundary [74]. Because these two classes of boundaries appear to be basic concepts in electromagnetics when expressed in terms of the four-dimensional formalism, they will be given special attention in this Chapter.

Conditions corresponding to those of the impedance boundary, (3.17) and (3.18), are obtained from (5.14) and (5.15) written in the form

$$\varepsilon_3 \wedge (\boldsymbol{\alpha}_{1s} \wedge \boldsymbol{\Phi} + \boldsymbol{\beta}_{1s} \wedge \eta_o \boldsymbol{\Psi}) = 0 \quad (5.20)$$
$$\varepsilon_3 \wedge (\boldsymbol{\alpha}_{2s} \wedge \boldsymbol{\Phi} + \boldsymbol{\beta}_{2s} \wedge \eta_o \boldsymbol{\Psi}) = 0, \quad (5.21)$$

where $\boldsymbol{\alpha}_{1s} \cdots \boldsymbol{\beta}_{2s}$ are four spatial one-forms, each defined by three, instead of four, scalar components. Actually, (5.20) and (5.21) are equivalent to the scalar equations

$$\mathbf{a}_{1s}|\mathbf{E} - \mathbf{b}_{1s}|\eta_o\mathbf{H} = 0 \quad (5.22)$$
$$\mathbf{a}_{2s}|\mathbf{E} - \mathbf{b}_{2s}|\eta_o\mathbf{H} = 0, \quad (5.23)$$

where the spatial vector $\mathbf{a}_{1s}$ is defined by the contraction operation as

$$\mathbf{a}_{1s} = \mathbf{e}_{123}\rfloor(\varepsilon_3 \wedge \boldsymbol{\alpha}_{1s}) = \mathbf{e}_{12}\rfloor\boldsymbol{\alpha}_{1s}. \quad (5.24)$$

The other three vectors $\mathbf{a}_{2s}, \mathbf{b}_{1s}, \mathbf{b}_{2s}$, are found through similar expressions. Similarity between (5.22) and (5.23) on one hand, and (3.17) and (3.18) on the other hand, is visible.

5.3 Decomposition of Plane Waves

One can show that plane waves can be decomposed in two parts each of which satisfies one of the GBC conditions (5.8) and (5.9) automatically. The form of the GBC conditions suggests a definition of two potential quantities, depending on the boundary parameters $\mathbf{a}_1 \cdots \mathbf{b}_2$, as

$$\phi_1 = k_o(\mathbf{a}_1 \cdot \mathbf{E} + \mathbf{b}_1 \cdot \eta_o\mathbf{H}), \quad (5.25)$$
$$\phi_2 = k_o(\mathbf{a}_2 \cdot \mathbf{E} + \mathbf{b}_2 \cdot \eta_o\mathbf{H}). \quad (5.26)$$

For a given plane wave with wave vector $\mathbf{k}$, the two potentials can be expressed in terms of the electric field and two vectors $\mathbf{c}_1(\mathbf{k}), \mathbf{c}_2(\mathbf{k})$ as

$$\phi_1 = -\mathbf{c}_1(\mathbf{k}) \cdot \mathbf{E}, \quad \phi_2 = -\mathbf{c}_2(\mathbf{k}) \cdot \mathbf{E}, \quad (5.27)$$

and, in terms of the magnetic field and two vectors $\mathbf{d}_1(\mathbf{k}), \mathbf{d}_2(\mathbf{k})$, as

$$\phi_1 = \mathbf{d}_1(\mathbf{k}) \cdot \eta_o\mathbf{H}, \quad \phi_2 = \mathbf{d}_2(\mathbf{k}) \cdot \eta_o\mathbf{H}, \quad (5.28)$$

when the four vector functions are defined by

$$\mathbf{c}_1(\mathbf{k}) = \mathbf{k} \times \mathbf{b}_1 - k_o\mathbf{a}_1, \quad (5.29)$$
$$\mathbf{c}_2(\mathbf{k}) = \mathbf{k} \times \mathbf{b}_2 - k_o\mathbf{a}_2, \quad (5.30)$$
$$\mathbf{d}_1(\mathbf{k}) = \mathbf{k} \times \mathbf{a}_1 + k_o\mathbf{b}_1, \quad (5.31)$$
$$\mathbf{d}_2(\mathbf{k}) = \mathbf{k} \times \mathbf{a}_2 + k_o\mathbf{b}_2. \quad (5.32)$$

One can show that the two pairs of vectors satisfy the relation

$$\begin{aligned} J(\mathbf{k}) &= \mathbf{k}\cdot\mathbf{c}_1(\mathbf{k})\times\mathbf{c}_2(\mathbf{k}) = \mathbf{k}\cdot\mathbf{d}_1(\mathbf{k})\times\mathbf{d}_2(\mathbf{k}) \\ &= k_o^3(\mathbf{a}_2\cdot\mathbf{b}_1 - \mathbf{b}_2\cdot\mathbf{a}_1) + k_o^2\mathbf{k}\cdot(\mathbf{b}_1\times\mathbf{b}_2 + \mathbf{a}_1\times\mathbf{a}_2) \\ &\quad + k_o\mathbf{k}\mathbf{k} : (\mathbf{a}_1\mathbf{b}_2 - \mathbf{b}_1\mathbf{a}_2), \end{aligned} \tag{5.33}$$

details of which are left as an exercise.

For $J(\mathbf{k}) \neq 0$, the vector triples $\{\mathbf{k}, \mathbf{c}_1(\mathbf{k}), \mathbf{c}_2(\mathbf{k})\}$ and $\{\mathbf{k}, \mathbf{d}_1(\mathbf{k}), \mathbf{d}_2(\mathbf{k})\}$ form two sets of basis vectors, whence the unit dyadic can be expanded in two forms as (see Appendix C)

$$\overline{\overline{\mathsf{I}}} = \frac{1}{J(\mathbf{k})}((\mathbf{c}_2(\mathbf{k})\times\mathbf{k})\mathbf{c}_1(\mathbf{k}) + (\mathbf{k}\times\mathbf{c}_1(\mathbf{k}))\mathbf{c}_2(\mathbf{k}) + (\mathbf{c}_1(\mathbf{k})\times\mathbf{c}_2(\mathbf{k}))\mathbf{k}(\mathbf{k})), \tag{5.34}$$

and

$$\overline{\overline{\mathsf{I}}} = \frac{1}{J(\mathbf{k})}((\mathbf{d}_2(\mathbf{k})\times\mathbf{k})\mathbf{d}_1(\mathbf{k}) + (\mathbf{k}\times\mathbf{d}_1(\mathbf{k}))\mathbf{d}_2(\mathbf{k}) + (\mathbf{d}_1(\mathbf{k})\times\mathbf{d}_2(\mathbf{k}))\mathbf{k}). \tag{5.35}$$

Assuming simple isotropic medium ("vacuum"), from $\mathbf{k}\cdot\mathbf{E} = 0$, the electric field vector can be expanded in terms of the two potentials from $\mathbf{E} = \overline{\overline{\mathsf{I}}}\cdot\mathbf{E}$, as

$$\begin{aligned} \mathbf{E} &= \frac{1}{J(\mathbf{k})}((\mathbf{c}_2(\mathbf{k})\times\mathbf{k})\mathbf{c}_1(\mathbf{k}) + (\mathbf{k}\times\mathbf{c}_1(\mathbf{k})))\mathbf{c}_2(\mathbf{k}))\cdot\mathbf{E} \\ &= \frac{1}{J(\mathbf{k})}\mathbf{k}\times(\mathbf{c}_2(\mathbf{k})\phi_1 - \mathbf{c}_1(\mathbf{k})\phi_2). \end{aligned} \tag{5.36}$$

The magnetic field vector can be expanded accordingly as

$$\begin{aligned} \eta_o\mathbf{H} &= \frac{1}{J(\mathbf{k})}((\mathbf{d}_2(\mathbf{k})\times\mathbf{k})\mathbf{d}_1(\mathbf{k}) + (\mathbf{k}\times\mathbf{d}_1(\mathbf{k})))\mathbf{d}_2(\mathbf{k}))\cdot\eta_o\mathbf{H} \\ &= -\frac{1}{J(\mathbf{k})}\mathbf{k}\times(\mathbf{d}_2(\mathbf{k})\phi_1 - \mathbf{d}_1(\mathbf{k})\phi_2). \end{aligned} \tag{5.37}$$

Because of the representations (5.36) and (5.37), any given plane wave can be decomposed in two components,

$$(\mathbf{E}, \mathbf{H}) = (\mathbf{E}_1, \mathbf{H}_1) + (\mathbf{E}_2, \mathbf{H}_2), \tag{5.38}$$

defined by

$$\phi_1 = 0 \;\Rightarrow\; \mathbf{c}_1(\mathbf{k})\cdot\mathbf{E}_1 = 0, \quad \mathbf{d}_1(\mathbf{k})\cdot\mathbf{H}_1 = 0 \quad \mathrm{TE}_{c1} \tag{5.39}$$

$$\phi_2 = 0 \;\Rightarrow\; \mathbf{c}_2(\mathbf{k})\cdot\mathbf{E}_2 = 0, \quad \mathbf{d}_2(\mathbf{k})\cdot\mathbf{H}_2 = 0 \quad \mathrm{TE}_{c2} \tag{5.40}$$

The two partial plane waves can be called either TE_{c1} and TE_{c2} waves, or TM_{d1} and TM_{d2} waves. For simplicity, let us choose the labels TE_{c1} and TE_{c2}. The corresponding fields can be represented by

$$\mathbf{E}_1 = -\frac{1}{J(\mathbf{k})}(\mathbf{k}\times\mathbf{c}_1(\mathbf{k}))\phi_2, \quad (5.41)$$

$$\eta_o\mathbf{H}_1 = \frac{1}{J(\mathbf{k})}(\mathbf{k}\times\mathbf{d}_1(\mathbf{k}))\phi_2, \quad (5.42)$$

$$\mathbf{E}_2 = \frac{1}{J(\mathbf{k})}(\mathbf{k}\times\mathbf{c}_2(\mathbf{k}))\phi_1, \quad (5.43)$$

$$\eta_o\mathbf{H}_2 = -\frac{1}{J(\mathbf{k})}(\mathbf{k}\times\mathbf{d}_2(\mathbf{k}))\phi_1. \quad (5.44)$$

It is left as an exercise to show that the TE_{c1} and TE_{c2} waves satisfy the plane-wave relations separately.

5.4 Reflection from GBC Boundary

Let us adopt the following shorthand notation related to waves incident to and reflected from the GBC boundary:

$$\mathbf{c}_1^{i,r} = \mathbf{c}_1(\mathbf{k}^{i,r}), \quad \mathbf{c}_2^{i,r} = \mathbf{c}_2(\mathbf{k}^{i,r}), \quad (5.45)$$

$$\mathbf{d}_1^{i,r} = \mathbf{d}_1(\mathbf{k}^{i,r}), \quad \mathbf{d}_2^{i,r} = \mathbf{d}_2(\mathbf{k}^{i,r}), \quad (5.46)$$

$$J^i = J(\mathbf{k}^i), \quad J^r = J(\mathbf{k}^r), \quad (5.47)$$

and

$$\phi_1^i = -\mathbf{c}_1^i\cdot\mathbf{E}^i = \mathbf{d}_1^i\cdot\eta_o\mathbf{H}^i, \qquad \phi_1^r = -\mathbf{c}_1^r\cdot\mathbf{E}^r = \mathbf{d}_1^r\cdot\eta_o\mathbf{H}^r, \quad (5.48)$$

$$\phi_2^i = -\mathbf{c}_2^i\cdot\mathbf{E}^i = \mathbf{d}_2^i\cdot\eta_o\mathbf{H}^i, \qquad \phi_2^r = -\mathbf{c}_2^r\cdot\mathbf{E}^r = \mathbf{d}_2^r\cdot\eta_o\mathbf{H}^r. \quad (5.49)$$

The GBC boundary conditions (5.8), (5.9) can now be expressed in terms of the potential quantities as

$$\phi_1^i + \phi_1^r = 0, \quad (5.50)$$

$$\phi_2^i + \phi_2^r = 0. \quad (5.51)$$

Applying (5.27) and (5.28) for the incident and reflected fields, the GBC boundary conditions take the form

$$\mathbf{c}_1^i\cdot\mathbf{E}^i + \mathbf{c}_1^r\cdot\mathbf{E}^r = 0, \quad (5.52)$$

$$\mathbf{c}_2^i\cdot\mathbf{E}^i + \mathbf{c}_2^r\cdot\mathbf{E}^r = 0, \quad (5.53)$$

or,

$$\mathbf{d}_1^i\cdot\mathbf{H}^i + \mathbf{d}_1^r\cdot\mathbf{H}^r = 0, \quad (5.54)$$

$$\mathbf{d}_2^i\cdot\mathbf{H}^i + \mathbf{d}_2^r\cdot\mathbf{H}^r = 0. \quad (5.55)$$

Fields of the incident and reflected TE_{c1} and TE_{c2} waves can be decomposed according to (5.41) - (5.44) as

$$\mathbf{E}_1^{i,r} = -\frac{1}{J^{i,r}}(\mathbf{k}^{i,r}\times\mathbf{c}_1^{i,r})\phi_2^{i,r}, \tag{5.56}$$

$$\eta_o\mathbf{H}_1^{i,r} = \frac{1}{J^{i,r}}(\mathbf{k}^{i,r}\times\mathbf{d}_1^{i,r})\phi_2^{i,r}, \tag{5.57}$$

$$\mathbf{E}_2^{i,r} = \frac{1}{J^{i,r}}(\mathbf{k}^{i,r}\times\mathbf{c}_2^{i,r})\phi_1^{i,r}, \tag{5.58}$$

$$\eta_o\mathbf{H}_2^{i,r} = -\frac{1}{J^{i,r}}(\mathbf{k}^{i,r}\times\mathbf{d}_2^{i,r})\phi_1^{i,r}. \tag{5.59}$$

From the GBC conditions (5.50) and (5.51) we see that $\phi_1^i = 0$ implies $\phi_1^r = 0$ and $\phi_2^i = 0$ implies $\phi_2^r = 0$. Thus, for a GBC boundary, there is no cross coupling between the TE_{c1} and TE_{c2} waves in reflection. This allows us to study reflection of both partial waves separately.

Assuming an incident TE_{c1} wave with $\phi_1^i = \phi_1^r = 0$, from (5.56) and (5.51) the reflected wave becomes

$$\mathbf{E}_1^r = \frac{1}{J^r}(\mathbf{k}^r\times\mathbf{c}_1^r)\phi_2^i = -\frac{1}{J^r}(\mathbf{k}^r\times\mathbf{c}_1^r)\mathbf{c}_2^i\cdot\mathbf{E}_1^i. \tag{5.60}$$

Similarly, for an incident TE_{c2} wave, from (5.58) and (5.50) we obtain

$$\mathbf{E}_2^r = -\frac{1}{J^r}(\mathbf{k}^r\times\mathbf{c}_2^r)\phi_1^i = \frac{1}{J^r}(\mathbf{k}^r\times\mathbf{c}_2^r)\mathbf{c}_1^i\cdot\mathbf{E}_2^i. \tag{5.61}$$

For an incident wave of any polarization, the reflection dyadic can be constructed by combining the two components as

$$\mathbf{E}^r = \mathbf{E}_1^r + \mathbf{E}_2^r = \overline{\overline{\mathsf{R}}}\cdot\mathbf{E}^i, \tag{5.62}$$

with the reflection dyadic defined by

$$\overline{\overline{\mathsf{R}}} = \frac{1}{J^r}\mathbf{k}^r\times(\mathbf{c}_2^r\mathbf{c}_1^i - \mathbf{c}_1^r\mathbf{c}_2^i). \tag{5.63}$$

The inverse of the reflection dyadic can be derived similarly and the result is

$$\overline{\overline{\mathsf{R}}}^{-1} = \frac{1}{J^i}\mathbf{k}^i\times(\mathbf{c}_2^i\mathbf{c}_1^r - \mathbf{c}_1^i\mathbf{c}_2^r). \tag{5.64}$$

Expanding the reflected magnetic field as

$$\begin{aligned}\eta_o\mathbf{H}^r &= \frac{1}{k_o}\mathbf{k}^r\times\mathbf{E}^r = \frac{1}{k_o}\mathbf{k}^r\times\overline{\overline{\mathsf{R}}}\cdot\mathbf{E}^i \\ &= -\frac{1}{k_o^2}\mathbf{k}^r\times\overline{\overline{\mathsf{R}}}\cdot(\mathbf{k}^i\times\eta_o\mathbf{H}^i),\end{aligned} \tag{5.65}$$

we obtain

$$\mathbf{H}^r = \frac{1}{k_o^2}(\mathbf{k}^r\mathbf{k}^i{}^\times_\times\overline{\overline{\mathsf{R}}})\cdot\mathbf{H}^i. \tag{5.66}$$

PEC Boundary

As a simple special case, let us check the reflection dyadic (5.63) for the PEC boundary, defined by

$$\mathbf{b}_1 = \mathbf{b}_2 = 0, \quad \mathbf{a}_1 = \mathbf{e}_1, \quad \mathbf{a}_2 = \mathbf{e}_2. \tag{5.67}$$

From (5.29) and (5.30) we obtain

$$\mathbf{c}_1(\mathbf{k}) = -k_o\mathbf{e}_1 = \mathbf{c}_1^i = \mathbf{c}_1^r, \tag{5.68}$$
$$\mathbf{c}_2(\mathbf{k}) = -k_o\mathbf{e}_2 = \mathbf{c}_2^i = \mathbf{c}_2^r, \tag{5.69}$$
$$J(\mathbf{k}) = k_o^2\mathbf{e}_1 \times \mathbf{e}_2 \cdot \mathbf{k} = k_o^2\mathbf{n}\cdot\mathbf{k}. \tag{5.70}$$

The reflection dyadic (5.63) can now be expanded as

$$\begin{aligned}\overline{\overline{\mathsf{R}}} &= \frac{1}{k_o^2\mathbf{n}\cdot\mathbf{k}^r}\mathbf{k}^r \times (\mathbf{c}_2\mathbf{c}_1 - \mathbf{c}_1\mathbf{c}_2)\\ &= \frac{1}{k_n}\mathbf{k}^r \times (\mathbf{e}_2\mathbf{e}_1 - \mathbf{e}_1\mathbf{e}_2)\\ &= \frac{1}{k_n}\mathbf{k}^r \times (\mathbf{n}\times\overline{\overline{\mathsf{I}}}) = -\overline{\overline{\mathsf{I}}}_t + \frac{\mathbf{n}\mathbf{k}_t}{k_n}.\end{aligned} \tag{5.71}$$

For an incident field $\mathbf{E}^i$ satisfying $\mathbf{k}^i\cdot\mathbf{E}^i = -k_nE_n^i + \mathbf{k}_t\cdot\mathbf{E}^i = 0$, we obtain

$$\mathbf{E}^r = -\mathbf{E}_t^i + \frac{\mathbf{n}}{k_n}\mathbf{k}_t\cdot\mathbf{E}^i = -\mathbf{E}_t^i + E_n^i\mathbf{n}, \tag{5.72}$$

whence the total field $\mathbf{E}^i + \mathbf{E}^r = 2E_n^i\mathbf{n}$ is seen to satisfy the PEC condition $\mathbf{n}\times\mathbf{E} = 0$.

It is left as an exercise to verify the reflection dyadic (5.63) for the special cases of the DB boundary and the GSH boundary.

5.5 Matched Waves

The expression (5.63) of the reflection dyadic fails for $J^r = J(\mathbf{k}^r) = 0$ and its inverse (5.64) fails for $J^i(\mathbf{k}^i) = 0$. In the former case there may exist a reflected wave for no incident wave and, in the latter case, the converse may be true. Actually, both of these define conditions for matched waves associated with the GBC boundary and since there is only a single wave, the attributes "incident" and "reflected" bear no meaning. Actually, they turn out to correspond to the same condition for a matched wave.

To verify this, let us consider a single plane wave with wave vector $\mathbf{k}$ and require that it satisfy the GBC conditions identically. Expressing (5.8) and (5.9) as

$$\phi_1 = k_o(\mathbf{a}_1\cdot\mathbf{E} + \mathbf{b}_1\cdot\eta_o\mathbf{H}) = -\mathbf{c}_1(\mathbf{k})\cdot\mathbf{E} = 0, \tag{5.73}$$
$$\phi_2 = k_o(\mathbf{a}_2\cdot\mathbf{E} + \mathbf{b}_2\cdot\eta_o\mathbf{H}) = -\mathbf{c}_2(\mathbf{k})\cdot\mathbf{E} = 0, \tag{5.74}$$

and applying $\mathbf{k}\cdot\mathbf{E}=0$, we conclude that, to have a solution $\mathbf{E}\neq 0$, the vectors $\mathbf{c}_1(\mathbf{k}), \mathbf{c}_2(\mathbf{k})$ and $\mathbf{k}$ must be linearly dependent. Thus, for a matched wave, the $\mathbf{k}$ vector must satisfy the dispersion equation

$$\begin{aligned} J(\mathbf{k}) &= \mathbf{c}_1(\mathbf{k})\times\mathbf{c}_2(\mathbf{k})\cdot\mathbf{k} \\ &= (\mathbf{k}\times\mathbf{b}_1-k_o\mathbf{a}_1)\times(\mathbf{k}\times\mathbf{b}_2-k_o\mathbf{a}_2)\cdot\mathbf{k}=0. \end{aligned} \quad (5.75)$$

In spite of its cubic appearance, the dispersion equation is actually quadratic in $\mathbf{k}$. This can be seen from its expansion

$$(\mathbf{a}_1\mathbf{b}_2-\mathbf{b}_1\mathbf{a}_2):\mathbf{kk}+k_o(\mathbf{a}_1\times\mathbf{a}_2+\mathbf{b}_1\times\mathbf{b}_2)\cdot\mathbf{k}+k_o^2(\mathbf{a}_2\cdot\mathbf{b}_1-\mathbf{a}_1\cdot\mathbf{b}_2)=0. \quad (5.76)$$

It can be given another form as

$$(\mathbf{a}_1\times\mathbf{k})\cdot(\mathbf{b}_2\times\mathbf{k})-(\mathbf{b}_1\times\mathbf{k})\cdot(\mathbf{a}_2\times\mathbf{k})=k_o(\mathbf{a}_1\times\mathbf{a}_2+\mathbf{b}_1\times\mathbf{b}_2)\cdot\mathbf{k}. \quad (5.77)$$

Because the GBC conditions (5.8) and (5.9) for a single plane wave can be expressed in the form (5.73) and (5.74) for the electric field, or in the form

$$\mathbf{d}_1(\mathbf{k})\cdot\mathbf{H}=0, \quad \mathbf{d}_2(\mathbf{k})\cdot\mathbf{H}=0, \quad (5.78)$$

for the magnetic field, the polarizations of a wave matched to the GBC boundary must be of the form

$$\mathbf{E} = A\mathbf{c}_1(\mathbf{k})\times\mathbf{c}_2(\mathbf{k}), \quad (5.79)$$

$$\mathbf{H} = B\mathbf{k}\times(\mathbf{c}_1(\mathbf{k})\times\mathbf{c}_2(\mathbf{k})). \quad (5.80)$$

or

$$\mathbf{H} = C\mathbf{d}_1(\mathbf{k})\times\mathbf{d}_2(\mathbf{k}), \quad (5.81)$$

$$\mathbf{E} = D\mathbf{k}\times(\mathbf{d}_1(\mathbf{k})\times\mathbf{d}_2(\mathbf{k})), \quad (5.82)$$

in terms of some coefficients $A\cdots D$. Actually, the dispersion equation (5.75) follows from $\mathbf{k}\cdot\mathbf{E}=0$ or $\mathbf{k}\cdot\mathbf{H}=0$.

Special Cases

Let us consider the dispersion equation (5.75) for some special cases of the GBC boundary.

- For the **DB boundary** with $\mathbf{a}_1=\mathbf{b}_2=\mathbf{n}$ and $\mathbf{a}_2=\mathbf{b}_1=0$, we have $\mathbf{c}_1(\mathbf{k})=-k_o\mathbf{n}$ and $\mathbf{c}_2(\mathbf{k})=\mathbf{k}\times\mathbf{n}$, whence the dispersion equation (5.75) becomes

$$(\mathbf{n}\times(\mathbf{k}\times\mathbf{n}))\cdot\mathbf{k}=\mathbf{k}_t\cdot\mathbf{k}_t=0. \quad (5.83)$$

  It is known that any plane wave with normal incidence, $\mathbf{k}_t=0$, is matched to the DB boundary [74, 75]. However, complex wave vectors of the form $\mathbf{k}_t=A\mathbf{v}_t\pm j\mathbf{n}\times\mathbf{v}_t$ define other possible matched waves for any tangential vector $\mathbf{v}_t$.

- For the **GSH boundary** with $\mathbf{b}_1 = \mathbf{a}_2 = 0$, $\mathbf{a}_1 = \mathbf{a}_t$ and $\mathbf{b}_2 = \mathbf{b}_t$, we have $\mathbf{c}_1(\mathbf{k}) = -k_o\mathbf{a}_t$ and $\mathbf{c}_2(\mathbf{k}) = \mathbf{k} \times \mathbf{b}_t$, whence the dispersion equation (5.77) takes the form

$$(\mathbf{a}_t \times \mathbf{k}) \cdot (\mathbf{b}_t \times \mathbf{k}) = k_o^2(\mathbf{a}_t \cdot \mathbf{b}_t) - (\mathbf{k}_t \cdot \mathbf{a}_t)(\mathbf{k}_t \cdot \mathbf{b}_t) = 0. \tag{5.84}$$

This coincides with the result (3.215).

- For the **simple-isotropic impedance boundary** with $\mathbf{a}_1 = \mathbf{e}_1$, $\mathbf{b}_1 = \mathbf{e}_2 Z_s/\eta_o$, $\mathbf{a}_2 = \mathbf{e}_2$ and $\mathbf{b}_2 = -\mathbf{e}_1 Z_s/\eta_o$, we have

$$\mathbf{c}_1(\mathbf{k}) = \frac{Z_s}{\eta_o}\mathbf{k} \times \mathbf{e}_2 - k_o\mathbf{e}_1, \qquad \mathbf{c}_2(\mathbf{k}) = -\frac{Z_s}{\eta_o}\mathbf{k} \times \mathbf{e}_1 - k_o\mathbf{e}_2. \tag{5.85}$$

The dispersion equation (5.76) is reduced to an equation for $k_3 = \mathbf{e}_3 \cdot \mathbf{k}$ as

$$k_3^2 + \frac{Z_s^2 + \eta_o^2}{Z_s\eta_o} k_o k_3 + k_o^2 = 0, \tag{5.86}$$

with the two solutions

$$k_3 = -k_o Z_s/\eta_o, \qquad k_3 = -k_o\eta_o/Z_s. \tag{5.87}$$

Because, for $\mathbf{k}^i$ and $\mathbf{k}^r$, k_3 respectively equals $-k_n$ and $+k_n$, the solutions coincide with those given in (3.90) and (3.91).

5.6 Eigenwaves

For a combination of incident and reflected plane waves, $\mathbf{E} = \mathbf{E}^i + \mathbf{E}^r$, $\mathbf{H} = \mathbf{H}^i + \mathbf{H}^r$, we can expand for $j = 1, 2$,

$$\mathbf{a}_j \cdot \mathbf{E} = \mathbf{a}_{jt} \cdot \mathbf{E}_t - \frac{a_{jn}}{k_o}(\mathbf{n} \times \mathbf{k}_t) \cdot \eta_o\mathbf{H}_t, \tag{5.88}$$

$$\mathbf{b}_j \cdot \eta_o\mathbf{H} = \mathbf{b}_{jt} \cdot \eta_o\mathbf{H}_t + \frac{b_{jn}}{k_o}(\mathbf{n} \times \mathbf{k}_t) \cdot \mathbf{E}_t. \tag{5.89}$$

Let us define

$$\mathbf{a}'_{jt} = \mathbf{a}_{jt} + \frac{b_{jn}}{k_o}\mathbf{n} \times \mathbf{k}_t, \tag{5.90}$$

$$\mathbf{b}'_{jt} = \mathbf{b}_{jt} - \frac{a_{jn}}{k_o}\mathbf{n} \times \mathbf{k}_t, \tag{5.91}$$

in terms of which the GBC conditions (5.8) and (5.9) can be expressed in the form of impedance-boundary conditions (3.17) and (3.18), involving the tangential field components, as

$$\mathbf{a}'_{1t} \cdot \mathbf{E}_t + \mathbf{b}'_{1t} \cdot \eta_o\mathbf{H}_t = 0, \tag{5.92}$$

$$\mathbf{a}'_{2t} \cdot \mathbf{E}_t + \mathbf{b}'_{2t} \cdot \eta_o\mathbf{H}_t = 0. \tag{5.93}$$

However, one must bear in mind that, in this form, the GBC conditions are valid for the fields of a sum of incident and reflected plane waves, only.

Eigenwaves are defined by requiring that the tangential field components of the reflected fields have the same polarizations as the corresponding incident fields. Considering an eigenwave reflecting from a GBC surface, we can substitute

$$\mathbf{E}_t^r = R\mathbf{E}_t^i, \tag{5.94}$$

and, applying $\eta_o\mathbf{H}_t^i = -\overline{\overline{\mathsf{J}}}_t \cdot \mathbf{E}_t^i$,

$$\eta_o\mathbf{H}_t^r = \overline{\overline{\mathsf{J}}}_t \cdot \mathbf{E}_t^r = R\overline{\overline{\mathsf{J}}}_t \cdot \mathbf{E}_t^i = -R\eta_o\mathbf{H}_t^i. \tag{5.95}$$

In terms of these, the GBC conditions (5.92) and (5.93) become

$$((1+R)\mathbf{a}'_{1t} - (1-R)\mathbf{b}'_{1t} \cdot \overline{\overline{\mathsf{J}}}_t) \cdot \mathbf{E}_t^i = 0, \tag{5.96}$$

$$((1+R)\mathbf{a}'_{2t} - (1-R)\mathbf{b}'_{2t} \cdot \overline{\overline{\mathsf{J}}}_t) \cdot \mathbf{E}_t^i = 0. \tag{5.97}$$

For solutions $\mathbf{E}_t^i \neq 0$, the two tangential vector expressions in brackets must be linearly dependent. This yields an equation for the eigenvalue R,

$$\mathbf{n} \cdot ((1+R)\mathbf{a}'_{1t} - (1-R)\mathbf{b}'_{1t} \cdot \overline{\overline{\mathsf{J}}}_t) \times ((1+R)\mathbf{a}'_{2t} - (1-R)\mathbf{b}'_{2t} \cdot \overline{\overline{\mathsf{J}}}_t) = 0. \tag{5.98}$$

Being quadratic in R, it can be straightforwardly solved in analytical form, after which the tangential component of the eigenfield corresponding to an eigenvalue R is obtained in the form

$$\mathbf{E}_t^i = A\mathbf{n} \times ((1+R)\mathbf{a}'_{1t} - (1-R)\mathbf{b}'_{1t} \cdot \overline{\overline{\mathsf{J}}}_t), \tag{5.99}$$

$$= B\mathbf{n} \times ((1+R)\mathbf{a}'_{2t} - (1-R)\mathbf{b}'_{2t} \cdot \overline{\overline{\mathsf{J}}}_t), \tag{5.100}$$

for some scalars A and B.

Simple-Isotropic Impedance Boundary

As an example, let us consider the simple-isotropic impedance boundary (3.74), for which we can set

$$\mathbf{a}_1 = \mathbf{e}_1 = \mathbf{a}'_{1t}, \qquad \mathbf{a}_2 = \mathbf{e}_2 = \mathbf{a}'_{2t}, \tag{5.101}$$

$$\mathbf{b}_1 = \mathbf{e}_2 Z_s/\eta_o = \mathbf{b}'_{1t}, \qquad \mathbf{b}_2 = -\mathbf{e}_1 Z_s/\eta_o = \mathbf{b}'_{2t}, \tag{5.102}$$

where $\mathbf{e}_1$ and $\mathbf{e}_2$ are two orthogonal tangential unit vectors. After some steps, the eigenvalue equation (5.98) can be given the form

$$\left(R - \frac{k_n Z_s - k_o\eta_o}{k_n Z_s + k_o\eta_o}\right)\left(R - \frac{k_o Z_s - k_n\eta_o}{k_o Z_s + k_n\eta_o}\right) = 0. \tag{5.103}$$

Details of the derivation are left as an exercise. The eigenvalues, and the respective eigenpolarizations found from (5.99), are

$$R = \frac{k_n Z_s - k_o\eta_o}{k_n Z_s + k_o\eta_o}, \quad \mathbf{E}_t^i = A\mathbf{n} \times \mathbf{k}_t, \tag{5.104}$$

$$R = \frac{k_o Z_s - k_n\eta_o}{k_o Z_s + k_n\eta_o}, \quad \mathbf{E}_t^i = A\mathbf{k}_t. \tag{5.105}$$

For the former eigenfield, from $\mathbf{k}^i \cdot \mathbf{E}^i = -k_n \mathbf{n} \cdot \mathbf{E}^i = 0$, the polarization is obviously TE$_n$. For the latter eigenfield, from $k_o \eta_o \mathbf{H}^i = \mathbf{k}^i \times \mathbf{E}^i \sim -k_n \mathbf{n} \times \mathbf{k}_t$ it is TM$_n$. Actually, the eigenvalues (5.104) and (5.105) reproduce the reflection coefficients (3.83) and (3.84), corresponding to waves incident with respective TE and TM polarizations.

Other special cases will be considered in subsequent Sections.

5.7 Duality Transformation

Applying the duality transformation of fields (1.10), with transformation parameters defined by (1.18) to leave the simple isotropic medium invariant, the GBC conditions (5.8), (5.9), written as

$$\begin{pmatrix} \mathbf{a}_1 & \mathbf{b}_1 \\ \mathbf{a}_2 & \mathbf{b}_2 \end{pmatrix} \cdot \begin{pmatrix} \mathbf{E} \\ \eta_o \mathbf{H} \end{pmatrix} = \begin{pmatrix} 0 \\ 0 \end{pmatrix}, \tag{5.106}$$

can be transformed to dual GBC conditions as

$$\begin{pmatrix} \mathbf{a}_{1d} & \mathbf{b}_{1d} \\ \mathbf{a}_{2d} & \mathbf{b}_{2d} \end{pmatrix} \cdot \begin{pmatrix} \mathbf{E}_d \\ \eta_o \mathbf{H}_d \end{pmatrix} = \begin{pmatrix} 0 \\ 0 \end{pmatrix}. \tag{5.107}$$

The dual boundary vectors are determined by

$$\begin{pmatrix} \mathbf{a}_{1d} & \mathbf{b}_{1d} \\ \mathbf{a}_{2d} & \mathbf{b}_{2d} \end{pmatrix} = \begin{pmatrix} \mathbf{a}_1 & \mathbf{b}_1 \\ \mathbf{a}_2 & \mathbf{b}_2 \end{pmatrix} \begin{pmatrix} \cos\varphi & -\sin\varphi \\ \sin\varphi & \cos\varphi \end{pmatrix}$$

$$= \begin{pmatrix} \mathbf{a}_1 \cos\varphi + \mathbf{b}_1 \sin\varphi & -\mathbf{a}_1 \sin\varphi + \mathbf{b}_1 \cos\varphi \\ \mathbf{a}_2 \cos\varphi + \mathbf{b}_2 \sin\varphi & -\mathbf{a}_2 \sin\varphi + \mathbf{b}_2 \cos\varphi \end{pmatrix}. \tag{5.108}$$

The dual of the set of vectors (5.29) - (5.32) are found as

$$\mathbf{c}_{1d}(\mathbf{k}) = \mathbf{c}_1(\mathbf{k}) \cos\varphi - \mathbf{d}_1(\mathbf{k}) \sin\varphi, \tag{5.109}$$

$$\mathbf{c}_{2d}(\mathbf{k}) = \mathbf{c}_2(\mathbf{k}) \cos\varphi - \mathbf{d}_2(\mathbf{k}) \sin\varphi, \tag{5.110}$$

$$\mathbf{d}_{1d}(\mathbf{k}) = \mathbf{d}_1(\mathbf{k}) \cos\varphi + \mathbf{c}_1(\mathbf{k}) \sin\varphi, \tag{5.111}$$

$$\mathbf{d}_{2d}(\mathbf{k}) = \mathbf{d}_2(\mathbf{k}) \cos\varphi + \mathbf{c}_2(\mathbf{k}) \sin\varphi. \tag{5.112}$$

Substituting these, we arrive at the property

$$\begin{aligned} J_d(\mathbf{k}) &= \mathbf{k} \cdot \mathbf{c}_{1d}(\mathbf{k}) \times \mathbf{c}_{2d}(\mathbf{k}) \\ &= \mathbf{k} \cdot \mathbf{d}_{1d}(\mathbf{k}) \times \mathbf{d}_{2d}(\mathbf{k}) \\ &= J(\mathbf{k}), \end{aligned} \tag{5.113}$$

details of which are left as an exercise. Because the function $J(\mathbf{k})$ is invariant in the duality transformation, so is the dispersion equation $J(\mathbf{k}) = 0$ of the matched wave. This can be compared with the known fact that the dispersion equation of a plane wave in a bi-anisotropic medium is invariant in a duality transformation of medium [72].

5.8 Soft-and-Hard/DB (SHDB) Boundary

In the four-dimensional formalism, the GBC conditions (5.14) and (5.15) contain the natural special case (5.18) and (5.19). In the corresponding three-dimensional formalism, (5.6) and (5.7) are reduced to conditions of the form

$$\alpha\mathbf{n}\cdot c\mathbf{B} + \mathbf{a}_t\cdot\mathbf{E}_t = 0, \tag{5.114}$$

$$\alpha\mathbf{n}\cdot c\mathbf{D} - \mathbf{a}_t\cdot\mathbf{H}_t = 0. \tag{5.115}$$

Because these conditions appear as a generalization of the soft-and-hard (SH) and DB boundary conditions, they have been called conditions of the soft-and-hard/DB (SHDB) boundary [69]. In fact, for $\mathbf{a}_t = 0$, (5.114) and (5.115) are reduced to the conditions of the DB boundary (4.1), while the SH-boundary conditions (3.179) are obtained for $\alpha = 0$. The minus sign in (5.115) is based on the minus sign in one of the Maxwell equations. Because in (5.18) and (5.19) the boundary depends on a single one-form $\boldsymbol{\alpha}$, this allows us to rank the SHDB conditions as one of the most fundamental boundary conditions in electromagnetics. Applying the Maxwell equations, for the simple-isotropic medium, the SHDB conditions (5.114), (5.115) can be expressed in operator form as

$$L(\nabla_t)\cdot\mathbf{E} = 0, \qquad L(\nabla_t)\cdot\mathbf{H} = 0, \tag{5.116}$$

with

$$L(\nabla_t) = \alpha\mathbf{n}\times\nabla_t - jk_o\mathbf{a}_t. \tag{5.117}$$

The SHDB conditions can be expressed for the simple-isotropic medium as

$$\mathbf{a}_t\cdot\mathbf{E} + \alpha\mathbf{n}\cdot\eta_o\mathbf{H} = 0 \tag{5.118}$$

$$\alpha\mathbf{n}\cdot\mathbf{E} - \mathbf{a}_t\cdot\eta_o\mathbf{H} = 0. \tag{5.119}$$

These can be obtained from the GBC conditions (5.8) and (5.9) through the substitutions $\mathbf{a}_1 = \mathbf{a}_t$, $\mathbf{a}_2 = \alpha\mathbf{n}$, $\mathbf{b}_1 = \alpha\mathbf{n}$ and $\mathbf{b}_2 = -\mathbf{a}_t$.

Considering a plane wave reflecting from an SHDB boundary, the vectors (5.29) - (5.32) take the form

$$\mathbf{c}_1(\mathbf{k}) = \alpha\mathbf{k}_t\times\mathbf{n} - k_o\mathbf{a}_t, \tag{5.120}$$

$$\mathbf{c}_2(\mathbf{k}) = -\mathbf{k}\times\mathbf{a}_t - k_o\alpha\mathbf{n}, \tag{5.121}$$

$$\mathbf{d}_1(\mathbf{k}) = \mathbf{k}\times\mathbf{a}_t + k_o\alpha\mathbf{n} \tag{5.122}$$

$$\mathbf{d}_2(\mathbf{k}) = \alpha\mathbf{k}_t\times\mathbf{n} - k_o\mathbf{a}_t. \tag{5.123}$$

Here we note that $\mathbf{c}_1^i = \mathbf{c}_1^r$ and $\mathbf{d}_2^i = \mathbf{d}_2^r$, and they have no normal components. Also, (5.33) now becomes

$$J(\mathbf{k}) = k_o((\mathbf{n}\times\mathbf{k}_t\cdot\mathbf{a}_t + \alpha k_o)^2 + (\mathbf{a}_t^2 - \alpha^2)k_n^2). \tag{5.124}$$

Because the right-hand side is independent of the sign of k_n, we conclude that, for the SHDB boundary, we can write

$$J(\mathbf{k}^i) = J(\mathbf{k}^r) = J. \tag{5.125}$$

If $\mathbf{a}_t$ is a real vector, we can set $\mathbf{a}_t = a\mathbf{e}_1$, in which case we have

$$J = k_o((ak_2 - \alpha k_o)^2 + (a^2 - \alpha^2)k_n^2). \tag{5.126}$$

Decomposing a plane wave in its TE_{c1} and TE_{c2} components, satisfying respectively $\mathbf{c}_1 \cdot \mathbf{E}_1 = 0$ and $\mathbf{c}_2 \cdot \mathbf{E}_2 = 0$, the fields can be shown to satisfy the orthogonality conditions

$$\mathbf{E}_1 \cdot \mathbf{E}_2 = 0, \qquad \mathbf{H}_1 \cdot \mathbf{H}_2 = 0. \tag{5.127}$$

Details of the proof are left as an exercise.

Because for the SHDB boundary (5.90) and (5.91) yield $\mathbf{b}'_{1t} = \mathbf{a}'_{2t} = 0$, the eigenvalue equation (5.98) is reduced to $(1 + R)(1 - R) = 0$. Thus, the eigenvalues of the reflection dyadic $\overline{\overline{\mathsf{R}}}_t$ are $R = \pm 1$.

Matched Waves

The wave vector $\mathbf{k}$ of a matched wave associated to the SHDB boundary [71] is required to satisfy the dispersion equation $J(\mathbf{k}){=}0$. Applying (5.126), it can be written in the form

$$(\alpha k_2 - ak_o)^2 + (\alpha^2 - a^2)k_1^2 = 0, \tag{5.128}$$

or

$$\alpha k_2 \pm \sqrt{a^2 - \alpha^2}k_1 - ak_o = 0. \tag{5.129}$$

This equals a set of two linear equations for $\mathbf{k}_t$. For real a, α and $\sqrt{a^2 - \alpha^2}$, the dispersion diagram consists of two straight lines on the plane of the boundary. They intersect at the point $k_1 = 0$, $k_2 = ak_o/\alpha$. The dispersion equations (5.129) can be compactly expressed as

$$\mathbf{k}_t \cdot \mathbf{u}_{t\pm} = k_o, \tag{5.130}$$

in terms of the (possibly complex) unit vectors

$$\mathbf{u}_{t\pm} = \mathbf{e}_2(\alpha/a) \pm \mathbf{e}_1\sqrt{1 - (\alpha/a)^2}, \quad \mathbf{u}_{t\pm} \cdot \mathbf{u}_{t\pm} = 1. \tag{5.131}$$

(5.129) is visualized by Figures 5.2 and 5.3.

- For the **DB boundary** with $a = 0$ the dispersion equation yields $k_o^2 - k_n^2 = \mathbf{k}_t \cdot \mathbf{k}_t = 0$ which includes the normally incident wave $\mathbf{k}_t = 0$.

- For the **SH boundary** with $\alpha = 0$ we obtain $k_1 = \pm k_o$, which corresponds to lateral waves propagating along the vectors $\pm\mathbf{e}_1$.

The dispersion properties for these two special cases coincide with those obtained in Chapters 3 and 4.

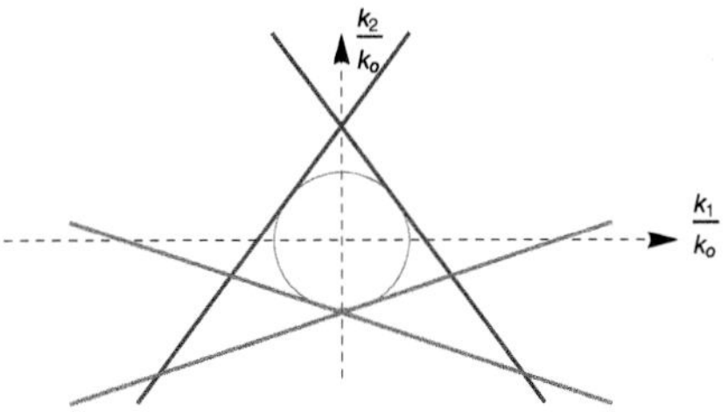

Figure 5.2: Visualization of the dispersion equation (5.129). The straight lines show the condition of a matched wave for the wave numbers k_1 and k_2 in the transverse wave plane $\mathbf{k}_t$. Blue lines: $a = 5, \alpha = 3$, red lines: $a = -20, \alpha = 19$. The circle corresponds to $k_1^2 + k_2^2 = k_o^2$.

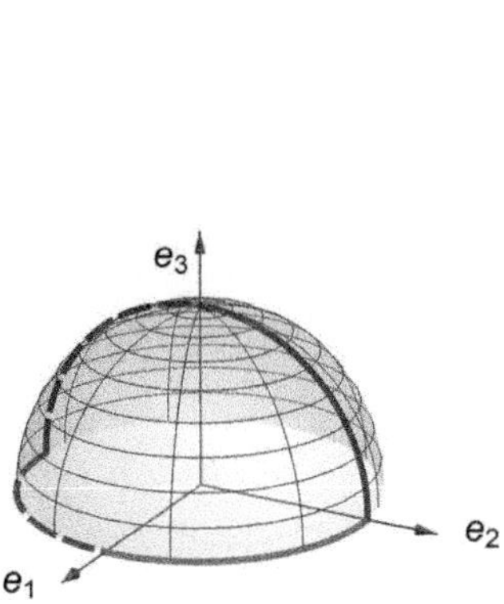

Figure 5.3: Directions of matched waves on the semisphere $|\mathbf{k}^i| = k_o$ for SHDB surface as function of a/α. For $\alpha = 0$, the wave vector is $\mathbf{e}_1 k_o$. When α increases, the wave direction turns in the transversal plane, reaching $\mathbf{e}_2$ when $\alpha = a$. For $\alpha < a$ the matched wave has a normal component, and finally has no tangential component when $a = 0$ (DB case). The dashed lines correspond to α/a decreasing from 0 to $-\infty$.

Reflection Dyadic

Substituting (5.120) and (5.121) in the reflection dyadic of the GBC boundary (5.63), for the SHDB boundary we can write

$$\begin{aligned}\overline{\overline{\mathsf{R}}} &= \frac{1}{J}\mathbf{k}^r \times (\mathbf{c}_2^r\mathbf{c}_1^i - \mathbf{c}_1^r\mathbf{c}_2^i) \\ &= \frac{1}{J}\mathbf{k}^r \times ((\mathbf{k}^r \times \mathbf{a}_t + k_o\alpha\mathbf{n})(\alpha\mathbf{n} \times \mathbf{k}_t + k_o\mathbf{a}_t) \\ &\quad -(\alpha\mathbf{n} \times \mathbf{k}_t + k_o\mathbf{a}_t)(\mathbf{k}^i \times \mathbf{a}_t + k_o\alpha\mathbf{n})). \end{aligned} \tag{5.132}$$

Expressing the reflection dyadic as $\overline{\overline{\mathsf{R}}} = \sum R_{ij}\mathbf{e}_i\mathbf{e}_j$, the coefficients R_{11} and R_{21} are represented for a certain special case in Fig. 5.4. The effect of the matched-wave condition for a certain $\mathbf{k}$ vector is clearly seen in the figures.

For the TE_{c1} and TE_{c2} polarizations (5.56), (5.58), the incident fields

$$\mathbf{E}_1^i = E_1\mathbf{k}^i \times \mathbf{c}_1^i, \tag{5.133}$$

$$\mathbf{E}_2^i = E_2\mathbf{k}^i \times \mathbf{c}_2^i, \tag{5.134}$$

create the reflected fields as

$$\mathbf{E}_1^r = -E_1\mathbf{k}^r \times \mathbf{c}_1^r, \tag{5.135}$$

$$\mathbf{E}_2^r = E_2\mathbf{k}^r \times \mathbf{c}_2^r, \tag{5.136}$$

when taking (5.125) and $\mathbf{c}_2^i \cdot \mathbf{E}_1^i = JE_1$ and $\mathbf{c}_1^i \cdot \mathbf{E}_2^i = JE_2$ into account.

Considering the components tangential to the boundary,

$$\mathbf{E}_{1t}^i = E_1(\mathbf{n}\cdot\mathbf{k}^i)(\alpha\mathbf{k}_t - k_o\mathbf{n}\times\mathbf{a}_t), \tag{5.137}$$

$$\mathbf{E}_{1t}^r = E_1(\mathbf{n}\cdot\mathbf{k}^r)(-\alpha\mathbf{k}_t + k_o\mathbf{n}\times\mathbf{a}_t) \tag{5.138}$$

$$\mathbf{E}_{2t}^i = E_2(-\mathbf{k}_t(\mathbf{k}_t\cdot\mathbf{a}_t) + k_o^2\mathbf{a}_t - k_o\alpha\mathbf{k}_t\times\mathbf{n}), \tag{5.139}$$

$$\mathbf{E}_{2t}^r = E_2(\mathbf{k}_t(\mathbf{k}_t\cdot\mathbf{a}_t) - k_o^2\mathbf{a}_t + k_o\alpha\mathbf{k}_t\times\mathbf{n}), \tag{5.140}$$

and $\mathbf{n}\cdot\mathbf{k}^r = -\mathbf{n}\cdot\mathbf{k}^i = k_n$, we find the relations

$$\mathbf{E}_{1t}^r = \mathbf{E}_{1t}^i, \qquad \mathbf{E}_{2t}^r = -\mathbf{E}_{2t}^i. \tag{5.141}$$

The corresponding relations for the magnetic fields are obtained from (1.47) and (1.48) as

$$\mathbf{H}_{1t}^r = -\mathbf{H}_{1t}^i, \qquad \mathbf{H}_{2t}^r = \mathbf{H}_{2t}^i. \tag{5.142}$$

Because we have $\mathbf{n}\times(\mathbf{E}_2^i+\mathbf{E}_2^r) = 0$, the TE_{c2} wave is reflected from the SHDB boundary as from the PEC boundary. Similarly, because of $\mathbf{n}\times(\mathbf{H}_1^i+\mathbf{H}_1^r) = 0$ the TE_{c1} wave is reflected as from the PMC boundary. One can state that the SHDB boundary has PEC/PMC equivalence, since any incident wave can be split in two parts one of which is reflected as from the PEC boundary, and the other one, as from the PMC boundary. This property follows from the fact that, for the SHDB boundary, the roots of the eigenvalue equation (5.98) are $R = \pm 1$.

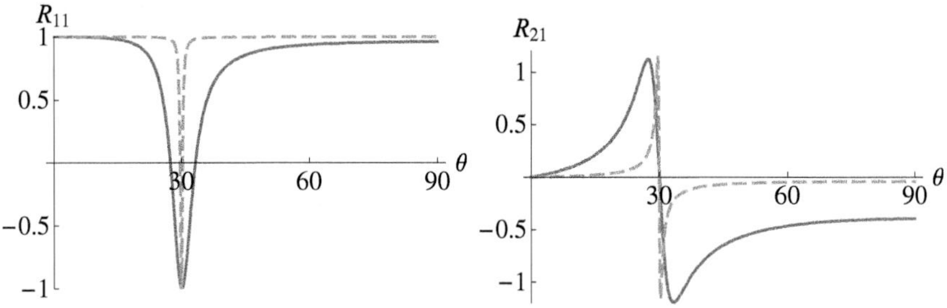

Figure 5.4: The co- and cross-polarized reflection coefficients R_{11} and R_{21} from a planar SHDB surface with parameters $\alpha = 2$ and $\mathbf{a}_t = \mathbf{e}_2$ as function of the angle $\theta = \arccos(k_n/k_o)$. The azimuth angle is $\varphi = \pi - 0.1 \approx 174.3°$ (solid blue) and $\pi - 0.01 \approx 179.4°$ (dashed orange). The matched wave at $\theta = 30°$ appears as a singularity in the figure.

Duality Transformation

Applying the duality transformation rules (5.108) to the SHDB conditions (5.114) and (5.115), the transformed boundary vectors can be shown to obey the rules

$$\mathbf{a}_{1d} = \mathbf{a}_t \cos\varphi + \mathbf{n}\alpha \sin\varphi, \tag{5.143}$$
$$\mathbf{b}_{1d} = -\mathbf{a}_t \sin\varphi + \mathbf{n}\alpha \cos\varphi, \tag{5.144}$$
$$\mathbf{a}_{2d} = \mathbf{n}\alpha \cos\varphi - \mathbf{a}_t \sin\varphi, \tag{5.145}$$
$$\mathbf{b}_{2d} = -\mathbf{n}\alpha \sin\varphi - \mathbf{a}_t \cos\varphi. \tag{5.146}$$

Actually, the transformed SHDB conditions can be written in the form

$$\begin{pmatrix} \cos\varphi & \sin\varphi \\ -\sin\varphi & \cos\varphi \end{pmatrix} \begin{pmatrix} \mathbf{a}_t \cdot \mathbf{E}_d + \alpha\mathbf{n} \cdot \eta_o \mathbf{H}_d \\ \alpha\mathbf{n} \cdot \mathbf{E}_d - \mathbf{a}_t \cdot \eta_o \mathbf{H}_d \end{pmatrix} = \begin{pmatrix} 0 \\ 0 \end{pmatrix}. \tag{5.147}$$

Multiplying this by the inverse of the first (rotation) matrix yields the original SHDB boundary conditions (5.114) and (5.115). Thus, the transformed SHDB conditions are equivalent to the original conditions. In other words, any SHDB boundary is self dual, i.e., invariant in the duality transformation (1.19). As a consequence, if we know the fields scattered by an object with SHDB boundary, the fields scattered by the same object for dual sources can be found by making the duality transformation for the scattered fields. Further, one can show that, because of the self-dual property, an object with SHDB boundary, and of certain symmetry, causes no back scattering. Such an object appears invisible for monostatic radar [62].

Realization by Pseudochiral Medium

Let us consider a possible realization of the SHDB boundary in terms of a special bi-anisotropic medium defined by conditions of the form [70]

$$\begin{pmatrix} \mathbf{D} \\ \mathbf{B} \end{pmatrix} = \begin{pmatrix} 0 & \xi\mathbf{e}_3\mathbf{e}_1 \\ -\xi\mathbf{e}_3\mathbf{e}_1 & 0 \end{pmatrix} \cdot \begin{pmatrix} \mathbf{E} \\ \mathbf{H} \end{pmatrix}, \tag{5.148}$$

which corresponds to the choice of parameter dyadics

$$\overline{\overline{\epsilon}} = 0, \quad \overline{\overline{\xi}} = \xi\mathbf{e}_3\mathbf{e}_1, \quad \overline{\overline{\zeta}} = -\xi\mathbf{e}_3\mathbf{e}_1, \quad \overline{\overline{\mu}} = 0. \tag{5.149}$$

In such a medium the fields satisfy the conditions

$$\mathbf{e}_3 \cdot \mathbf{B} + \xi\mathbf{e}_1 \cdot \mathbf{E} = 0, \tag{5.150}$$

$$\mathbf{e}_3 \cdot \mathbf{D} - \xi\mathbf{e}_1 \cdot \mathbf{H} = 0. \tag{5.151}$$

Due to continuity, the same conditions are valid at a planar interface with normal vector $\mathbf{n} = \mathbf{e}_3$. Thus, the interface of such a medium acts as an SHDB boundary with $\mathbf{a}_t/\alpha = \xi\mathbf{e}_1$ in the conditions (5.114) and (5.115). In other words, the SHDB boundary can be realized by an interface of a medium defined by the medium dyadics (5.149).

Because a medium of this kind may appear somewhat strange, let us add a few terms to the medium conditions:

$$\begin{pmatrix} \mathbf{D} \\ \mathbf{B} \end{pmatrix} = \begin{pmatrix} \overline{\overline{\epsilon}} & \xi\mathbf{e}_3\mathbf{e}_1 + \overline{\overline{\xi}}' \\ -\xi\mathbf{e}_3\mathbf{e}_1 + \overline{\overline{\zeta}}' & \overline{\overline{\mu}} \end{pmatrix} \cdot \begin{pmatrix} \mathbf{E} \\ \mathbf{H} \end{pmatrix}, \tag{5.152}$$

whence the fields at the interface are related by

$$\mathbf{e}_3 \cdot \mathbf{B} + \xi\mathbf{e}_1 \cdot \mathbf{E} - \mathbf{e}_3 \cdot \overline{\overline{\mu}} \cdot \mathbf{H} - \mathbf{e}_3 \cdot \overline{\overline{\zeta}}' \cdot \mathbf{E} = 0, \tag{5.153}$$

$$\mathbf{e}_3 \cdot \mathbf{D} - \xi\mathbf{e}_1 \cdot \mathbf{H} - \mathbf{e}_3 \cdot \overline{\overline{\epsilon}} \cdot \mathbf{E} - \mathbf{e}_3 \cdot \overline{\overline{\xi}}' \cdot \mathbf{H} = 0. \tag{5.154}$$

They can be reduced to the form of (5.150) and (5.151) by requiring that the medium dyadics satisfy

$$\mathbf{e}_3 \cdot \overline{\overline{\mu}} = \mathbf{e}_3 \cdot \overline{\overline{\epsilon}} = 0, \quad \mathbf{e}_3 \cdot \overline{\overline{\xi}}' = \mathbf{e}_3 \cdot \overline{\overline{\zeta}}' = 0. \tag{5.155}$$

To further improve the realizability, let us assume that the medium is reciprocal, whence the medium dyadics are required to satisfy [28]

$$\overline{\overline{\epsilon}}^T = \overline{\overline{\epsilon}}, \quad \overline{\overline{\mu}}^T = \overline{\overline{\mu}}, \quad \overline{\overline{\xi}}^T = -\overline{\overline{\zeta}}. \tag{5.156}$$

These conditions will be met if we choose

$$\overline{\overline{\epsilon}} = \epsilon\overline{\overline{\mathsf{I}}}_t, \quad \overline{\overline{\mu}} = \mu\overline{\overline{\mathsf{I}}}_t, \quad \overline{\overline{\xi}}' = -\overline{\overline{\zeta}}' = \xi\mathbf{e}_1\mathbf{e}_3. \tag{5.157}$$

Defining now

$$\overline{\overline{\xi}} = -\overline{\overline{\zeta}} = \xi\overline{\overline{\mathsf{S}}}, \tag{5.158}$$

where $\overline{\overline{\mathsf{S}}}$ is the symmetric dyadic

$$\overline{\overline{\mathsf{S}}} = \mathbf{e}_3\mathbf{e}_1 + \mathbf{e}_1\mathbf{e}_3, \tag{5.159}$$

the medium equations (5.152) have the form

$$\begin{pmatrix} \mathbf{D} \\ \mathbf{B} \end{pmatrix} = \begin{pmatrix} \epsilon\overline{\overline{\mathsf{I}}}_t & \xi\overline{\overline{\mathsf{S}}} \\ -\xi\overline{\overline{\mathsf{S}}} & \mu\overline{\overline{\mathsf{I}}}_t \end{pmatrix} \cdot \begin{pmatrix} \mathbf{E} \\ \mathbf{H} \end{pmatrix}. \tag{5.160}$$

Since (5.150) and (5.151) are obviously satisfied by fields in such a medium, the conditions (5.160) for $\mathbf{a}_t/\alpha = \xi\mathbf{e}_1$ define a possible bi-anisotropic medium yielding the SHDB boundary conditions (5.114), (5.115) at its interface.

The system (5.160) defined above is algebraically more complete than (5.149), because the dyadic matrix of (5.160) can be inverted so that **E** and **H** can be expressed in terms of **D** and **B**. In fact, one can verify through multiplication that the inverse of the dyadic matrix in (5.160) can be expressed as

$$\frac{1}{\xi^2}\begin{pmatrix} \mu\mathbf{e}_3\mathbf{e}_3 + \xi^2\mathbf{e}_2\mathbf{e}_2/\epsilon & -\xi\overline{\overline{\mathsf{S}}} \\ \xi\overline{\overline{\mathsf{S}}} & \epsilon\mathbf{e}_3\mathbf{e}_3 + \xi^2\mathbf{e}_2\mathbf{e}_2/\mu \end{pmatrix}. \tag{5.161}$$

It is obvious that the magnetoelectric dyadic $\overline{\overline{\xi}} = -\overline{\overline{\zeta}} = \xi\overline{\overline{\mathsf{S}}}$ plays a central role in the medium producing the SHDB boundary conditions at its interface. Because of $\mathrm{tr}\overline{\overline{\mathsf{S}}} = 0$, the trace of the dyadics $\overline{\overline{\xi}}$ and $\overline{\overline{\zeta}}$ vanishes. The form of the dyadic $\overline{\overline{\xi}}$ implies that the medium belongs to the class of pseudochiral media. The medium is lossless when the parameter ξ is purely imaginary [89, 108]. To find a realization for the medium, let us rewrite

$$\overline{\overline{\xi}} = (\mathbf{e}_3\mathbf{e}_1 + \mathbf{e}_1\mathbf{e}_3)\xi = (\mathbf{e}_3'\mathbf{e}_3' - \mathbf{e}_1'\mathbf{e}_1')\xi, \tag{5.162}$$

where

$$\mathbf{e}_3' = \frac{\mathbf{e}_1 + \mathbf{e}_3}{\sqrt{2}} \quad \text{and} \quad \mathbf{e}_1' = \frac{\mathbf{e}_1 - \mathbf{e}_3}{\sqrt{2}} \tag{5.163}$$

respectively denote the unit vectors $\mathbf{e}_3$ and $\mathbf{e}_1$ rotated by $\pi/4$.

Because dyadics of the form $\overline{\overline{\xi}} = -\overline{\overline{\zeta}} = \xi\mathbf{u}\mathbf{u}$ correspond to a uniaxial chiral medium which can be realized by similar helices parallel to the unit vector **u**, the medium defined by (5.162) consists of two sets of helices, one set parallel to the vector $\mathbf{e}_3'$ and the other set of similar helices but of opposite handedness, parallel to the orthogonal vector $\mathbf{e}_1'$, Figure 5.8 [70]. The final medium is not chiral because its mirror image rotated by $\pi/2$ equals the original one. This is the reason why it has been called pseudochiral in the past.

To obtain a medium with isotropic permittivity and permeability one can add a third set of helices parallel to the x_2 axis. To avoid chirality caused by these helices, half of the amount should be right handed and the other half left handed.

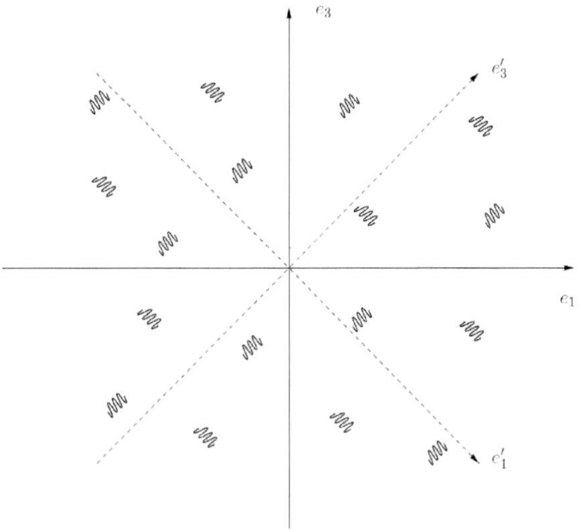

Figure 5.5: The magnetoelectric effect of the bianisotropic medium is realized by two sets of helices. The axes of the helices are parallel to two orthogonal directions in the x_1x_3 plane. The helices with orthogonal axes have opposite handedness.

Realization by Skewon-Axion Medium

As another possibility for the realization of the SHDB boundary, let us consider the interface of a bianisotropic medium which has been called by the name skewon-axion medium. The medium is most naturally defined in terms of four-dimensional formalism, arising from linear medium conditions of the form (see Appendix A)

$$\mathbf{\Psi} = \overline{\overline{\mathsf{M}}}|\mathbf{\Phi}, \tag{5.164}$$

relating the electromagnetic two-forms $\mathbf{\Psi}$ and $\mathbf{\Phi}$ through a dyadic $\overline{\overline{\mathsf{M}}}$ (actually a bidyadic [72]), involving 36 parameters. When $\overline{\overline{\mathsf{M}}}$ is a multiple of the 4D unit dyadic (1 parameter), it defines an axion medium. When $\overline{\overline{\mathsf{M}}}$ can be associated to an antisymmetric dyadic (15 parameters), it defines a skewon medium [20]. A medium defined by a dyadic $\overline{\overline{\mathsf{M}}}$ consisting of a linear combination of these is called a skewon-axion medium [72].

Applying Gibbsian 3D formalism, the most general skewon-axion medium can be defined by medium equations of the form [59, 69, 72]

$$\begin{pmatrix} \mathbf{D} \\ \mathbf{H} \end{pmatrix} = \begin{pmatrix} \overline{\overline{\epsilon}}' & \overline{\overline{\alpha}} \\ \overline{\overline{\beta}} & \overline{\overline{\mu}}^{-1} \end{pmatrix} \cdot \begin{pmatrix} \mathbf{E} \\ \mathbf{B} \end{pmatrix}, \tag{5.165}$$

when the four dyadics are defined by

$$\overline{\overline{\alpha}} = \overline{\overline{\mathsf{A}}} - \mathrm{tr}\overline{\overline{\mathsf{A}}}\,\overline{\overline{\mathsf{I}}}, \tag{5.166}$$

$$\overline{\overline{\epsilon}}' = \mathbf{c}\times\overline{\overline{\mathsf{I}}}, \tag{5.167}$$

$$\overline{\overline{\mu}}^{-1} = \mathbf{g}\times\overline{\overline{\mathsf{I}}}, \tag{5.168}$$

$$\overline{\overline{\beta}} = \overline{\overline{\mathsf{A}}}^T + a\overline{\overline{\mathsf{I}}}. \tag{5.169}$$

Here, $\overline{\overline{\mathsf{A}}}$ is a 3D dyadic, $\mathbf{c}, \mathbf{g}$ are two vectors and a (the axion parameter) is a scalar. Together, they correspond to $9 + 3 + 3 + 1 = 16$ scalar parameters. Because $\overline{\overline{\epsilon}}'$ and $\overline{\overline{\mu}}^{-1}$ are antisymmetric, they do not have inverses. Thus, it is not possible to express the skewon-axion medium equations in the form (1.3). Actually, (5.165), with (5.166) - (5.169) inserted, yields the most natural 3D representation of the skewon-axion medium.

One can show that, denoting $\mathbf{k} = \omega\mathbf{p}$, the electric field of a plane wave in a skewon-axion medium satisfies an equation of the form

$$\mathbf{q}(\mathbf{p})\times\mathbf{E} = 0, \tag{5.170}$$

in terms of the vector function

$$\mathbf{q}(\mathbf{p}) = (\mathbf{g}\cdot\mathbf{p} - a)\mathbf{p} + \mathbf{p}\cdot\overline{\overline{\mathsf{A}}} - \mathbf{c}. \tag{5.171}$$

Details of the derivation are left as an exercise.

Because of the form (5.170), (5.171), there is no restriction to the wave vector $\mathbf{k}$ by a skewon-axion medium and $\mathbf{E}$ must be a multiple of $\mathbf{q}(\mathbf{p})$ for any choice of $\mathbf{p} = \mathbf{k}/\omega$. Since there is no dispersion equation for a plane wave in a skewon-axion medium, it serves as an example of a NDE medium (No Dispersion Equation medium). Other examples of NDE media have been discussed in [72]. Boundary conditions generalizing SH and D'B' boundaries have been introduced in the past together with possible realization [68].

To verify that an interface of a skewon-axion medium may act as an SHDB boundary, let us consider a special case of the skewon-axion medium [67]. In terms of an orthonormal vector basis $\mathbf{e}_1, \mathbf{e}_2, \mathbf{e}_3 = \mathbf{n}$, the medium under study is defined in terms of dyadic parameters (5.166) – (5.169) in which we substitute

$$\overline{\overline{\mathsf{A}}} = B\overline{\overline{\mathsf{I}}} + \mathbf{a}_1\mathbf{e}_1, \tag{5.172}$$

$$\mathbf{c} = c\mathbf{e}_1, \tag{5.173}$$

$$\mathbf{g} = A\mathbf{a}_1, \tag{5.174}$$

$$a = B + Ac \tag{5.175}$$

The number of free parameters of the medium is now reduced from 16 to $3(\mathbf{c}) + 4(\overline{\overline{\mathsf{A}}}) + 1(\mathbf{g}) + 1(a) = 8$ and the medium conditions (5.165) can be rewritten as

$$\mathbf{D} = \mathbf{a}_1(\mathbf{e}_1\cdot\mathbf{B}) - (2B + \mathbf{e}_1\cdot\mathbf{a}_1)\mathbf{B} + c\mathbf{e}_1\times\mathbf{E}, \tag{5.176}$$

$$\mathbf{H} = A\mathbf{a}_1\times\mathbf{B} + (2B + Ac)\mathbf{E} + \mathbf{e}_1(\mathbf{a}_1\cdot\mathbf{E}). \tag{5.177}$$

Considering a plane wave in the special skewon-axion medium defined by (5.176) and (5.177), it can be shown that the fields of the plane wave satisfy the conditions

$$A\mathbf{e}_3 \cdot \mathbf{B} + \mathbf{e}_2 \cdot \mathbf{E} = 0, \tag{5.178}$$

$$A\mathbf{e}_3 \cdot \mathbf{D} - \mathbf{e}_2 \cdot \mathbf{H} = 0. \tag{5.179}$$

Details of the analysis are left as an exercise.

Let us assume a planar boundary $\mathbf{e}_3 \cdot \mathbf{r} = 0$ separating the skewon-axion medium half space $\mathbf{e}_3 \cdot \mathbf{r} > 0$ from an isotropic half space $\mathbf{e}_3 \cdot \mathbf{r} < 0$. A plane wave incident from the isotropic half space is transmitted into the skewon-axion half space and satisfies the conditions (5.178) and (5.179) for its fields. Because of continuity of the components $\mathbf{e}_3 \cdot \mathbf{D}$, $\mathbf{e}_3 \cdot \mathbf{B}$ and $\mathbf{e}_2 \cdot \mathbf{H}$, $\mathbf{e}_2 \cdot \mathbf{E}$ across the interface, for $\mathbf{a}_t/\alpha = \mathbf{e}_2/A$, (5.178) and (5.179) can be identified as the SHDB conditions (5.114) and (5.115) at the interface of the special skewon-axion medium. Since they are linear and valid for any plane wave, they are actually independent of how the fields are produced.

5.9 Generalized Soft-and-Hard/DB (GSHDB) Boundary

Like the SHDB boundary, the GSHDB boundary can be regarded as a fundamental concept in electromagnetics, because its four-dimensional definition (see Appendix A) takes the simple form

$$\boldsymbol{\varepsilon}_3 \wedge \boldsymbol{\alpha} \wedge \boldsymbol{\Phi} = 0, \qquad \boldsymbol{\varepsilon}_3 \wedge \boldsymbol{\beta} \wedge \boldsymbol{\Psi} = 0, \tag{5.180}$$

in terms of two one-forms, $\boldsymbol{\alpha}$ and $\boldsymbol{\beta}$. Because the SHDB boundary conditions (5.18) and (5.19) correspond to the special case $\boldsymbol{\alpha} = \boldsymbol{\beta}$, the GBC boundary defined by the conditions (5.180) has been called the Generalized Soft-and-Hard/DB (GSHDB) boundary [74].

In the three-dimensional formalism, the GSHDB boundary can be defined by

$$\alpha c\mathbf{n} \cdot \mathbf{B} + \mathbf{a}_t \cdot \mathbf{E} = 0, \tag{5.181}$$

$$\beta c\mathbf{n} \cdot \mathbf{D} + \mathbf{b}_t \cdot \mathbf{H} = 0. \tag{5.182}$$

In the simple isotropic medium the conditions can be written as

$$\mathbf{a}_t \cdot \mathbf{E} + \alpha\mathbf{n} \cdot \eta_o\mathbf{H} = 0, \tag{5.183}$$

$$\beta\mathbf{n} \cdot \mathbf{E} + \mathbf{b}_t \cdot \eta_o\mathbf{H} = 0. \tag{5.184}$$

Of the number of parameters is $2(\mathbf{a}_t) + 1(\alpha) + 1(\beta) + 2(\mathbf{b}_t) = 6$, two of which can be chosen at will. This leaves $\mathbf{a}_t/\alpha$ and $\mathbf{b}_t/\beta$ as two independent vector parameters.

Applying the Maxwell equations, the GSHDB conditions (5.181) and (5.182) can be replaced by operator conditions

$$\mathbf{L}_a(\nabla_t)\cdot\mathbf{E}_t = 0, \qquad \mathbf{L}_b(\nabla_t)\cdot\mathbf{H}_t = 0, \tag{5.185}$$

where we define

$$\mathbf{L}_a(\nabla_t) = \alpha\mathbf{n}\times\nabla_t - jk_o\mathbf{a}_t, \tag{5.186}$$

$$\mathbf{L}_b(\nabla_t) = \beta\mathbf{n}\times\nabla_t + jk_o\mathbf{b}_t. \tag{5.187}$$

For the GSHDB boundary, the vectors (5.29) - (5.32) are defined by

$$\mathbf{c}_1(\mathbf{k}) = \alpha\mathbf{k}\times\mathbf{n} - k_o\mathbf{a}_t, \tag{5.188}$$

$$\mathbf{c}_2(\mathbf{k}) = \mathbf{k}\times\mathbf{b}_t - k_o\beta\mathbf{n}, \tag{5.189}$$

$$\mathbf{d}_1(\mathbf{k}) = \mathbf{k}\times\mathbf{a}_t + k_o\alpha\mathbf{n}, \tag{5.190}$$

$$\mathbf{d}_2(\mathbf{k}) = \beta\mathbf{k}\times\mathbf{n} + k_o\mathbf{b}_t, \tag{5.191}$$

whence we have

$$\mathbf{c}_1^i = \mathbf{c}_1^r = \alpha\mathbf{k}_t\times\mathbf{n} - k_o\mathbf{a}_t, \tag{5.192}$$

$$\mathbf{d}_2^i = \mathbf{d}_2^r = \beta\mathbf{k}_t\times\mathbf{n} + k_o\mathbf{b}_t. \tag{5.193}$$

Expanding (5.33) we obtain

$$J(\mathbf{k}) = \mathbf{c}_1(\mathbf{k})\times\mathbf{c}_2(\mathbf{k})\cdot\mathbf{k} \tag{5.194}$$

$$= k_o\mathbf{k}_t\mathbf{k}_t : (\mathbf{a}_t\mathbf{b}_t + \alpha\beta\overline{\overline{\mathsf{I}}}_t) + k_o^2\mathbf{k}_t\cdot(\mathbf{n}\times(\alpha\mathbf{b}_t - \beta\mathbf{a}_t)) - k_o^3\mathbf{a}_t\cdot\mathbf{b}_t. \tag{5.195}$$

Since the expression (5.195) does not contain $\mathbf{n}\cdot\mathbf{k}$, it satisfies the condition

$$J^r = J^i = J. \tag{5.196}$$

Applying this and (5.56) - (5.59), we can expand

$$\eta_o J(\mathbf{k})(\mathbf{H}_1^i + \mathbf{H}_1^r) = (\mathbf{k}^i\times\mathbf{d}_1^i - \mathbf{k}^r\times\mathbf{d}_1^r)\phi_1^i$$

$$= (\mathbf{k}^i\times(\mathbf{k}^i\times\mathbf{a}_t + k_o\alpha\mathbf{n}) - \mathbf{k}^r\times(\mathbf{k}^r\times\mathbf{a}_t + k_o\alpha\mathbf{n}))\phi_1^i$$

$$= 2\mathbf{n}(\mathbf{n}\cdot\mathbf{k}^i)(\mathbf{k}_t\cdot\mathbf{a}_t)\phi_1^i, \tag{5.197}$$

whence the total field $\mathbf{H}_1^i + \mathbf{H}_1^r$ has no tangential component. This means that the TE_{c1} wave is reflected from the GSHDB boundary as from the PMC boundary. After similar reasoning, we can find that the TE_{c2} wave is reflected as from the PEC boundary. Thus, we can state that the GSHDB boundaries have PEC/PMC equivalence: any plane wave can be split in two waves, one of which is reflected as from the PEC boundary and, the other one, as from the PMC boundary.

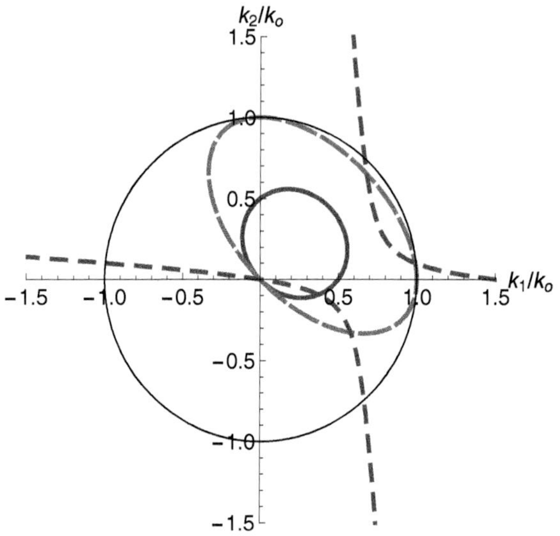

Figure 5.6: Solutions for the dispersion equation (5.202) depicting $\mathbf{k}_t(\varphi)/k_o$ for plane waves matched to a GSHDB boundary for parameters $\alpha = 2a_1, \beta = 2b_2$ (solid blue), $\alpha = a_1, \beta = b_2$ (dashed red), and $a_1 = 10\alpha, b_2 = 1.5\beta$ (short-dashed green).

Matched Waves

To find the dispersion equation governing matched waves at the GSHDB boundary, we can substitute $\mathbf{a}_1 = \mathbf{a}_t$, $\mathbf{b}_1 = \alpha\mathbf{n}$, $\mathbf{a}_2 = \beta\mathbf{n}$ and $\mathbf{b}_2 = \mathbf{b}_t$ in (5.76), valid for GBC boundaries. The resulting condition equals $J(\mathbf{k}) = 0$, which reduces to a quadratic equation for $\mathbf{k}_t$. Applying (5.195), it can be expressed as

$$\mathbf{k}_t\mathbf{k}_t : (\mathbf{a}_t\mathbf{b}_t + \alpha\beta\overline{\overline{\mathsf{I}}}_t) + k_o\mathbf{k}_t \cdot (\mathbf{n} \times (\alpha\mathbf{b}_t - \beta\mathbf{a}_t)) = k_o^2\mathbf{a}_t \cdot \mathbf{b}_t, \tag{5.198}$$

which can be written in another form as

$$(\mathbf{a}_t \cdot \mathbf{k}_t)(\mathbf{b}_t \cdot \mathbf{k}_t) + (\alpha\mathbf{k}_t - k_o\mathbf{n} \times \mathbf{a}_t) \cdot (\beta\mathbf{k}_t + k_o\mathbf{n} \times \mathbf{b}_t) = 0. \tag{5.199}$$

For the GSH boundary with $\alpha = \beta = 0$, (5.198) coincides with (3.215). For the SHDB boundary with $\alpha = \beta$ and $\mathbf{a}_t = -\mathbf{b}_t = a\mathbf{e}_1$, (5.199) reduces to

$$(\alpha^2 - a^2)k_1^2 + (\alpha k_2 - k_o a)^2 = 0, \tag{5.200}$$

which coincides with (5.128).

As an example, let us assume orthogonal vectors $\mathbf{a}_t = a_1\mathbf{e}_1$ and $\mathbf{b}_t = b_2\mathbf{e}_2$ and $\mathbf{k}_t = k_t(\mathbf{e}_1\cos\varphi + \mathbf{e}_2\sin\varphi)$. In this case, (5.198) becomes

$$a_1b_2k_t^2\sin\varphi\cos\varphi + \alpha\beta k_t^2 - k_ok_t(\alpha b_2\cos\varphi + \beta a_1\sin\varphi) = 0. \tag{5.201}$$

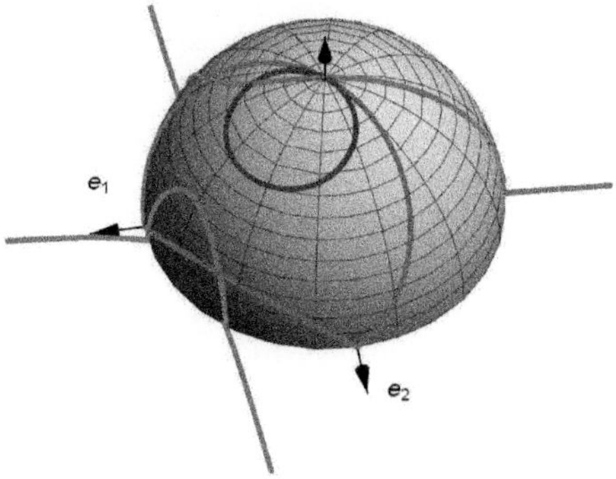

Figure 5.7: Paths on the unit sphere corresponding to solutions of (5.202) define directions of the normalized wave vector k_t/k_o of a wave matched to the GSHDB boundary with the parameters of Fig. 5.6. The curve outside the sphere corresponds to the real part of the wave vector.

Obviously, one of the solutions is $k_t = 0$, or $\mathbf{k} = \mathbf{n}k_o$, which corresponds to a wave with normal incidence. The second solution is

$$k_t(\varphi) = k_o \frac{\alpha b_2 \cos\varphi + \beta a_1 \sin\varphi}{a_1 b_2 \sin\varphi\cos\varphi + \alpha\beta}, \tag{5.202}$$

with

$$\mathbf{n}\cdot\mathbf{k} = \sqrt{k_o^2 - k_t^2(\varphi)}. \tag{5.203}$$

Figure 5.6 depicts solutions $\mathbf{k}_t(\varphi)/k_o$ for the dispersion equation (5.202) for certain values of the relative parameters α/a_1 and β/b_2. The corresponding path of the vector $\mathbf{k}/k_o$ is seen in Figure 5.7. Outside the unit sphere the $\mathbf{k}$ vector has an imaginary normal component, which is not shown in the figure.

Eigenfields

To find eigensolutions for the reflection from the GSHDB boundary, i.e., solutions R for the condition (5.94), we can apply the condition (5.98) for the GSHDB boundary by substituting

$$\mathbf{a}'_{1t} = \mathbf{a}_t + (\alpha/k_o)\mathbf{n}\times\mathbf{k}_t, \quad \mathbf{a}'_{2t} = 0, \tag{5.204}$$

$$\mathbf{b}'_{1t} = 0, \quad \mathbf{b}'_{2t} = \mathbf{b}_t - (\beta/k_o)\mathbf{n}\times\mathbf{k}_t. \tag{5.205}$$

Because in this case the eigenequation (5.98) is reduced to

$$\mathbf{n}\cdot((1+R)\mathbf{a}'_{1t})\times((1-R)\mathbf{b}'_{2t}\cdot\overline{\overline{\mathsf{J}}}_t)=0, \tag{5.206}$$

the possible eigenvalues are

$$R_\pm=\pm 1. \tag{5.207}$$

Applying these to the total tangential fields at the boundary,

$$\mathbf{E}_t \;=\; \mathbf{E}^i_t+\mathbf{E}^r_t=(1+R)\mathbf{E}^i_t, \tag{5.208}$$

$$\eta_o\mathbf{H}_t \;=\; -\overline{\overline{\mathsf{J}}}_t\cdot\mathbf{E}^i_t+\overline{\overline{\mathsf{J}}}_t\cdot\mathbf{E}^r_t=-(1-R)\overline{\overline{\mathsf{J}}}_t\cdot\mathbf{E}^i_t, \tag{5.209}$$

the eigenwaves corresponding to $R=-1$ and $R=+1$ are respectively reflected as from the PEC boundary and the PMC boundary.

As a conclusion we can state that the GSHDB boundary together with all of its special cases (SHDB, SH, DB etc) has PEC/PMC equivalence in plane-wave reflection. Thus, any plane wave can be split in two partial fields, one of which is reflected as from the PEC boundary, and the other one as from the PMC boundary. Because the eigenvalues are independent of the **k** vector of the wave, the same property is valid for any sum of plane waves.

Duality Transformation

Applying the duality transformation rules (5.108) to the GSHDB conditions (5.183) and (5.184), the transformed boundary vectors become

$$\mathbf{a}_{1d} \;=\; \mathbf{n}\alpha\sin\varphi+\mathbf{a}_t\cos\varphi, \tag{5.210}$$

$$\mathbf{b}_{1d} \;=\; \mathbf{n}\alpha\cos\varphi-\mathbf{a}_t\sin\varphi, \tag{5.211}$$

$$\mathbf{a}_{2d} \;=\; \mathbf{n}\beta\cos\varphi+\mathbf{b}_t\sin\varphi, \tag{5.212}$$

$$\mathbf{b}_{2d} \;=\; -\mathbf{n}\beta\sin\varphi+\mathbf{b}_t\cos\varphi. \tag{5.213}$$

In the general case, the transformed GSHDB boundary is not a GSHDB boundary. Because the tangential parts of the four vectors are restricted to be multiples of either $\mathbf{a}_t$ or $\mathbf{b}_t$, the transformed vector parameters define another subclass of GBC boundaries.

Eliminating successively the terms $\mathbf{n}\cdot\mathbf{E}_d$ and $\mathbf{n}\cdot\mathbf{H}_d$ from the two equations, the transformed GSHDB boundary conditions can be written in the form

$$\begin{pmatrix} \beta\mathbf{a}_t\cos^2\varphi-\alpha\mathbf{b}_t\sin^2\varphi & \alpha\beta\mathbf{n}-\mathbf{g}_t\sin\varphi\cos\varphi \\ \alpha\beta\mathbf{n}+\mathbf{g}_t\sin\varphi\cos\varphi & \alpha\mathbf{b}_t\cos^2\varphi-\beta\mathbf{a}_t\sin^2\varphi \end{pmatrix}\cdot\begin{pmatrix}\mathbf{E}_d\\ \eta_o\mathbf{H}_d\end{pmatrix}=\begin{pmatrix}0\\0\end{pmatrix}, \tag{5.214}$$

with

$$\mathbf{g}_t=\beta\mathbf{a}_t+\alpha\mathbf{b}_t. \tag{5.215}$$

One can immediately check that, for $\varphi=0$, the original GSHDB conditions are recovered from (5.214). For $\varphi=\pi/2$, the GSHDB boundary is transformed to another GSHDB boundary defined by

$$\begin{pmatrix} -\mathbf{b}_t & \beta\mathbf{n} \\ \alpha\mathbf{n} & -\mathbf{a}_n \end{pmatrix}\begin{pmatrix}\mathbf{E}_d\\ \eta_o\mathbf{H}_d\end{pmatrix}=\begin{pmatrix}0\\0\end{pmatrix}. \tag{5.216}$$

This is obtained by substituting $\mathbf{a}_t \leftrightarrow -\mathbf{b}_t$ and $\alpha \leftrightarrow \beta$ in the original GSHDB conditions (5.183), (5.184). For $\sin\varphi\cos\varphi \neq 0$ the GSHDB boundary is transformed to a more general GBC boundary.

Realization

A possible realization for the GSHDB boundary in terms of an interface of a bi-anisotropic medium can be found by requiring that the GSHDB conditions (5.181) and (5.182) be satisfied everywhere in the medium,

$$\alpha c\mathbf{e}_3 \cdot (\overline{\overline{\zeta}} \cdot \mathbf{E} + \overline{\overline{\mu}} \cdot \mathbf{H}) + \mathbf{a}_t \cdot \mathbf{E} = 0, \tag{5.217}$$

$$\beta c\mathbf{e}_3 \cdot (\overline{\overline{\epsilon}} \cdot \mathbf{E} + \overline{\overline{\xi}} \cdot \mathbf{H}) + \mathbf{b}_t \cdot \mathbf{H} = 0, \tag{5.218}$$

where we denote $\mathbf{n} = \mathbf{e}_3$. Obviously, these will be satisfied for any fields when the medium dyadics are restricted by the conditions

$$\mathbf{e}_3 \cdot \overline{\overline{\epsilon}} = 0, \quad \mathbf{e}_3 \cdot \overline{\overline{\mu}} = 0, \tag{5.219}$$

$$\alpha c\mathbf{e}_3 \cdot \overline{\overline{\zeta}} + \mathbf{a}_t = 0, \quad \beta c\mathbf{e}_3 \cdot \overline{\overline{\xi}} + \mathbf{b}_t = 0. \tag{5.220}$$

In this case, the medium dyadics must be of the form

$$\overline{\overline{\epsilon}} = \mathbf{e}_3 \times \overline{\overline{\epsilon}}', \tag{5.221}$$

$$\overline{\overline{\xi}} = \mathbf{e}_3 \times \overline{\overline{\xi}}' - \mathbf{e}_3\mathbf{b}_t/\beta c, \tag{5.222}$$

$$\overline{\overline{\zeta}} = \mathbf{e}_3 \times \overline{\overline{\zeta}}' - \mathbf{e}_3\mathbf{a}_t/\alpha c \tag{5.223}$$

$$\overline{\overline{\mu}} = \mathbf{e}_3 \times \overline{\overline{\mu}}', \tag{5.224}$$

where the primed dyadics bear no restrictions. Because of continuity, the fields at the interface of this medium will satisfy the GSHDB boundary conditions.

5.10 GBC Boundaries with PEC/PMC Equivalence

Because the SHDB and GSHDB boundaries have eigenvalues ± 1, both of these boundaries are equivalent to PEC and PMC boundaries for their proper eigenfields. It is interesting to find the most general GBC boundary sharing this property [73].

Requiring that the eigenvalue equation (5.98) be satisfied by $R = +1$ leads to the condition

$$\mathbf{n} \cdot (\mathbf{a}'_{1t} \times \mathbf{a}'_{2t}) = 0, \tag{5.225}$$

while for $R = -1$ it is reduced to

$$\mathbf{n} \cdot (\mathbf{b}'_{1t} \cdot \overline{\overline{\mathsf{J}}}_t) \times (\mathbf{b}'_{2t} \cdot \overline{\overline{\mathsf{J}}}_t) = \mathbf{n} \cdot (\mathbf{b}'_{1t} \times \mathbf{b}'_{2t}) = 0. \tag{5.226}$$

Here we have applied the rule

$$(\mathbf{b}'_{1t}\cdot\overline{\overline{\mathsf{J}}}_t)\times(\mathbf{b}'_{2t}\cdot\overline{\overline{\mathsf{J}}}_t)=(\mathbf{b}'_{1t}\times\mathbf{b}'_{2t})\cdot\overline{\overline{\mathsf{J}}}_t^{(2)}=(\mathbf{b}'_{1t}\times\mathbf{b}'_{2t})\cdot\mathbf{nn}. \tag{5.227}$$

From (5.225) and (5.226) it follows that the tangential vectors $\mathbf{a}'_{1t}$ and $\mathbf{a}'_{2t}$ on one hand, and $\mathbf{b}'_{1t}$ and $\mathbf{b}'_{2t}$ on the other hand, must be linearly dependent, whence they satisfy conditions of the form

$$A\mathbf{a}'_{1t}+C\mathbf{a}'_{2t} = 0, \tag{5.228}$$

$$B\mathbf{b}'_{1t}+D\mathbf{b}'_{2t} = 0, \tag{5.229}$$

for some scalars $A\cdots D$.

Under the assumption that the GBC medium have eigenvalues ± 1, the boundary conditions (5.92) and (5.93) can be written as

$$\begin{pmatrix} \mathbf{a}'_{1t} & \mathbf{b}'_{1t} \\ \mathbf{a}'_{2t} & \mathbf{b}'_{2t} \end{pmatrix}\cdot\begin{pmatrix} \mathbf{E} \\ \eta_o\mathbf{H} \end{pmatrix}=\begin{pmatrix} 0 \\ 0 \end{pmatrix}. \tag{5.230}$$

These are of the form of the GSH-boundary conditions, (3.180), except that the vectors $\mathbf{a}'_{1t}\cdots\mathbf{b}'_{2t}$ are not constant but depend on $\mathbf{k}_t$. Applying (5.228) and (5.229), (5.230) equal the conditions

$$\begin{pmatrix} A & C \\ B & D \end{pmatrix}\begin{pmatrix} \mathbf{a}'_{1t} & \mathbf{b}'_{1t} \\ \mathbf{a}'_{2t} & \mathbf{b}'_{2t} \end{pmatrix}\cdot\begin{pmatrix} \mathbf{E} \\ \eta_o\mathbf{H} \end{pmatrix}=\begin{pmatrix} (A\mathbf{b}'_{1t}+C\mathbf{b}'_{2t})\cdot\eta_o\mathbf{H} \\ (B\mathbf{a}'_{1t}+D\mathbf{a}'_{2t})\cdot\mathbf{E} \end{pmatrix}=\begin{pmatrix} 0 \\ 0 \end{pmatrix}. \tag{5.231}$$

Substituting (5.90) and (5.91), (5.231) can be rewritten as

$$0 = (B(\mathbf{a}_{1t}+\frac{b_{1n}}{k_o}\mathbf{n}\times\mathbf{k}_t)+D(\mathbf{a}_{2t}+\frac{b_{2n}}{k_o}\mathbf{n}\times\mathbf{k}_t))\cdot\mathbf{E} \tag{5.232}$$

$$= (B\mathbf{a}_{1t}+D\mathbf{a}_{2t})\cdot\mathbf{E}+(Bb_{1n}+Db_{2n})\mathbf{n}\cdot\eta_o\mathbf{H}, \tag{5.233}$$

$$0 = (A(\mathbf{b}_{1t}-\frac{a_{1n}}{k_o}\mathbf{n}\times\mathbf{k}_t)+C(\mathbf{b}_{2t}-\frac{a_{2n}}{k_o}\mathbf{n}\times\mathbf{k}_t))\cdot\eta_o\mathbf{H} \tag{5.234}$$

$$= (A\mathbf{b}_{1t}+C\mathbf{b}_{2t})\cdot\eta_o\mathbf{H}+(Aa_{1n}+Ca_{2n})\mathbf{n}\cdot\mathbf{E}, \tag{5.235}$$

where we have applied the plane-wave field relations. Since these conditions have the form (5.183) and (5.184), one may conclude that the most general GBC boundary with the PEC/PMC equivalence is actually the GSHDB boundary.

It is left as an exercise to show that the most general impedance boundary with PEC/PMC equivalence is the GSH boundary.

5.11 Some Special GBC Boundaries

Let us consider a few obvious special cases of the GBC conditions (5.8) and (5.9).

The E Boundary

As a first example, the E boundary is defined in terms of the conditions

$$\mathbf{a}_1 \cdot \mathbf{E} = 0 \quad (5.236)$$

$$\mathbf{a}_2 \cdot \mathbf{E} = 0, \quad (5.237)$$

where the two vectors $\mathbf{a}_1$ and $\mathbf{a}_2$ are assumed linearly independent but, otherwise, they can be freely chosen [75]. Denoting

$$\mathbf{a}_{12} = \mathbf{a}_1 \times \mathbf{a}_2 \neq 0, \quad (5.238)$$

(5.236) and (5.237) can be replaced by the vector condition

$$\mathbf{a}_{12} \times \mathbf{E} = 0, \quad (5.239)$$

which requires that the electric field at the boundary be parallel to the vector $\mathbf{a}_{12}$. For tangential vectors $\mathbf{a}_1$ and $\mathbf{a}_2$, i.e., for the normal vector $\mathbf{a}_{12} = a_{12}\mathbf{n}$, the E boundary equals the PEC boundary. Otherwise, we can choose either $\mathbf{a}_1$ or $\mathbf{a}_2$ tangential without losing generality.

To find plane-wave reflection from the E boundary, we can write from (5.29) – (5.32),

$$\mathbf{c}_1(\mathbf{k}) = -k_o\mathbf{a}_1, \quad \mathbf{c}_2(\mathbf{k}) = -k_o\mathbf{a}_2 \quad (5.240)$$

$$\mathbf{d}_1(\mathbf{k}) = \mathbf{k} \times \mathbf{a}_1, \quad \mathbf{d}_2(\mathbf{k}) = \mathbf{k} \times \mathbf{a}_2, \quad (5.241)$$

whence

$$J(\mathbf{k}) = \mathbf{c}_1(\mathbf{k}) \times \mathbf{c}_2(\mathbf{k}) \cdot \mathbf{k} = \mathbf{d}_1(\mathbf{k}) \times \mathbf{d}_2(\mathbf{k}) \cdot \mathbf{k} = k_o^2 \mathbf{a}_{12} \cdot \mathbf{k}. \quad (5.242)$$

Thus, the reflection dyadic (5.63) takes the form

$$\overline{\overline{\mathsf{R}}} = \frac{1}{\mathbf{a}_{12} \cdot \mathbf{k}^r} \mathbf{k}^r \times (\mathbf{a}_2\mathbf{a}_1 - \mathbf{a}_1\mathbf{a}_2) \quad (5.243)$$

$$= \frac{1}{\mathbf{a}_{12} \cdot \mathbf{k}^r} \mathbf{k}^r \times (\mathbf{a}_{12} \times \overline{\overline{\mathsf{I}}}) \quad (5.244)$$

$$= -\overline{\overline{\mathsf{I}}} + \frac{\mathbf{a}_{12}\mathbf{k}^r}{\mathbf{a}_{12} \cdot \mathbf{k}^r}. \quad (5.245)$$

In the special case $\mathbf{a}_{12} = \mathbf{n}(\mathbf{n} \cdot \mathbf{a}_{12})$, we have

$$\overline{\overline{\mathsf{R}}} = -\overline{\overline{\mathsf{I}}} + \frac{\mathbf{n}\mathbf{k}^r}{\mathbf{n} \cdot \mathbf{k}^r}, \quad (5.246)$$

which equals the reflection dyadic (5.71) of the PEC boundary.

For the E boundary we can arrive at the following conclusions.

- Because from (5.245) we have $\overline{\overline{\mathsf{R}}} \cdot \mathbf{a}_{12} = 0$, there is no reflection for the polarization $\mathbf{E}^i = A\mathbf{a}_{12}$. The same follows from the boundary condition (5.239), which is identically satisfied for $\mathbf{E} = \mathbf{E}^i = A\mathbf{a}_{12}$. Thus, an incident wave with such a polarization is matched to the E boundary for any $\mathbf{k}$ vector satisfying $\mathbf{k}^i \cdot \mathbf{E}^i = 0$. Assuming real $\mathbf{a}_{12}$, the possible solutions for $J(\mathbf{k}^i) = 0$ make a circular disk of radius k_o on the plane orthogonal to $\mathbf{a}_{12}$, and the dispersion curve for $\mathbf{k}_t$ equals its projection ellipse on the boundary plane. For the PEC boundary with $\mathbf{a}_{12} = a_{12}\mathbf{n}$, the matched waves are any lateral waves satisfying $\mathbf{k}^i \cdot \mathbf{n} = 0$.

- Considering an incident TE$_n$ wave satisfying $\mathbf{n} \cdot \mathbf{E}^i = 0$, from $\mathbf{k}^r \cdot \mathbf{E}^i = \mathbf{k}^i \cdot \mathbf{E}^i = 0$ and (5.245) we have $\mathbf{E}^r = \overline{\overline{\mathsf{R}}} \cdot \mathbf{E}^i = -\mathbf{E}^i$, whence the reflected wave is also a TE$_n$ wave. Since the total field is tangential and satisfies $\mathbf{E} = \mathbf{E}^i + \mathbf{E}^r = 0$, the E boundary can be replaced by the PEC boundary for any TE$_n$ wave.

- One can show that (5.245) satisfies $\overline{\overline{\mathsf{R}}} \cdot (\overline{\overline{\mathsf{R}}} + \overline{\overline{\mathsf{I}}}) = 0$, whence $\overline{\overline{\mathsf{R}}}$ has two eigenvalues, -1 and 0. The former corresponds to the PEC boundary case and, the latter, to the matched-wave case. Any given incident field can be decomposed in the two eigenfields as

$$\mathbf{E}^i = \frac{1}{\mathbf{n} \cdot \mathbf{a}_{12}}(\mathbf{a}_{12}(\mathbf{n} \cdot \mathbf{E}^i) - \mathbf{n} \times (\mathbf{a}_{12} \times \mathbf{E}^i)), \tag{5.247}$$

  when $\mathbf{n} \cdot \mathbf{a}_{12} \neq 0$. The latter (TE$_n$) component reflects from the E boundary as from the PEC boundary, while the former component creates no reflected field.

Similar properties can be found for the H boundary, defined by the dual conditions

$$\mathbf{b}_1 \cdot \mathbf{H} = 0, \qquad \mathbf{b}_2 \cdot \mathbf{H} = 0, \tag{5.248}$$

or, by

$$\mathbf{b}_{12} \times \mathbf{H} = 0, \qquad \mathbf{b}_{12} = \mathbf{b}_1 \times \mathbf{b}_2 \neq 0. \tag{5.249}$$

EH Boundary

Conditions of the form

$$\mathbf{a}_1 \cdot \mathbf{E} = 0, \qquad \mathbf{b}_2 \cdot \mathbf{H} = 0, \tag{5.250}$$

define another special case of the GBC boundary which has been called the EH boundary in the past [75]. Here, $\mathbf{a}_1$ and $\mathbf{b}_2$ may be any two vectors. Expanding (5.29) – (5.32) as

$$\mathbf{c}_1(\mathbf{k}) = -k_o\mathbf{a}_1, \qquad \mathbf{c}_2(\mathbf{k}) = \mathbf{k} \times \mathbf{b}_2, \tag{5.251}$$

$$\mathbf{d}_1(\mathbf{k}) = \mathbf{k} \times \mathbf{a}_1, \qquad \mathbf{d}_2(\mathbf{k}) = k_o\mathbf{b}_2, \tag{5.252}$$

we obtain

$$J(\mathbf{k}) = -k_o(\mathbf{k} \times \mathbf{a}_1) \cdot (\mathbf{k} \times \mathbf{b}_2). \tag{5.253}$$

In this case, the reflection dyadic (5.63) takes the form

$$\overline{\overline{\mathsf{R}}} = -\frac{k_o}{J^r}\mathbf{k}^r \times ((\mathbf{k}^r \times \mathbf{b}_2)\mathbf{a}_1 - \mathbf{a}_1(\mathbf{k}^i \times \mathbf{b}_2)). \tag{5.254}$$

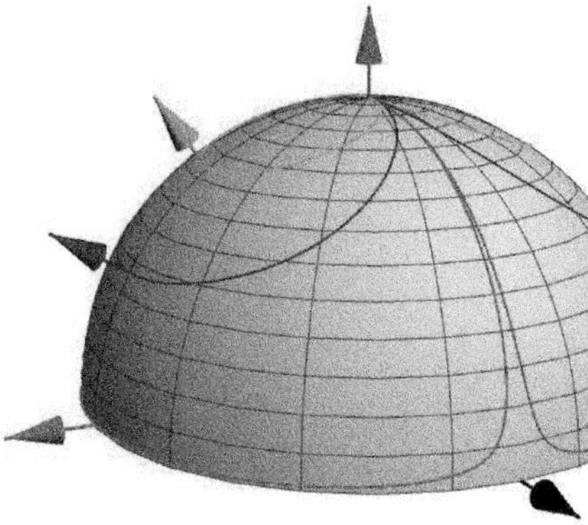

Figure 5.8: Paths of the normalized matched-wave vector $\mathbf{k}^i/k_o$ on the unit half sphere above the EH boundary for $\mathbf{b}_2 = \mathbf{n}$ and $\mathbf{a}_1$ making angles $\psi = \pi/5, \pi/3$ and $\pi/2 - .01$ with the normal direction $\mathbf{n} = \mathbf{e}_3$ (arrows in descending order). The arrow on the right is parallel to $-\mathbf{a}_1 \times \mathbf{b}_2 \sim \mathbf{e}_2$.

One can show that when $J(\mathbf{k}^i) \neq 0$, any incident field can be expanded as

$$\mathbf{E}^i = \frac{k_o}{J^i}\mathbf{k}^i \times ((\mathbf{k}^i \times \mathbf{b}_2)(\mathbf{a}_1 \cdot \mathbf{E}^i) + k_o\mathbf{a}_1(\mathbf{b}_2 \cdot \eta_o\mathbf{H}^i)), \tag{5.255}$$

$$\eta_o\mathbf{H}^i = \frac{k_o}{J^i}\mathbf{k}^i \times ((\mathbf{k}^i \times \mathbf{a}_1)(\mathbf{b}_2 \cdot \eta_o\mathbf{H}^i) - k_o\mathbf{b}_2(\mathbf{a}_1 \cdot \mathbf{E}^i)), \tag{5.256}$$

details of which are left as an exercise. Thus, the incident plane wave can be decomposed in two parts, the TE$_a$ component, satisfying $\mathbf{a}_1 \cdot \mathbf{E}^i = 0$, and the TM$_b$ component, satisfying $\mathbf{b}_2 \cdot \mathbf{H}^i = 0$. The reflected fields are obtained from (5.254) as

$$\mathbf{E}^r = -\frac{k_o}{J^r}\mathbf{k}^r \times (\mathbf{k}^r \times \mathbf{b}_2(\mathbf{a}_1 \cdot \mathbf{E}^i) + k_o\mathbf{a}_1(\mathbf{b}_2 \cdot \eta_o\mathbf{H}^i)), \tag{5.257}$$

$$\eta_o\mathbf{H}^r = -\frac{k_o}{J^r}\mathbf{k}^r \times (\mathbf{k}^r \times \mathbf{a}_1(\mathbf{b}_2 \cdot \eta_o\mathbf{H}^i) - k_o\mathbf{b}_2(\mathbf{a}_1 \cdot \mathbf{E}^i)). \tag{5.258}$$

They obey a similar decomposition and the two components do not couple in reflection. Actually, TE$_a$ and TM$_b$ could be respectively relabeled as TE$_{c1}$ and TE$_{c2}$.

Let us summarize some properties of the EH boundary.

- For $\mathbf{n} \cdot \mathbf{a}_1 = \mathbf{n} \cdot \mathbf{b}_2 = 0$ the EH boundary equals the GSH boundary. For $\mathbf{n} \times \mathbf{a}_1 = \mathbf{n} \times \mathbf{b}_2 = 0$ it equals the DB boundary.
- The condition for a wave matched to the EH boundary becomes

$$J(\mathbf{k}) = 0 \quad \Rightarrow \quad (\mathbf{k} \times \mathbf{a}_1) \cdot (\mathbf{k} \times \mathbf{b}_2) = 0. \tag{5.259}$$

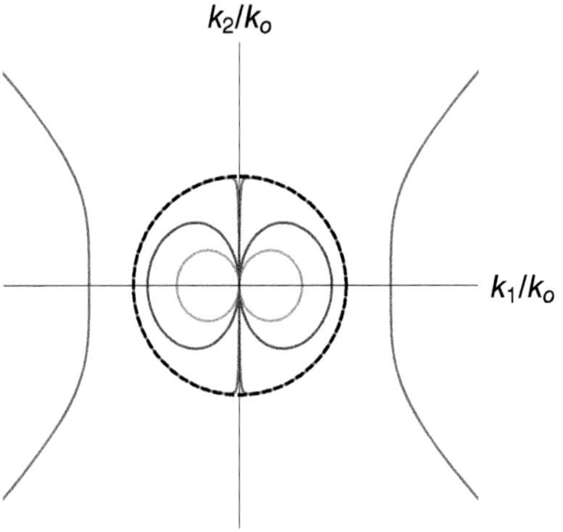

Figure 5.9: Projections of the wave-vector paths on the boundary plane make the corresponding dispersion diagrams for the matched waves on different EH boundaries of Figure 5.8. The lines outside the unit circle represent waves with imaginary normal component of $\mathbf{k}^i$.

Because from (5.254) we obtain

$$\overline{\overline{\mathsf{R}}}\cdot(\mathbf{a}_1\times(\mathbf{k}^i\times\mathbf{b}_2))=0, \tag{5.260}$$

the matched wave is polarized as $\mathbf{E}=A\mathbf{a}_1\times(\mathbf{k}\times\mathbf{b}_2)$ and $\mathbf{H}=B\mathbf{k}\times\mathbf{b}_2$. It is easy to check that the results for the GSH and DB boundaries are obtained as two special cases.

- For the special EH conditions

$$\mathbf{a}_{1t}\cdot\mathbf{E}=0,\qquad \mathbf{n}\cdot\mathbf{H}=0, \tag{5.261}$$

and

$$\mathbf{b}_{2t}\cdot\mathbf{H}=0,\qquad \mathbf{n}\cdot\mathbf{E}=0, \tag{5.262}$$

the boundary can be replaced by the respective PEC and PMC boundaries. The proof is left as an exercise.

As a numerical example, let us assume $\mathbf{b}_2=\mathbf{n}=\mathbf{e}_3$ and $\mathbf{a}_1=\mathbf{e}_3\cos\psi+\mathbf{e}_1\sin\psi$, whence $\mathbf{a}_1\times\mathbf{b}_2=-\mathbf{e}_2\sin\psi$. Substituting these, and

$$\mathbf{k}=k_o(\mathbf{e}_1\sin\theta\cos\varphi+\mathbf{e}_2\sin\theta\sin\varphi+\mathbf{e}_3\cos\theta), \tag{5.263}$$

in (5.259), yields the relation

$$\sin\theta(\tan\psi-\frac{\tan\theta}{\cos\varphi})=0. \tag{5.264}$$

One solution of (5.259) is $\mathbf{k}\times\mathbf{b}_2 = \mathbf{k}\times\mathbf{n} = 0$, or $\sin\theta = 0$, which corresponds to the normally incident wave. Requiring $\tan\theta = \tan\psi\cos\varphi$ we can expand

$$\begin{aligned}\mathbf{k}_t &= k_o\sin\theta(\mathbf{e}_1\cos\varphi + \mathbf{e}_2\sin\varphi)\\ &= k_o\frac{\tan\theta}{\sqrt{1+\tan^2\theta}}(\mathbf{e}_1\cos\varphi + \mathbf{e}_2\sin\varphi)\\ &= k_o\frac{\sin\psi\cos\varphi}{\sqrt{\cos^2\psi + \sin^2\psi\cos^2\varphi}}(\mathbf{e}_1\cos\varphi + \mathbf{e}_2\sin\varphi). \qquad (5.265)\end{aligned}$$

Figure 5.8 depicts the path of the direction of the normalized wave vector $\mathbf{k}/k_o$ of a matched wave on the unit sphere for three different vectors $\mathbf{a}_1$ at three angles ψ. Because of the square root, there is another set of similar paths on the other half sphere. Figure 5.9 shows the projection of both sets of paths on the plane $x_3 = 0$ of the EH boundary. There is also a solution $\theta = 0$ (origin in Figure 5.9) of (5.264) for any ψ. For $\psi \to \pi/2$, $\mathbf{a}_1$ becomes tangential to the boundary and the dispersion diagram approaches a circle corresponding to the PEC boundary. In Figure 5.9 dispersion curves for the complex vector $\mathbf{a}_1 = j\mathbf{e}_3 + \sqrt{2}\mathbf{e}_1$ have been added. Since the real part of k_t/k_o stays outside the unit circle, k_3 is imaginary corresponding to either a surface wave or a leaky wave.

5.12 Summary of GBC Conditions

In the present Chapter, the most general linear and local form of boundary conditions (GBC conditions) have been introduced, along with a number of special cases. They can be represented by (5.10), (5.11), in the form

$$a_{1n}\mathbf{n}\cdot\mathbf{E} + \mathbf{a}_{1t}\cdot\mathbf{E} + \mathbf{b}_{1t}\cdot\eta_o\mathbf{H}_t = 0, \qquad (5.266)$$

$$\mathbf{a}_{2t}\cdot\mathbf{E} + b_{2n}\mathbf{n}\cdot\eta_o\mathbf{H} + \mathbf{b}_{2t}\cdot\eta_o\mathbf{H} = 0, \qquad (5.267)$$

In the following table, the scalar and vector parameters of different special cases of the GBC boundary have been summarized for comparison.

GBC	a_{1n}/b_{2n}	$\mathbf{a}_{1t}/\mathbf{a}_{2t}$	$\mathbf{b}_{1t}/\mathbf{b}_{2t}$
Impedance	$0/0$	$\mathbf{a}_{1t}/\mathbf{a}_{2t}$	$\mathbf{b}_{1t}/\mathbf{b}_{2t}$
DB	a_{1n}/b_{2n}	$0/0$	$0/0$
SH	$0/0$	$\mathbf{a}_{1t}/0$	$0/\mathbf{a}_{1t}$
SHDB	a_{1n}/a_{1n}	$\mathbf{a}_{1t}/0$	$0/-\mathbf{a}_{1t}$
GSH	$0/0$	$\mathbf{a}_{1t}/0$	$0/\mathbf{b}_{2t}$
GSHDB	a_{1n}/b_{2n}	$\mathbf{a}_{1t}/0$	$0/\mathbf{b}_{2t}$
E	$a_{1n}/0$	$\mathbf{a}_{1t}/\mathbf{a}_{2t}$	$0/0$
H	$0/b_{2n}$	$0/0$	$\mathbf{b}_{1t}/\mathbf{b}_{2t}$
EH	a_{1n}/b_{2n}	$0/\mathbf{a}_{2t}$	$\mathbf{b}_{1t}/0$

5.13 Reciprocity of GBC Boundaries

According to [87], reaction between sources a and fields from sources b is defined as

$$< ab >= \int (\mathbf{J}_{ea} \cdot \mathbf{E}_b - \mathbf{J}_{ma} \cdot \mathbf{H}_b) dV, \qquad (5.268)$$

where the integral is over a volume containing the sources a. Similar expression can be written for the reaction $< ba >$ when the volume contains the sources b. When the integral contains both sources a and b, the condition of reciprocity can be expressed as

$$< ab > - < ba >= 0. \qquad (5.269)$$

Let us denote the two pairs of time-harmonic sources by $\mathbf{J}'_e, \mathbf{J}'_m$ and $\mathbf{J}''_e, \mathbf{J}''_m$ and the corresponding fields by $\mathbf{E}', \mathbf{H}'$ and $\mathbf{E}'', \mathbf{H}''$ and assume them located at finite distance from the origin. The condition of reciprocity (5.269) can be expressed as

$$\int_V (\mathbf{E}' \cdot \mathbf{J}''_e - \mathbf{H}' \cdot \mathbf{J}''_m) dV = \int_V (\mathbf{E}'' \cdot \mathbf{J}'_e - \mathbf{H}'' \cdot \mathbf{J}'_m) dV, \qquad (5.270)$$

where V is the volume containing all sources. Applying the Maxwell equations, (5.270) can be reduced to the integral over the boundary of V,

$$\oint \mathbf{n} \cdot (\mathbf{E}' \times \mathbf{H}'' - \mathbf{E}'' \times \mathbf{H}') dS = 0, \qquad (5.271)$$

which is known as the Lorentz reciprocity theorem [29, 31]. The integration surface is formed by the plane $\mathbf{n} \cdot \mathbf{r} = 0$ with GBC boundary conditions and the half-sphere in infinity. Since the fields in an isotropic medium satisfy the radiation condition, the integral over the half-sphere vanishes [2], whence the integration surface consists of the GBC boundary plane S. A boundary is nonreciprocal unless it is reciprocal for all sources.

Let us consider special cases of the GBC boundary for which we can find conditions of reciprocity for the boundary parameters. Here one may note that reciprocity is not a property which is preserved in duality transformation. In fact, the PEC boundary, which is known to be reciprocal, can be transformed to a PEMC boundary, known to be nonreciprocal.

Impedance Boundary

For the impedance boundary defined by the conditions (3.1) as

$$\mathbf{E}_t = \overline{\overline{\mathsf{Z}}}_t \cdot (\mathbf{n} \times \mathbf{H}_t), \qquad (5.272)$$

in terms of the impedance dyadic $\overline{\overline{\mathsf{Z}}}_t$, (5.271) yields

$$\int_S (\mathbf{n} \times \mathbf{H}'_t) \cdot (\overline{\overline{\mathsf{Z}}}_t - \overline{\overline{\mathsf{Z}}}{}_t^T) \cdot (\mathbf{n} \times \mathbf{H}''_t) dS = 0. \qquad (5.273)$$

Since, for reciprocity, this must be valid for any $\mathbf{H}'$ and $\mathbf{H}''$, the impedance dyadic must satisfy

$$\overline{\overline{\mathsf{Z}}}_t = \overline{\overline{\mathsf{Z}}}_t^T, \tag{5.274}$$

i.e., it must be symmetric. As a conclusion, any impedance boundary which contains the antisymmetric (PEMC) component is nonreciprocal.

GSH Boundary

The generalized soft-and-hard (GSH) boundary defined by conditions (3.180) as

$$\mathbf{a}_t \cdot \mathbf{E} = 0, \qquad \mathbf{b}_t \cdot \mathbf{H} = 0, \tag{5.275}$$

is a special case of the impedance boundary. Since the conditions (5.275) cannot be expressed simply in terms of the impedance dyadic $\overline{\overline{\mathsf{Z}}}_t$, let us consider the reciprocity property separately. The tangential fields are of the form

$$\mathbf{E}'_t = E'\mathbf{n} \times \mathbf{a}_t, \qquad \mathbf{E}''_t = E''\mathbf{n} \times \mathbf{a}_t, \tag{5.276}$$

$$\mathbf{H}'_t = H'\mathbf{n} \times \mathbf{b}_t, \qquad \mathbf{H}''_t = H''\mathbf{n} \times \mathbf{b}_t, \tag{5.277}$$

whence the condition (5.271) becomes

$$\int_S \mathbf{n} \cdot (E'H'' + E''H')(\mathbf{n} \cdot \mathbf{a}_t \times \mathbf{b}_t) dS = 0. \tag{5.278}$$

To be valid for all values $E' \cdots H''$, the boundary parameters must satisfy

$$\mathbf{n} \cdot \mathbf{a}_t \times \mathbf{b}_t = 0 \quad \Rightarrow \quad \mathbf{b}_t = \alpha \mathbf{a}_t. \tag{5.279}$$

In conclusion, the GSH boundary is reciprocal only when it actually equals the SH boundary.

DB Boundary

For the DB boundary, defined by the conditions (4.1) as

$$\mathbf{n} \cdot \mathbf{D} = 0, \qquad \mathbf{n} \cdot \mathbf{B} = 0, \tag{5.280}$$

the fields at the boundary satisfy

$$-j\omega \mathbf{n} \cdot \mathbf{B} = \mathbf{n} \cdot (\nabla \times \mathbf{E}) = 0, \quad \Rightarrow \quad \nabla_t \times \mathbf{E}_t = 0, \tag{5.281}$$

$$j\omega \mathbf{n} \cdot \mathbf{D} = \mathbf{n} \cdot (\nabla \times \mathbf{H}) = 0, \quad \Rightarrow \quad \nabla_t \times \mathbf{H}_t = 0. \tag{5.282}$$

From conditions of this form we conclude that the tangential fields at the DB boundary can be expressed in terms of four scalar potential quantities as

$$\mathbf{E}'_t = \nabla_t \phi', \qquad \mathbf{E}''_t = \nabla_t \phi'', \tag{5.283}$$

$$\mathbf{H}'_t = \nabla_t\psi', \quad \mathbf{H}''_t = \nabla_t\psi''. \tag{5.284}$$

In the integrand of (5.271) we can expand

$$\mathbf{E}'_t \times \mathbf{H}''_t - \mathbf{E}''_t \times \mathbf{H}'_t = (\nabla_t\phi') \times (\nabla_t\psi'') - (\nabla_t\phi'') \times (\nabla_t\psi') \tag{5.285}$$

$$= \nabla_t \times (\phi'\nabla_t\psi'' - \phi''\nabla_t\psi'), \tag{5.286}$$

which is of the form $\nabla_t \times \mathbf{f}_t$, whence the integrand of (5.271) has the form $\mathbf{n} \cdot (\nabla_t \times \mathbf{f}_t)$. Due to Stokes' theorem, the integral of a planar curl of a vector reduces to a line integral in infinity,

$$\int_S \nabla_t \times \mathbf{f}_t \cdot \mathbf{n} dS = \oint_C \mathbf{f}_t \cdot d\mathbf{c}. \tag{5.287}$$

The integral vanishes because of the $1/r$ dependence of the potentials. In conclusion, the DB boundary is reciprocal.

D'B' Boundary

Being non-local, the D'B' boundary, defined by conditions of the form

$$\nabla \cdot (\mathbf{nn} \cdot \mathbf{D}) = 0, \quad \nabla \cdot (\mathbf{nn} \cdot \mathbf{B}) = 0, \tag{5.288}$$

is not an example of the GBC boundary. However, the reciprocity property can be studied in a manner similar to that of the DB boundary. Because fields outside sources satisfy

$$\nabla \cdot \mathbf{D} = 0, \quad \nabla \cdot \mathbf{B} = 0, \tag{5.289}$$

the D'B' boundary conditions can be alternatively expressed as

$$\nabla \cdot (\mathbf{D} - \mathbf{nn} \cdot \mathbf{D}) = \nabla_t \cdot \mathbf{D}_t = 0, \tag{5.290}$$

$$\nabla \cdot (\mathbf{B} - \mathbf{nn} \cdot \mathbf{B}) = \nabla_t \cdot \mathbf{B}_t = 0. \tag{5.291}$$

Thus, in an isotropic medium the tangential fields at the planar D'B' boundary can be represented as

$$\mathbf{E}'_t = \nabla_t \times \mathbf{n}\phi', \quad \mathbf{E}''_t = \nabla_t \times \mathbf{n}\phi'', \tag{5.292}$$

$$\mathbf{H}'_t = \nabla_t \times \mathbf{n}\psi', \quad \mathbf{H}''_t = \nabla_t \times \mathbf{n}\psi'', \tag{5.293}$$

in terms of four scalar potentials.

Expanding

$$\begin{aligned}\mathbf{n} \cdot (\mathbf{E}'_t \times \mathbf{H}''_t - \mathbf{E}''_t \times \mathbf{H}'_t) &= \mathbf{n} \cdot \big[(\nabla_t\phi' \times \mathbf{n}) \times (\nabla_t\psi'' \times \mathbf{n}) \\ &\quad -(\nabla_t\phi'' \times \mathbf{n}) \times (\nabla_t\psi' \times \mathbf{n})\big] \\ &= \mathbf{n} \cdot (\nabla_t\phi' \times \nabla_t\psi'' - \nabla_t\phi'' \times \nabla_t\psi') \\ &= \mathbf{n} \cdot (\nabla_t \times (\phi'\nabla_t\psi'' - \phi''\nabla_t\psi')),\end{aligned} \tag{5.294}$$

the integrand of (5.271) is seen to be of the form $\mathbf{n} \cdot \nabla_t \times \mathbf{f}_t$, whence, again, its integral over the surface S yields zero. Thus, the D'B' boundary is reciprocal.

Finding the reciprocity condition for the mixed-impedance (DB/D'B') boundary can be found along similar lines. Details are left as an exercise.

SHDB Boundary

The SHDB boundary is defined by the conditions (5.114) and (5.115) as

$$\alpha\mathbf{n}\cdot c\mathbf{B} + \mathbf{a}_t\cdot\mathbf{E} = 0, \qquad (5.295)$$

$$\alpha\mathbf{n}\cdot c\mathbf{D} - \mathbf{a}_t\cdot\mathbf{H} = 0. \qquad (5.296)$$

Substituting **B** and **D** from the Maxwell equations, the conditions can be expressed in the form (5.116)

$$\mathbf{L}(\nabla_t)\cdot\mathbf{E}_t = 0, \qquad \mathbf{L}(\nabla_t)\cdot\mathbf{H}_t = 0, \qquad (5.297)$$

in terms of the vector operator

$$\mathbf{L}(\nabla_t) = \alpha\mathbf{n}\times\nabla_t - jk_o\mathbf{a}_t. \qquad (5.298)$$

Let us define another planar operator as

$$\mathbf{G}(\nabla_t) = \mathbf{L}(\nabla_t)\times\mathbf{n} = \alpha\nabla_t + jk_o\mathbf{n}\times\mathbf{a}_t. \qquad (5.299)$$

Because the two operators satisfy

$$\mathbf{L}(\nabla_t)\cdot\mathbf{G}(\nabla_t) = 0, \qquad (5.300)$$

the tangential fields can be represented as

$$\mathbf{E}_t = \mathbf{G}(\nabla_t)\phi, \qquad \mathbf{H}_t = \mathbf{G}(\nabla_t)\psi, \qquad (5.301)$$

in terms of two scalar functions. The fields so defined satisfy the SHDB conditions (5.297) for any scalar functions ϕ and ψ.

Expanding

$$\begin{aligned}\mathbf{n}\cdot(\mathbf{E}_t\times\mathbf{H}_t) &= \mathbf{n}\cdot(\mathbf{G}(\nabla_t)\phi\times\mathbf{G}(\nabla_t)\psi)\\ &= \mathbf{n}\cdot(\alpha\nabla_t\phi + jk_o\mathbf{n}\times\mathbf{a}_t\phi)\times(\alpha\nabla_t\psi + jk_o\mathbf{n}\times\mathbf{a}_t\psi)\\ &= \alpha^2\mathbf{n}\cdot\nabla_t\times(\phi\nabla_t\psi) + jk_o\alpha\mathbf{a}_t\cdot(\psi\nabla_t\phi - \phi\nabla_t\psi),\end{aligned} \qquad (5.302)$$

and substituting in (5.271), the integrand due to the first term of the form $\mathbf{n}\cdot\nabla_t\times\mathbf{f}_t$ in (5.302) is seen to vanish in the integration. The remaining part of the integrand due to last term of (5.302) equals

$$jk_o\alpha\mathbf{a}_t\cdot(\psi''\nabla_t\phi' - \phi'\nabla_t\psi'' - \psi'\nabla_t\phi'' + \phi''\nabla_t\psi'). \qquad (5.303)$$

For a reciprocal SHDB boundary, this must be zero for any potentials $\phi'\cdots\psi''$. Obviously, this is valid when either $\alpha = 0$ or $\mathbf{a}_t = 0$, which correspond to the respective special cases of the SH boundary and the DB boundary. Thus, in conclusion, the SHDB boundary is reciprocal only in its two special cases, the SH boundary and the DB boundary.

GSHDB boundary

The generalization of the SHDB boundary, the GSHDB boundary, is defined by conditions (5.181) and (5.182) as

$$\alpha c\mathbf{n}\cdot\mathbf{B}+\mathbf{a}_t\cdot\mathbf{E} = 0, \tag{5.304}$$

$$\beta c\mathbf{n}\cdot\mathbf{D}+\mathbf{b}_t\cdot\mathbf{H} = 0. \tag{5.305}$$

After substituting **B** and **D** from the Maxwell equations we obtain equations of the form (5.185),

$$\mathbf{L}_a(\nabla_t)\cdot\mathbf{E}_t=0,\quad \mathbf{L}_b(\nabla_t)\cdot\mathbf{H}_t=0, \tag{5.306}$$

with

$$\mathbf{L}_a(\nabla_t) = \alpha\mathbf{n}\times\nabla_t - jk_o\mathbf{a}_t, \tag{5.307}$$

$$\mathbf{L}_b(\nabla_t) = \beta\mathbf{n}\times\nabla_t + jk_o\mathbf{b}_t. \tag{5.308}$$

Thus, the fields transverse to the boundary can be expressed in terms of two scalar functions ϕ and ψ as

$$\mathbf{E}_t=\mathbf{n}\times\mathbf{L}_a(\nabla_t)\phi,\qquad \mathbf{H}_t=\mathbf{n}\times\mathbf{L}_b(\nabla_t)\psi. \tag{5.309}$$

Expanding the field expression

$$\begin{aligned}\mathbf{n}\cdot\mathbf{E}_t\times\mathbf{H}_t &= \mathbf{n}\cdot(\mathbf{n}\times\mathbf{L}_a(\nabla_t)\phi)\times(\mathbf{n}\times\mathbf{L}_b(\nabla_t)\psi)\\ &= \mathbf{n}\cdot(\mathbf{L}_a(\nabla_t)\phi\times\mathbf{L}_b(\nabla_t)\psi)\\ &= \mathbf{n}\cdot(\alpha\mathbf{n}\times\nabla_t\phi-jk_o\mathbf{a}_t\phi)\times(\beta\mathbf{n}\times\nabla_t\psi+jk_o\mathbf{b}_t\psi)\\ &= \alpha\beta\mathbf{n}\cdot(\nabla_t\phi)\times(\nabla_t\psi)-jk_o\alpha\psi\mathbf{b}_t\cdot(\nabla_t\phi)\\ &\quad -jk_o\beta\phi\mathbf{a}_t\cdot(\nabla_t\psi)+k_o^2(\mathbf{n}\cdot\mathbf{a}_t\times\mathbf{b}_t)\phi\psi,\end{aligned} \tag{5.310}$$

the term $\mathbf{n}\cdot(\nabla_t\phi)\times(\nabla_t\psi)=\mathbf{n}\cdot\nabla\times(\phi\nabla_t\psi)$ can be omitted because it corresponds to terms vanishing in the integration (5.271). The integrable part of the expression $\mathbf{n}\cdot(\mathbf{E}'\times\mathbf{H}''-\mathbf{E}''\times\mathbf{H}')$ equals

$$\begin{aligned}\mathbf{n}\cdot(\mathbf{E}'\times\mathbf{H}''-\mathbf{E}''\times\mathbf{H}') &= -jk_o\alpha\mathbf{b}_t\cdot((\nabla_t\phi')\psi''-(\nabla_t\phi'')\psi')\\ &\quad -jk_o\beta\mathbf{a}_t\cdot((\nabla_t\psi'')\phi'-(\nabla_t\psi')\phi'')\\ &\quad +k_o^2(\mathbf{n}\cdot\mathbf{a}_t\times\mathbf{b}_t)(\phi'\psi''-\phi''\psi').\end{aligned} \tag{5.311}$$

To be reciprocal, the parameters $\alpha,\beta,\mathbf{a}_t$ and $\mathbf{b}_t$ of the GSHDB boundary must make this expression vanish for any possible potentials $\phi'\cdots\psi''$. This requires satisfaction of the three conditions

$$\alpha\mathbf{b}_t=0,\qquad \beta\mathbf{a}_t=0,\qquad \mathbf{a}_t\times\mathbf{b}_t=0, \tag{5.312}$$

between the four parameters $\alpha,\beta,\mathbf{a}_t$ and $\mathbf{b}_t$. Of these we must exclude the two possibilities $\alpha=0,\mathbf{a}_t=0$ and $\beta=0,\mathbf{b}_t=0$, because each of the two leads to an incomplete set of boundary conditions.

Actually, the conditions (5.312) leave us with two possibilities

- $\alpha \neq 0,\ \mathbf{b}_t = 0,\ \beta \neq 0,\ \mathbf{a}_t = 0$. This corresponds to the DB boundary.
- $\alpha = 0,\ \mathbf{b}_t \neq 0,\ \beta = 0,\ \mathbf{a}_t \neq 0$. Because of $\mathbf{a}_t \times \mathbf{b}_t = 0$, this corresponds to the SH boundary.

As the conclusion, the GSHDB boundary is reciprocal when it actually equals the reciprocal SHDB boundary, i.e., it coincides either with the SH boundary or the DB boundary.

Although there is no proof yet, one may conjecture that the GBC boundary is reciprocal only when it equals the impedance boundary with symmetric impedance dyadic $\overline{\overline{\mathsf{Z}}}_t$, or the DB boundary.

E Boundary and H Boundary

The E boundary and the H boundary are defined by the respective conditions

$$\mathbf{a}_1 \cdot \mathbf{E} = 0, \quad \mathbf{a}_2 \cdot \mathbf{E} = 0, \quad \Rightarrow \quad \mathbf{E} = E\mathbf{a}_1 \times \mathbf{a}_2 \neq 0, \tag{5.313}$$

$$\mathbf{b}_1 \cdot \mathbf{H} = 0, \quad \mathbf{b}_2 \cdot \mathbf{H} = 0, \quad \Rightarrow \quad \mathbf{H} = H\mathbf{b}_1 \times \mathbf{b}_2 \neq 0. \tag{5.314}$$

We can assume $\mathbf{n} \cdot \mathbf{a}_2 = 0$ and $\mathbf{n} \cdot \mathbf{b}_1 = 0$ without loss of generality.

Expanding for the E boundary

$$\mathbf{n} \cdot (\mathbf{E}' \times \mathbf{H}'' - \mathbf{E}'' \times \mathbf{H}') = (\mathbf{n} \times (\mathbf{a}_1 \times \mathbf{a}_2)) \cdot (E'\mathbf{H}'' - E''\mathbf{H}'), \tag{5.315}$$

and for the H boundary,

$$\mathbf{n} \cdot (\mathbf{E}' \times \mathbf{H}'' - \mathbf{E}'' \times \mathbf{H}') = (\mathbf{n} \times (\mathbf{b}_1 \times \mathbf{b}_2)) \cdot (H'\mathbf{E}'' - H''\mathbf{E}'), \tag{5.316}$$

and requiring these to be zero for all $\mathbf{E}', \mathbf{E}''$ and $\mathbf{H}', \mathbf{H}''$, leads to the conditions

$$\mathbf{n} \times (\mathbf{a}_1 \times \mathbf{a}_2) = 0, \quad \Rightarrow \quad \mathbf{n} \times \mathbf{E} = 0, \tag{5.317}$$

$$\mathbf{n} \times (\mathbf{b}_1 \times \mathbf{b}_2) = 0, \quad \Rightarrow \quad \mathbf{n} \times \mathbf{H} = 0. \tag{5.318}$$

Thus, the reciprocal E-boundary equals the PEC boundary and, the reciprocal H boundary equals the PMC boundary.

5.14 Realization of the GBC Boundary

To find a realization of the GBC boundary in terms of an interface of a medium, let us start from the conditions of the most general bi-anisotropic medium (1.3). Because of the continuity of $\mathbf{e}_3 \cdot \mathbf{B}$ and $\mathbf{e}_3 \cdot \mathbf{D}$ through the interface, the fields below the interface must satisfy the GBC conditions (5.6) and (5.7) as

$$\alpha_1 c\mathbf{e}_3 \cdot (\overline{\overline{\zeta}} \cdot \mathbf{E} + \overline{\overline{\mu}} \cdot \mathbf{H}) + \frac{\beta_1}{\epsilon_o}\mathbf{e}_3 \cdot (\overline{\overline{\epsilon}} \cdot \mathbf{E} + \overline{\overline{\xi}} \cdot \mathbf{H}) + \mathbf{a}_{1t} \cdot \mathbf{E} + \eta_o \mathbf{b}_{1t} \cdot \mathbf{H} = 0, \tag{5.319}$$

$$\alpha_2 c\mathbf{e}_3 \cdot (\overline{\overline{\zeta}} \cdot \mathbf{E} + \overline{\overline{\mu}} \cdot \mathbf{H}) + \frac{\beta_2}{\epsilon_o}\mathbf{e}_3 \cdot (\overline{\overline{\epsilon}} \cdot \mathbf{E} + \overline{\overline{\xi}} \cdot \mathbf{H}) + \mathbf{a}_{2t} \cdot \mathbf{E} + \eta_o \mathbf{b}_{2t} \cdot \mathbf{H} = 0. \tag{5.320}$$

While it is known that there is no unique way to define a medium corresponding to given boundary conditions, let us consider a simple scenario by requiring that the conditions (5.319) and (5.320) are actually valid everywhere in the medium for any possible fields. In such a case, the medium dyadics are required to satisfy the conditions

$$\mathbf{e}_3\cdot(c\alpha_1\overline{\overline{\zeta}}+\beta_1\overline{\overline{\epsilon}}/\epsilon_o)+\mathbf{a}_{1t} = 0, \tag{5.321}$$

$$\mathbf{e}_3\cdot(c\alpha_2\overline{\overline{\zeta}}+\beta_2\overline{\overline{\epsilon}}/\epsilon_o)+\mathbf{a}_{2t} = 0, \tag{5.322}$$

$$\mathbf{e}_3\cdot(c\alpha_1\overline{\overline{\mu}}+\beta_1\overline{\overline{\xi}}/\epsilon_o)+\eta_o\mathbf{b}_{1t} = 0, \tag{5.323}$$

$$\mathbf{e}_3\cdot(c\alpha_2\overline{\overline{\mu}}+\beta_2\overline{\overline{\xi}}/\epsilon_o)+\eta_o\mathbf{b}_{2t} = 0. \tag{5.324}$$

Assuming $\Delta=\alpha_1\beta_2-\alpha_2\beta_1\neq 0$, we obtain the following restrictions to the medium dyadics,

$$\Delta\mathbf{e}_3\cdot\overline{\overline{\epsilon}} = \epsilon_o(\alpha_2\mathbf{a}_{1t}-\alpha_1\mathbf{a}_{2t}), \tag{5.325}$$

$$\Delta\mathbf{e}_3\cdot\overline{\overline{\xi}} = \epsilon_o\eta_o(\alpha_2\mathbf{b}_{1t}-\alpha_1\mathbf{b}_{2t}), \tag{5.326}$$

$$\Delta\mathbf{e}_3\cdot\overline{\overline{\zeta}} = \frac{1}{c}(\beta_1\mathbf{a}_{2t}-\beta_2\mathbf{a}_{1t}), \tag{5.327}$$

$$\Delta\mathbf{e}_3\cdot\overline{\overline{\mu}} = \frac{\eta_o}{c}(\beta_1\mathbf{b}_{2t}-\beta_2\mathbf{b}_{1t}), \tag{5.328}$$

whence they can be expressed as

$$\overline{\overline{\epsilon}} = \frac{\epsilon_o}{\Delta}\mathbf{e}_3(\alpha_2\mathbf{a}_{1t}-\alpha_1\mathbf{a}_{2t})+\mathbf{e}_3\times\overline{\overline{\epsilon}}', \tag{5.329}$$

$$\overline{\overline{\xi}} = \frac{1}{c\Delta}\mathbf{e}_3(\alpha_2\mathbf{b}_{1t}-\alpha_1\mathbf{b}_{2t})+\mathbf{e}_3\times\overline{\overline{\xi}}', \tag{5.330}$$

$$\overline{\overline{\zeta}} = \frac{1}{c\Delta}\mathbf{e}_3(\beta_1\mathbf{a}_{2t}-\beta_2\mathbf{a}_{1t})+\mathbf{e}_3\times\overline{\overline{\zeta}}' \tag{5.331}$$

$$\overline{\overline{\mu}} = \frac{\mu_o}{\Delta}\mathbf{e}_3(\beta_1\mathbf{b}_{2t}-\beta_2\mathbf{b}_{1t})+\mathbf{e}_3\times\overline{\overline{\mu}}'. \tag{5.332}$$

Here, $\overline{\overline{\epsilon}}',\overline{\overline{\xi}}',\overline{\overline{\zeta}}',\overline{\overline{\mu}}'$ may be any four dyadics.

After substituting (5.329) – (5.332), one can verify that (5.319) and (5.320) are satisfied. Thus, the interface really acts as a boundary, defined by the conditions (5.8) and (5.9). The question on how the medium dyadics (5.329) – (5.332) can be realized by some physical material is, however, out of scope of this treatise. Present-day advances in metasurface research [78, 1, 123, 124] may eventually provide practical fabrication possibilities.

As a special case, for the GSHDB boundary conditions, the medium dyadics (5.329) – (5.332) are reduced to the form (5.221) – (5.224).

5.15 Problems

5.1 Check the reflection dyadic of the GBC boundary (5.63) for the special case of the DB boundary and find the solutions for the eigenvalue equation (5.98).

5.2 Check the reflection dyadic of the GBC boundary (5.63) for the special case of the GSH boundary and find the solutions for the eigenvalue equation (5.98).

5.3 Derive the form (5.103) of the eigenvalue equation (5.98) corresponding to the simple-isotropic boundary.

5.4 Derive (5.124).

5.5 Show that the solutions of the eigenvalue equation (5.98) are $R = \pm 1$ for the SHDB boundary.

5.6 Derive the expressions (5.29) - (5.32) and verify the relation (5.33)

5.7 Derive (5.34) and (5.35). Hint: start by expanding the expression $\mathbf{k}\times((\mathbf{c}_1 \times \mathbf{c}_2) \times \overline{\overline{\mathsf{I}}})$.

5.8 Check that the field expressions (5.41) - (5.44) satisfy the plane-wave conditions.

5.9 Show that any incident field can be expanded in the form (5.255).

5.10 Show that the special EH boundaries defined by (5.261) and (5.262) can be replaced by the respective PEC and PMC boundaries.

5.11 Derive the property (5.113) for the dual of the function $J(\mathbf{k})$.

5.12 Derive the equation for the electric field of a plane wave in a skewon-axion medium defined by (5.166) - (5.169) and show that it is satisfied by any wave vector $\mathbf{k}$.

5.13 Show that a plane wave in a skewon-axion medium, defined by medium conditions (5.165) with medium dyadics of the form (5.166) - (5.169), satisfies SHDB conditions (5.178), (5.179) when the parameters are chosen as (5.172) - (5.175).

5.14 Decomposing a plane wave in its TE_{c1} and TE_{c2} components with respect to vectors $\mathbf{c}_1(\mathbf{k}) \cdots \mathbf{d}_2(\mathbf{k})$ in (5.120) - (5.123) associated to the SHDB boundary, show that the fields satisfy the orthogonality conditions

$$\mathbf{E}_1 \cdot \mathbf{E}_2 = 0, \qquad \mathbf{H}_1 \cdot \mathbf{H}_2 = 0.$$

5.15 Show that the most general impedance boundary with PEC/PMC equivalence is the SGH boundary.

5.16 Find the reciprocity condition for the mixed-impedance (DB/D'B') boundary.

Chapter 6

Sesquilinear Boundary Conditions

A class of boundary conditions can be defined by requiring that the complex Poynting vector of any electromagnetic field have no normal component at the boundary surface,

$$\mathbf{n} \cdot \mathbf{E} \times \mathbf{H}^* = 0, \tag{6.1}$$

whence the tangential components of $\mathbf{E}$ and $\mathbf{H}^*$ must be linearly dependent at the boundary,

$$A\mathbf{E}_t + B\mathbf{H}_t^* = 0. \tag{6.2}$$

The condition (6.1) is not a linear one like the impedance-boundary condition (3.1). Actually, the condition (6.1) involves a function of two vector variables $f(\mathbf{E}_t, \mathbf{H}_t) = \mathbf{n} \cdot \mathbf{E}_t \times \mathbf{H}_t^*$, called sesquilinear.

The impedance-boundary condition (3.1) is linear, since it can be defined in terms of a bilinear function $f(\mathbf{x}, \mathbf{y}) = \mathbf{n} \cdot (\mathbf{x} \times \overline{\overline{\mathsf{Z}}}_t \cdot (\mathbf{n} \times \mathbf{y}))$ as a condition of the form $f(\mathbf{E}_t, \mathbf{H}_t) = 0$. By definition, a bilinear function is required to satisfy

$$\begin{aligned} f(\alpha\mathbf{x}_1 + \beta\mathbf{x}_2, \mathbf{y}) &= \alpha f(\mathbf{x}_1, \mathbf{y}) + \beta f(\mathbf{x}_2, \mathbf{y}), & (6.3)\\ f(\mathbf{x}, \alpha\mathbf{y}_1 + \beta\mathbf{y}_2) &= \alpha f(\mathbf{x}, \mathbf{y}_1) + \beta f(\mathbf{x}, \mathbf{y}_2), & (6.4) \end{aligned}$$

for any complex vectors $\mathbf{x}, \mathbf{x}_1, \mathbf{x}_2, \mathbf{y}, \mathbf{y}_1, \mathbf{y}_2$ and complex scalars α, β.

In contrast, the function $f(\mathbf{x}, \mathbf{y}) = \mathbf{n} \cdot \mathbf{x} \times \mathbf{y}^*$ is sesquilinear ("one-and-half linear") because it satisfies conditions of the form [15]

$$\begin{aligned} f(\alpha\mathbf{x}_1 + \beta\mathbf{x}_2, \mathbf{y}) &= \alpha f(\mathbf{x}_1, \mathbf{y}) + \beta f(\mathbf{x}_2, \mathbf{y}), & (6.5)\\ f(\mathbf{x}, \alpha\mathbf{y}_1 + \beta\mathbf{y}_2) &= \alpha^* f(\mathbf{x}, \mathbf{y}_1) + \beta^* f(\mathbf{x}, \mathbf{y}_2). & (6.6) \end{aligned}$$

Let us call conditions satisfying $f(\mathbf{E}_t, \mathbf{H}_t) = 0$ for any sesquilinear function as sesquilinear (SQL) boundary conditions, and the corresponding boundaries as the SQL boundaries.

Because of (6.1), the real part of the Poynting vector cannot have a normal component, whence no power may pass through the surface of the SQL boundary. This is why conditions of the form (6.1) have been labeled as "ideal boundary conditions" [39, 40, 41]. The SQL boundary can neither be active nor lossy. However, a given passive and lossless boundary is not necessarily an SQL boundary. PEC and PMC boundaries serve as two simple examples of the SQL boundary.

6.1 Isotropic and Anisotropic SQL Boundaries

The conditions (6.1) can be generalized by considering sesquilinear functions of the form $f(\mathbf{x},\mathbf{y}) = \mathbf{n}\cdot\mathbf{x}_t \times (\overline{\overline{\mathsf{X}}}_t \cdot \mathbf{y}_t^*)$, whence the SQL boundary condition can be expressed as

$$\mathbf{E}_t = e^{j\Theta}\overline{\overline{\mathsf{X}}}_t \cdot \eta_o \mathbf{H}_t^*. \tag{6.7}$$

Here, $\overline{\overline{\mathsf{X}}}_t$ is assumed to be a real 2D dyadic and Θ is a real scalar. The class of SQL boundaries can be split in two classes depending on the dyadic $\overline{\overline{\mathsf{X}}}_t$:

- Isotropic SQL boundaries

$$\mathbf{n}\mathbf{n}{}^\times_\times\overline{\overline{\mathsf{X}}}_t = \overline{\overline{\mathsf{X}}}_t, \tag{6.8}$$

- Anisotropic SQL boundaries

$$\mathbf{n}\mathbf{n}{}^\times_\times\overline{\overline{\mathsf{X}}}_t = -\overline{\overline{\mathsf{X}}}_t. \tag{6.9}$$

The general isotropic SQL boundary is defined by a condition of the form

$$\mathbf{E}_t = X_1\mathbf{H}_t^* + X_2\mathbf{n}\times\mathbf{H}_t^*, \tag{6.10}$$

where X_1 and X_2 are two complex scalars. The boundary is isotropic since there is no preferred tangential vector involved in the condition.

To define an example of anisotropic SQL boundaries let us first write (6.7) in the equivalent form

$$\overline{\overline{\mathsf{A}}}_t \cdot \mathbf{E}_t = \overline{\overline{\mathsf{B}}}_t \cdot \mathbf{H}_t^*, \tag{6.11}$$

where, $\overline{\overline{\mathsf{A}}}_t$ and $\overline{\overline{\mathsf{B}}}_t$ are some 2D dyadics. One can show that, for the choice of $\overline{\overline{\mathsf{A}}}_t = \mathbf{c}_t^*\mathbf{a}_t$ and $\overline{\overline{\mathsf{B}}}_t = (\mathbf{n}\times\mathbf{c}_t)\mathbf{b}_t^*$, the condition (6.11) is equivalent to

$$\mathbf{a}_t\cdot\mathbf{E}_t = 0, \qquad \mathbf{b}_t\cdot\mathbf{H}_t = 0. \tag{6.12}$$

Since this coincides with the GSH condition (3.180), one can conclude that the generalized soft-and-hard boundary serves as an example of an anisotropic SQL boundary. In the case $\mathbf{b}_t = \mathbf{a}_t^*$, the fields satisfy the condition $\mathbf{n}\cdot\mathbf{E}\times\mathbf{H}^* = 0$ at the boundary. In fact, assuming $\mathbf{a}_t\times\mathbf{a}_t^* = \mathbf{n}(\mathbf{n}\cdot\mathbf{a}_t\times\mathbf{a}_t^*) \neq 0$, from

$$\begin{aligned}(\mathbf{a}_t\times\mathbf{a}_t^*)\cdot(\mathbf{E}_t\times\mathbf{H}_t^*) &= (\mathbf{a}_t\cdot\mathbf{E}_t)(\mathbf{a}_t\cdot\mathbf{H}_t)^* - (\mathbf{a}_t^*\cdot\mathbf{E}_t)(\mathbf{a}_t\cdot\mathbf{H}_t^*)\\ &= (\mathbf{a}_t\cdot\mathbf{E}_t)(\mathbf{a}_t\cdot\mathbf{H}_t)^* - (\mathbf{a}_t^*\cdot\mathbf{E}_t)(\mathbf{b}_t\cdot\mathbf{H}_t)^*\\ &= (\mathbf{n}\cdot\mathbf{a}_t\times\mathbf{a}_t^*)\mathbf{n}\cdot(\mathbf{E}_t\times\mathbf{H}_t^*) = 0,\end{aligned} \tag{6.13}$$

it follows that the fields satisfy the SQL conditions (6.1).

Let us, however, concentrate on the isotropic SQL boundary defined by (6.10), and, in particular, on its two special cases, $X_1 = 0$ and $X_2 = 0$. In the latter case, the relation (6.7) can be expressed in the form

$$\mathbf{E}_t = e^{j\Theta} \tan\psi \ \eta_o \mathbf{H}_t^*, \tag{6.14}$$

where the scalars Θ and ψ are assumed real. In the case $X_1 = 0$, (6.7) takes the form

$$\mathbf{E}_t = e^{j\Theta} \tan\psi \ \eta_o \mathbf{n} \times \mathbf{H}_t^*. \tag{6.15}$$

In both of these cases, for $\psi = 0$ and $\psi = \pi/2$, the isotropic SQL boundary respectively equals the PEC and PMC boundary.

6.2 Reflection from Isotropic SQL Boundary

Leaving the case (6.15) as the topic of an exercise, let us concentrate on the SQL boundary obeying the condition (6.14) and consider plane-wave reflection from such a boundary.

Assuming that the wave vector $\mathbf{k}^i$ of the incident wave is a real vector, the incident plane-wave field

$$\mathbf{E}^i(\mathbf{r}) = \mathbf{E}^i e^{-j\mathbf{k}^i\cdot\mathbf{r}}, \quad \mathbf{H}^i(\mathbf{r}) = \mathbf{H}^i e^{-j\mathbf{k}^i\cdot\mathbf{r}}, \tag{6.16}$$

has the phase dependence $\exp(-j\mathbf{k}_t \cdot \mathbf{r})$ along the SQL boundary. Because of the conjugation sign in (6.14), the reflected field must contain a component with the dependence $\exp(j\mathbf{k}_t \cdot \mathbf{r})$, which is associated to a retroreflected wave propagating with the wave vector $-\mathbf{k}^i$. For the boundary condition (6.14) to be valid for the total field, it must split in two conditions corresponding to waves propagating in the two opposite directions along the boundary plane. Thus, the total reflected field must be composed of a forward (mirror) reflected $(\mathbf{E}', \mathbf{H}')$ and a retroreflected $(\mathbf{E}'', \mathbf{H}'')$ component as

$$\mathbf{E}^r(\mathbf{r}) = \mathbf{E}' e^{-j\mathbf{k}'\cdot\mathbf{r}} + \mathbf{E}'' e^{-j\mathbf{k}''\cdot\mathbf{r}}, \tag{6.17}$$

$$\mathbf{H}^r(\mathbf{r}) = \mathbf{H}' e^{-j\mathbf{k}'\cdot\mathbf{r}} + \mathbf{H}'' e^{-j\mathbf{k}''\cdot\mathbf{r}}, \tag{6.18}$$

with

$$\mathbf{k}' = \mathbf{k}^i - 2\mathbf{n}\mathbf{n}\cdot\mathbf{k}^i, \quad \mathbf{k}'' = -\mathbf{k}^i. \tag{6.19}$$

Assuming, more generally, that $\mathbf{k}^i$ may be a complex vector, one can show that the relations (6.19) must actually be replaced by

$$\mathbf{k}' = \mathbf{k}^i - 2\mathbf{n}\mathbf{n}\cdot\mathbf{k}^i, \quad \mathbf{k}'' = -\mathbf{k}^{i*}. \tag{6.20}$$

Inserting the total fields in (6.14), the wave components with the same spatial dependence at the boundary can be separated in two conditions as (proof left as an exercise)

$$(\mathbf{E}^i + \mathbf{E}')_t = e^{j\Theta} \tan\psi \ \eta_o \mathbf{H}_t''^*, \tag{6.21}$$

$$\mathbf{E}_t'' = e^{j\Theta} \tan\psi \ \eta_o (\mathbf{H}^i + \mathbf{H}')_t^*. \tag{6.22}$$

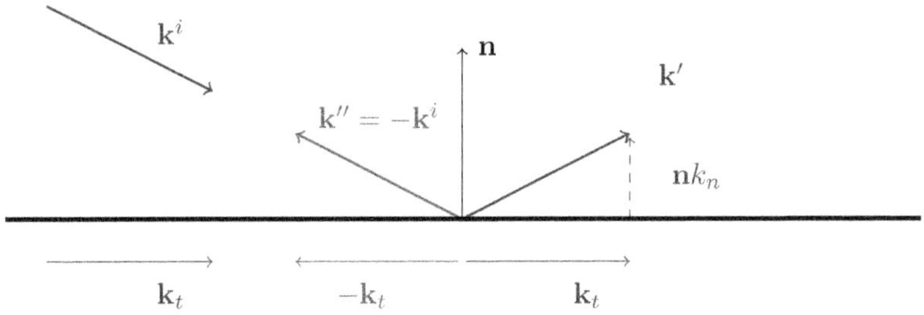

Figure 6.1: Wave incident to an SQL boundary with real $\mathbf{k}^i$ vector is bireflected as two waves with two $\mathbf{k}$ vectors. The vector $\mathbf{k}'' = -\mathbf{k}^i$ corresponds to a retroreflected wave.

We can apply the relations (1.47) and (1.48) between the tangential components of plane-wave fields, which in the present case can be expressed as

$$\eta_o \mathbf{H}_t^i = -\overline{\overline{\mathsf{J}}}_t \cdot \mathbf{E}_t^i, \quad \eta_o \mathbf{H}_t' = \overline{\overline{\mathsf{J}}}_t \cdot \mathbf{E}_t', \quad \eta_o \mathbf{H}_t'' = \overline{\overline{\mathsf{J}}}_t^{\,*} \cdot \mathbf{E}_t''. \tag{6.23}$$

The dyadic $\overline{\overline{\mathsf{J}}}_t$ is defined by (1.43) and it satisfies

$$\overline{\overline{\mathsf{J}}}_t = \frac{1}{k_n k_o} \mathbf{n} \times (k_n^2 \overline{\overline{\mathsf{I}}}_t + \mathbf{k}_t \mathbf{k}_t), \quad \overline{\overline{\mathsf{J}}}_t^2 = -\overline{\overline{\mathsf{I}}}_t. \tag{6.24}$$

After eliminating the magnetic field components, the conditions (6.21) and (6.22) take the form

$$(\mathbf{E}^i + \mathbf{E}')_t = e^{j\Theta} \tan\psi \, \overline{\overline{\mathsf{J}}}_t \cdot \mathbf{E}_t''^*, \tag{6.25}$$

$$\mathbf{E}_t''^* = -e^{-j\Theta} \tan\psi \, \overline{\overline{\mathsf{J}}}_t \cdot (\mathbf{E}^i - \mathbf{E}')_t, \tag{6.26}$$

from which the reflected fields can be solved as

$$\mathbf{E}_t' = -\cos 2\psi \, \mathbf{E}_t^i, \tag{6.27}$$

$$\mathbf{E}_t'' = -e^{j\Theta} \sin 2\psi \, (\overline{\overline{\mathsf{J}}}_t \cdot \mathbf{E}_t^i)^* \tag{6.28}$$

$$= e^{j\Theta} \sin 2\psi \, \eta_o \mathbf{H}_t^{i*}. \tag{6.29}$$

The corresponding magnetic field components are obtained as

$$\eta_o \mathbf{H}_t' = \overline{\overline{\mathsf{J}}}_t \cdot \mathbf{E}_t' = -\cos 2\psi \, \overline{\overline{\mathsf{J}}}_t \cdot \mathbf{E}_t^i$$

$$= \cos 2\psi \, \eta_o \mathbf{H}_t^i \tag{6.30}$$

$$\eta_o \mathbf{H}_t'' = \overline{\overline{\mathsf{J}}}_t \cdot \mathbf{E}_t'' = e^{j\Theta} \sin 2\psi \, \mathbf{E}_t^{i*}$$

$$= e^{j\Theta} \sin 2\psi \, \overline{\overline{\mathsf{J}}}_t \cdot \eta_o \mathbf{H}_t^{i*}. \tag{6.31}$$

It is left as an exercise to verify that the total plane-wave fields $\mathbf{E} = \mathbf{E}^i + \mathbf{E}' + \mathbf{E}''$ and $\mathbf{H} = \mathbf{H}^i + \mathbf{H}' + \mathbf{H}''$ satisfy the condition (6.14) of the SQL boundary.

The total reflected fields can be constructed as

$$\begin{aligned} \mathbf{E}' &= (\bar{\bar{\mathsf{I}}}_t - \frac{\mathbf{n}}{k_n}\mathbf{k}_t)\cdot\mathbf{E}'_t \\ &= -\cos 2\psi\, \bar{\bar{\mathsf{C}}}\cdot\mathbf{E}^i, \quad (6.32)\\ \mathbf{H}' &= \cos 2\psi\, \bar{\bar{\mathsf{C}}}\cdot\mathbf{H}^i, \quad (6.33)\\ \mathbf{E}'' &= (\bar{\bar{\mathsf{I}}}_t + \frac{\mathbf{n}}{k_n}\mathbf{k}_t)^*\cdot\mathbf{E}''_t \\ &= e^{j\Theta}\sin 2\psi\, \eta_o\mathbf{H}^{i*}, \quad (6.34)\\ \eta_o\mathbf{H}'' &= e^{j\Theta}\sin 2\psi\, \mathbf{E}^{i*}. \quad (6.35) \end{aligned}$$

Here, $\bar{\bar{\mathsf{C}}}$ is the mirror reflection dyadic

$$\bar{\bar{\mathsf{C}}} = \bar{\bar{\mathsf{I}}}_t - \mathbf{nn} = \bar{\bar{\mathsf{I}}} - 2\mathbf{nn}, \quad (6.36)$$

satisfying

$$\bar{\bar{\mathsf{C}}}^T = \bar{\bar{\mathsf{C}}}, \quad \bar{\bar{\mathsf{C}}}^2 = \bar{\bar{\mathsf{I}}}, \quad \bar{\bar{\mathsf{C}}}^{(2)} = -\bar{\bar{\mathsf{C}}}, \quad (6.37)$$

$$(\bar{\bar{\mathsf{C}}}\cdot\mathbf{a})\times(\bar{\bar{\mathsf{C}}}\cdot\mathbf{b}) = \bar{\bar{\mathsf{C}}}^{(2)}\cdot(\mathbf{a}\times\mathbf{b}) = -\bar{\bar{\mathsf{C}}}\cdot(\mathbf{a}\times\mathbf{b}). \quad (6.38)$$

Let us consider some special cases.

- There is no retroreflected component $\mathbf{E}''_t$ when $\sin 2\psi = 0$. This is satisfied for $\psi = \pm n\pi/2$, i.e., when either $\tan\psi = 0$ or $\cot\psi = 0$. In these two cases the SQL boundary equals the respective PEC and PMC boundaries.

- There is only the retroreflected component when $\mathbf{E}' = 0$. This requires $\cos 2\psi = 0$ (or $\tan\psi = \pm 1$, or $\psi = \pm\pi/4$), which corresponds to the SQL-boundary condition of the form

$$\mathbf{E}_t = e^{j\Theta}\eta_o\mathbf{H}_t^*. \quad (6.39)$$

  The double sign is absorbed in Θ.

- For normal incidence, $\mathbf{k}_t = 0$, $k_n = k_o$, both components are retroreflected because of $\mathbf{k}' = \mathbf{k}'' = -\mathbf{k}^i = \mathbf{n}k_o$. In this case, (6.27) is unchanged but (6.28) is reduced to

$$\mathbf{E}''_t = -e^{j\Theta}\sin 2\psi\, \mathbf{n}\times\mathbf{E}_t^{i*}. \quad (6.40)$$

- There does not exist any matched waves for the SQL boundary. In fact, the primed and double-primed reflected waves cannot vanish simultaneously for any $\mathbf{k}^i$, as can be seen from (6.32) – (6.35).

Polarization Properties

To study the polarization of plane-wave fields, let us apply the polarization vector of a complex vector $\mathbf{a}$ defined by (see Appendix B)

$$\mathbf{p}(\mathbf{a}) = \frac{\mathbf{a} \times \mathbf{a}^*}{j\mathbf{a} \cdot \mathbf{a}^*}. \tag{6.41}$$

Applied to the reflected field $\mathbf{E}'$, after some algebra we obtain

$$\mathbf{p}(\mathbf{E}') = -\overline{\overline{\mathsf{C}}} \cdot \mathbf{p}(\mathbf{E}^i). \tag{6.42}$$

From this it follows that $|\mathbf{p}(\mathbf{E}')| = |\mathbf{p}(\mathbf{E}^i)|$, whence the ellipticity of $\mathbf{E}'$ equals that of $\mathbf{E}^i$. Similarly, we have

$$\mathbf{p}(\mathbf{E}'') = \mathbf{p}(\mathbf{H}^{i*}) = -\mathbf{p}(\mathbf{H}^i), \tag{6.43}$$

whence $|\mathbf{p}(\mathbf{E}'')| = |\mathbf{p}(\mathbf{H}^i)| = |\mathbf{p}(\mathbf{E}^i)|$ and the ellipticity of the retroreflected field also equals that of the incident field.

Assuming real $\mathbf{k}^i$, from $\mathbf{k}^i \times (\mathbf{E}^i \times \mathbf{E}^{i*}) = 0$ it follows that $\mathbf{p}(\mathbf{E}^i)$ is parallel to $\mathbf{k}^i$. Similarly, $\mathbf{k}' = \overline{\overline{\mathsf{C}}} \cdot \mathbf{k}^i$ is parallel to $\mathbf{p}(\mathbf{E}')$ and $\mathbf{k}'' = -\mathbf{k}^i$ is parallel to $\mathbf{p}(\mathbf{E}'')$. From $\mathbf{k}' = \overline{\overline{\mathsf{C}}} \cdot \mathbf{k}^i$ we have

$$\mathbf{k}' \cdot \mathbf{p}(\mathbf{E}') = -\mathbf{k}^i \cdot \overline{\overline{\mathsf{C}}}^2 \cdot \mathbf{p}(\mathbf{E}^i) = -\mathbf{k}^i \cdot \mathbf{p}(\mathbf{E}^i), \tag{6.44}$$

whence the handedness of the reflected field $\mathbf{E}'$ is opposite to that of the incident field $\mathbf{E}^i$ when looking in the direction of propagation.

Applying the polarization vector to the retroreflected field, for real $\mathbf{k}^i$, we obtain

$$\mathbf{p}(\mathbf{E}'') = -\mathbf{p}(\mathbf{H}^i) = -\mathbf{p}(\mathbf{k}^i \times \mathbf{E}^i) \tag{6.45}$$

$$= -\frac{\mathbf{k}^i\mathbf{k}^i \cdot (\mathbf{E}^i \times \mathbf{E}^{i*})}{jk_o^2|\mathbf{E}^i|^2} = -\frac{\mathbf{k}^i\mathbf{k}^i}{k_o^2} \cdot \mathbf{p}(\mathbf{E}^i). \tag{6.46}$$

From this it follows that

$$\mathbf{k}'' \cdot \mathbf{p}(\mathbf{E}'') = \mathbf{k}^i \cdot \mathbf{p}(\mathbf{E}^i), \tag{6.47}$$

whence handedness is preserved in retroreflection from such an isotropic SQL boundary.

In conclusion, if $\mathbf{E}^i$ is elliptically polarized, the reflected wave component $\mathbf{E}'$ has similar polarization but opposite handedness. The major and minor axes of the $\mathbf{E}'_t$ component are parallel to those of the $\mathbf{E}^i_t$ component. In contrast, while the major and minor axes of the component $\mathbf{E}''_t$ are parallel to those of the $\mathbf{H}^i_t$ component, they appear rotated by $\pi/2$ from those of the $\mathbf{E}^i_t$ component. However, the handedness equals that of the incident wave.

A planar SQL boundary defined by (6.39), which acts as a retroreflector, can be compared with the phase-conjugate mirror [6] or a plane filled with corner reflectors, both of which are famous for their retroreflecting property. It is known that, in reflection from the PEC surface, handedness of the polarization is changed while it remains the same for the SH surface. Thus, realizing a retroreflecting SQL boundary in terms of corner reflectors, it must consist of SH surface (e.g., tuned corrugated surface) instead of solid metal [38].

6.3 Eigenfields

The eigenvectors of the dyadic $\overline{\overline{\mathsf{J}}}_t$ from (2.12), (2.13), satisfy

$$\overline{\overline{\mathsf{J}}}_t \cdot \mathbf{x}_{t\pm} = \pm j\mathbf{x}_{t\pm}, \quad \mathbf{x}_{t\pm} = k_n\mathbf{k}_t \mp jk_o\mathbf{n} \times \mathbf{k}_t. \tag{6.48}$$

For real $\mathbf{k}$, we have

$$\mathbf{x}_{t\pm}^* = \mathbf{x}_{t\mp}. \tag{6.49}$$

Assuming

$$X = \mathbf{n} \cdot \mathbf{x}_{t+} \times \mathbf{x}_{t-} = 2jk_ok_nk_t^2 \neq 0, \tag{6.50}$$

and applying (2.16), we can expand the incident field in terms of the eigenvectors as

$$\begin{aligned} \mathbf{E}_t^i &= E_+^i\mathbf{x}_{t+} + E_-^i\mathbf{x}_{t-}, & (6.51) \\ \eta_o\mathbf{H}_t^i &= -j(E_+^i\mathbf{x}_{t+} - E_-^i\mathbf{x}_{t-}), & (6.52) \end{aligned}$$

with

$$E_\pm = \pm\frac{1}{X}(\mathbf{x}_{t\mp} \times \mathbf{n}) \cdot \mathbf{E}_t^i. \tag{6.53}$$

Expansions of the reflected-field components

$$\begin{aligned} \mathbf{E}_t' &= E_+'\mathbf{x}_{t+} + E_-'\mathbf{x}_{t-}, & (6.54) \\ \mathbf{E}_t'' &= E_+''\mathbf{x}_{t+} + E_-''\mathbf{x}_{t-}, & (6.55) \end{aligned}$$

are related to the incident field by

$$\begin{aligned} \begin{pmatrix} E_+' \\ E_-' \end{pmatrix} &= -\cos 2\psi \begin{pmatrix} E_+^i \\ E_-^i \end{pmatrix} & (6.56) \\ \begin{pmatrix} E_+'' \\ E_-'' \end{pmatrix} &= je^{j\Theta}\sin 2\psi \begin{pmatrix} 0 & -1 \\ 1 & 0 \end{pmatrix} \begin{pmatrix} E_+^i \\ E_-^i \end{pmatrix}, & (6.57) \end{aligned}$$

as can be seen from (6.27) - (6.31). The expressions for the transverse fields are, thus,

$$\begin{aligned} \mathbf{E}_t' &= -\cos 2\psi(E_+^i\mathbf{x}_{t+} + E_-^i\mathbf{x}_{t-}), & (6.58) \\ \eta_o\mathbf{H}_t' &= -j\cos 2\psi(E_+^i\mathbf{x}_{t+} - E_-^i\mathbf{x}_{t-}), & (6.59) \\ \mathbf{E}_t'' &= je^{j\Theta}\sin 2\psi(E_+^i\mathbf{x}_{t-} - E_-^i\mathbf{x}_{t+}), & (6.60) \\ \eta_o\mathbf{H}_{t\pm}'' &= e^{j\Theta}\sin 2\psi(E_+^i\mathbf{x}_{t-} + E_-^i\mathbf{x}_{t+}). & (6.61) \end{aligned}$$

Normal Incidence

In the special case of normal incidence, $\mathbf{k}^i = -\mathbf{n}k_o$, both reflected components appear retroreflected, $\mathbf{k}' = \mathbf{k}'' = -\mathbf{k}^i$. Because we now have $\overline{\overline{\mathsf{J}}}_t = \mathbf{n}\times\overline{\overline{\mathsf{I}}}$, the eigenvectors $\mathbf{x}_{t\pm}$ can be replaced by

$$\mathbf{e}_\pm = \mathbf{e}_1 \mp j\mathbf{e}_2, \qquad \mathbf{e}_\pm^* = \mathbf{e}_\mp, \tag{6.62}$$

where $\mathbf{e}_1, \mathbf{e}_2, \mathbf{e}_3 = \mathbf{n}$ make a real orthonormal vector basis and $\mathbf{e}_\pm$ are two circularly polarized vectors satisfying

$$\mathbf{n}\times\mathbf{e}_\pm = \pm j\mathbf{e}_\pm, \tag{6.63}$$

$$\mathbf{e}_+\times\mathbf{e}_- = -2j\mathbf{n}, \tag{6.64}$$

$$\mathbf{e}_\pm\cdot\mathbf{e}_\pm = 0, \tag{6.65}$$

$$\mathbf{e}_+\cdot\mathbf{e}_- = 2. \tag{6.66}$$

The total tangential fields at the boundary are

$$\begin{aligned} \mathbf{E}_t &= \mathbf{E}_t^i + \mathbf{E}_t' + \mathbf{E}_t'' && (6.67)\\ &= (1-\cos 2\psi)\mathbf{E}_t^i - e^{j\Theta}\sin 2\psi\mathbf{n}\times\mathbf{E}_t^{i*}, && (6.68)\\ \eta_o\mathbf{H}_t &= \mathbf{n}\times(-\mathbf{E}_t^i + \mathbf{E}_t' + \mathbf{E}_t''), && (6.69)\\ &= -(1+\cos 2\psi)\mathbf{n}\times\mathbf{E}_t^i + e^{j\Theta}\sin 2\psi\mathbf{E}_t^{i*}. && (6.70) \end{aligned}$$

For $\psi = 0$ we obtain

$$\mathbf{E}_t = 0, \quad \eta_o\mathbf{H}_t = -2\mathbf{n}\times\mathbf{E}_t^i = 2\eta_o\mathbf{H}_t^i, \tag{6.71}$$

which equals the PEC condition. For $\psi = \pi/2$ we have

$$\mathbf{E}_t = 2\mathbf{E}_t^i, \quad \eta_o\mathbf{H}_t = 0, \tag{6.72}$$

which equals the PMC condition.

Assuming the most general elliptic polarization of the incident field in the form

$$\mathbf{E}^i = \mathbf{E}_t^i = \mathbf{e}_+E_+ + \mathbf{e}_-E_-, \tag{6.73}$$

where E_+ and E_- are two real quantities, the reflected components become

$$\begin{aligned} \mathbf{E}_t' &= -\cos 2\psi(\mathbf{e}_+E_+ + \mathbf{e}_-E_-), && (6.74)\\ \eta_o\mathbf{H}_t' &= -j\cos 2\psi(\mathbf{e}_+E_+ - \mathbf{e}_-E_-), && (6.75)\\ \mathbf{E}_t'' &= je^{j\Theta}\sin 2\psi(\mathbf{e}_-E_+ - \mathbf{e}_+E_-), && (6.76)\\ \eta_o\mathbf{H}_{t\pm}'' &= e^{j\Theta}\sin 2\psi(\mathbf{e}_-E_+ + \mathbf{e}_+E_-). && (6.77) \end{aligned}$$

We can construct the polarization vectors as

$$\begin{aligned} \mathbf{p}(\mathbf{E}_t^i) &= -\mathbf{n}\frac{E_+^2 - E_-^2}{E_+^2 + E_-^2}, && (6.78)\\ \mathbf{p}(\mathbf{E}_t') &= \mathbf{p}(\mathbf{E}^i), && (6.79)\\ \mathbf{p}(\mathbf{E}_t'') &= -\mathbf{p}(\mathbf{E}^i), && (6.80) \end{aligned}$$

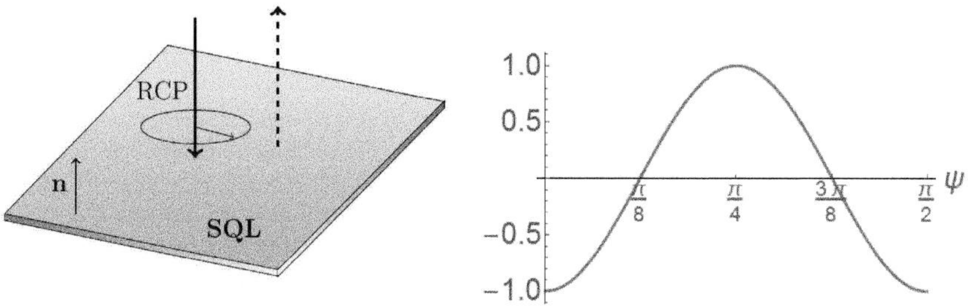

Figure 6.2: A normally incident right-hand circularly-polarized wave is reflected from an isotropic SQL plane. Dependence of the polarization vector $\mathbf{n}\cdot\mathbf{p}(\mathbf{E}'+\mathbf{E}'')$ on the parameter ψ is shown in the figure. For $\psi = \pi/8$ and $\psi = 3\pi/8$ the polarization of the reflected wave is linear and right handed for $\mathbf{n}\cdot\mathbf{p} > 0$.

which shows us that the ellipticity of the three fields is the same. The handedness of $\mathbf{E}''$ equals that of $\mathbf{E}^i$, while for $\mathbf{E}'$ it is the opposite.

As an example, let us consider right-hand circularly-polarized wave incident to the isotropic SQL boundary. It is left as an exercise to show that the polarization vector of the total reflected field can be found to equal

$$\mathbf{p}(\mathbf{E}' + \mathbf{E}'') = -\mathbf{n}\cos 4\psi. \tag{6.81}$$

The reflected total field is elliptically polarized, in general. For $\psi = \pi/8$ and $\psi = 3\pi/8$ it is linearly polarized, for $\psi = \pi/4$ it is right-hand circularly polarized, and for $\psi = 0$ and $\pi/2$ it is left-hand circularly polarized, see Figure 6.2.

6.4 Power Balance

From (6.27) - (6.31) we obtain the orthogonality conditions for fields associated to the general plane-wave reflection from an isotropic SQL boundary,

$$\mathbf{E}_t^i \times \eta_o\mathbf{H}_t^{\prime\prime *} = 0, \tag{6.82}$$
$$\mathbf{E}_t^{\prime\prime} \times \eta_o\mathbf{H}_t^{i*} = 0, \tag{6.83}$$
$$\mathbf{E}_t^{\prime\prime} \times \eta_o\mathbf{H}_t^{\prime *} = 0, \tag{6.84}$$
$$\mathbf{E}_t^{\prime} \times \eta_o\mathbf{H}_t^{\prime\prime *} = 0. \tag{6.85}$$

Applying these properties, the boundary condition (6.1) can be expressed for the plane-wave fields in the form

$$\mathbf{n}\cdot(\mathbf{E}^i \times \mathbf{H}^{i*} + \mathbf{E}' \times \mathbf{H}'^* + \mathbf{E}'' \times \mathbf{H}''^*) = 0. \tag{6.86}$$

The real part of this equation can be interpreted so that the power incident to the boundary equals the sum of powers of the two reflected waves. This verifies the fact that the SQL boundary is lossless.

Applying the eigenexpansions (6.51) - (6.61), we can write

$$\mathbf{n}\cdot\mathbf{E}^i\times\eta_o\mathbf{H}^{i*} = -P^i, \tag{6.87}$$
$$\mathbf{n}\cdot\mathbf{E}'\times\eta_o\mathbf{H}'^* = \cos^2 2\psi\ P^i = P', \tag{6.88}$$
$$\mathbf{n}\cdot\mathbf{E}''\times\eta_o\mathbf{H}''^* = \sin^2 2\psi\ P^i = P'', \tag{6.89}$$

with

$$P^i = 2k_ok_nk_t^2(|E_+^i|^2+|E_-^i|^2) = P_+^i + P_-^i. \tag{6.90}$$

The power of the eigencomponents is reflected as

$$P_+' = \cos^2 2\psi P_+^i, \qquad P_-' = \cos^2 2\psi P_-^i, \tag{6.91}$$
$$P_+'' = \sin^2 2\psi P_-^i, \qquad P_-'' = \sin^2 2\psi P_+^i. \tag{6.92}$$

In conclusion, the power of the eigenfields E_+ and E_- do not couple in mirror reflection from an isotropic SQL boundary, while in retroreflection they become cross coupled.

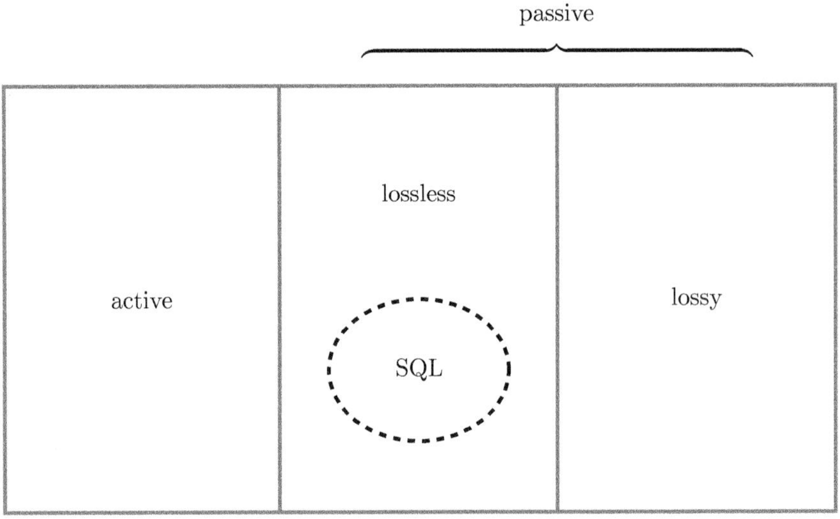

Figure 6.3: SQL boundary appears as a subset of boundaries satisfying lossless conditions.

6.5 Image theory

The expressions for the reflected field vectors (6.32) - (6.35) can be written for the plane-wave field functions as

$$\mathbf{E}^r(\mathbf{r}) = -\cos 2\psi\ \overline{\overline{\mathsf{C}}}\cdot\mathbf{E}^i(\overline{\overline{\mathsf{C}}}\cdot\mathbf{r}), \tag{6.93}$$

$$\mathbf{H}^r(\mathbf{r}) = \cos 2\psi\ \overline{\overline{\mathsf{C}}}\cdot\mathbf{H}^i(\overline{\overline{\mathsf{C}}}\cdot\mathbf{r}), \tag{6.94}$$

$$\mathbf{E}''(\mathbf{r}) = e^{j\Theta}\sin 2\psi\ \eta_o\mathbf{H}^{i*}(\mathbf{r}), \tag{6.95}$$

$$\mathbf{H}''(\mathbf{r}) = (1/\eta_o)e^{j\Theta}\sin 2\psi\ \mathbf{E}^{i*}(\mathbf{r})). \tag{6.96}$$

Because the expressions on the right are independent of the wave vector $\mathbf{k}^i$ and represent linear relations between the reflected and incident fields, they are equally valid for an incident field consisting of any sum or an integral of plane waves.

Let us consider a combination of electric and magnetic current sources $\mathbf{J}_e(\mathbf{r})$ and $\mathbf{J}_m(\mathbf{r})$ giving rise to the incident field $\mathbf{E}^i, \mathbf{H}^i$. The relation between the fields and the sources is governed by the Maxwell equations which can be written in the form

$$\begin{pmatrix} \mathbf{J}_e(\mathbf{r}) \\ \mathbf{J}_m(\mathbf{r}) \end{pmatrix} = \mathcal{M}(\nabla)\cdot\begin{pmatrix} \mathbf{E}^i(\mathbf{r}) \\ \mathbf{H}^i(\mathbf{r}) \end{pmatrix}, \tag{6.97}$$

where the dyadic-matrix operator $\mathcal{M}(\nabla)$ is defined by

$$\mathcal{M}(\nabla) = \begin{pmatrix} -j(k_o/\eta_o)\overline{\overline{\mathsf{I}}} & \nabla\times\overline{\overline{\mathsf{I}}} \\ -\nabla\times\overline{\overline{\mathsf{I}}} & -jk_o\eta_o\overline{\overline{\mathsf{I}}} \end{pmatrix}. \tag{6.98}$$

If the incident field is originally given, its source in the homogeneous space can be recovered from (6.97). Similarly, if the reflected fields are known, their apparent source in the homogeneous space can be found by applying the operator $\mathcal{M}(\nabla)$ to the reflected field vectors. The apparent sources of the reflected fields are called reflection images of the original sources.

Applying the expressions (6.93) – (6.96), the reflection-image sources can be recovered from

$$\begin{aligned}
\begin{pmatrix} \mathbf{J}'_e(\mathbf{r}) \\ \mathbf{J}'_m(\mathbf{r}) \end{pmatrix} &= \mathcal{M}(\nabla)\cdot\begin{pmatrix} \mathbf{E}'(\mathbf{r}) \\ \mathbf{H}'(\mathbf{r}) \end{pmatrix} \\
&= \cos 2\psi\mathcal{M}(\nabla)\cdot\begin{pmatrix} -\overline{\overline{\mathsf{C}}}\cdot\mathbf{E}^i(\overline{\overline{\mathsf{C}}}\cdot\mathbf{r}) \\ \overline{\overline{\mathsf{C}}}\cdot\mathbf{H}^i(\overline{\overline{\mathsf{C}}}\cdot\mathbf{r}) \end{pmatrix}, \\
\begin{pmatrix} \mathbf{J}''_e(\mathbf{r}) \\ \mathbf{J}''_m(\mathbf{r}) \end{pmatrix} &= \mathcal{M}(\nabla)\cdot\begin{pmatrix} \mathbf{E}''(\mathbf{r}) \\ \mathbf{H}''(\mathbf{r}) \end{pmatrix} \\
&= e^{j\Theta}\sin 2\psi\mathcal{M}(\nabla)\cdot\begin{pmatrix} \eta_o\mathbf{H}^i(\mathbf{r}) \\ \mathbf{E}^i(\mathbf{r})/\eta_o \end{pmatrix}^*.
\end{aligned} \tag{6.99}$$

Inserting the expressions (6.32) - (6.35) and applying the rule (6.38), we obtain

$$\mathbf{J}'_e(\mathbf{r}) = \cos 2\psi(j(k_o/\eta_o)\overline{\overline{\mathsf{C}}}\cdot\mathbf{E}^i(\overline{\overline{\mathsf{C}}}\cdot\mathbf{r}) + \nabla\times(\overline{\overline{\mathsf{C}}}\cdot\mathbf{H}^i(\overline{\overline{\mathsf{C}}}\cdot\mathbf{r}))), \quad (6.100)$$

$$\mathbf{J}'_m(\mathbf{r}) = \cos 2\psi(\nabla\times(\overline{\overline{\mathsf{C}}}\cdot\mathbf{E}^i(\overline{\overline{\mathsf{C}}}\cdot\mathbf{r})) - jk_o\eta_o\mathbf{H}^i(\overline{\overline{\mathsf{C}}}\cdot\mathbf{r})), \quad (6.101)$$

$$\mathbf{J}''_e(\mathbf{r}) = e^{j\Theta}\sin 2\psi(-jk_o\mathbf{H}^i(\mathbf{r}) + \nabla\times\mathbf{E}^i(\mathbf{r})/\eta_o)^*, \quad (6.102)$$

$$\mathbf{J}''_m(\mathbf{r}) = e^{j\Theta}\sin 2\psi(-\eta_o\nabla\times\mathbf{H}^i(\mathbf{r}) - jk_o\mathbf{E}^i(\mathbf{r}))^*. \quad (6.103)$$

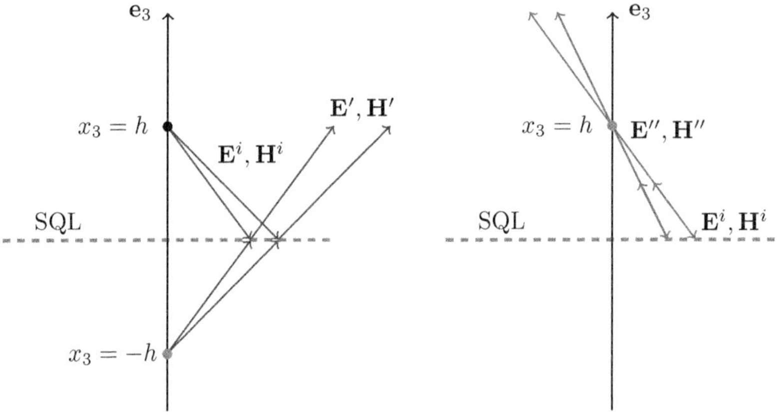

Figure 6.4: Ray diagrams corresponding to the mirror-reflected field component $\mathbf{E}', \mathbf{H}'$, and the retroreflected field component $\mathbf{E}'', \mathbf{H}''$, due to an incident field starting from a point source at $x_3 = h$. The mirror-reflected rays appear to arise from a point source at the mirror image point of the original source. The retroreflected rays appear to start from a point source $z = h$ for the half space $x_3 > h$, while for the region below $z = h$ they appear to be created by a sink at $x_3 = h$.

Finally, applying the Maxwell operator (6.97), the reflection image sources take the form

$$\mathbf{J}'_e(\mathbf{r}) = -\cos 2\psi\ \overline{\overline{\mathsf{C}}}\cdot\mathbf{J}_e(\overline{\overline{\mathsf{C}}}\cdot\mathbf{r}), \quad (6.104)$$

$$\mathbf{J}'_m(\mathbf{r}) = \cos 2\psi\ \overline{\overline{\mathsf{C}}}\cdot\mathbf{J}_m(\overline{\overline{\mathsf{C}}}\cdot\mathbf{r}), \quad (6.105)$$

$$\mathbf{J}''_e(\mathbf{r}) = e^{j\Theta}\sin 2\psi\ \eta_o\mathbf{J}^*_m(\mathbf{r}) \quad (6.106)$$

$$\mathbf{J}''_m(\mathbf{r}) = e^{j\Theta}\sin 2\psi\ \mathbf{J}^*_e(\mathbf{r})/\eta_o. \quad (6.107)$$

As an example, let us consider a vertical electric-current dipole at the distance h above the isotropic SQL boundary,

$$\mathbf{J}_e(\mathbf{r}) = \mathbf{e}_3 IL\delta(\mathbf{r} - \mathbf{e}_3 h). \quad (6.108)$$

The image source consists of the electric and magnetic dipoles

$$\mathbf{J}'_e(\mathbf{r}) = \mathbf{e}_3 IL\cos 2\psi\; \delta(\mathbf{r}+\mathbf{e}_3 h), \tag{6.109}$$

$$\mathbf{J}''_m(\mathbf{r}) = \mathbf{e}_3 \frac{IL}{\eta_o} e^{j\Theta} \sin 2\psi\; \delta(\mathbf{r}-\mathbf{e}_3 h). \tag{6.110}$$

The image component $\mathbf{J}'_e(\mathbf{r})$ responsible for the reflected field $\mathbf{E}', \mathbf{H}'$ appears to be in the mirror image point $x_3 = -h$ and it is similar to that of the PEC boundary, except for the coefficient $\cos 2\psi$. The image source responsible for the retroreflected field $\mathbf{E}'', \mathbf{H}''$ appears to be at the point of the original source. This means that the retroreflected field has a singularity at the point of the original source. In the region $\mathbf{e}_3 \cdot \mathbf{r} = x_3 > h$ the reflected field can be computed in normal manner. However, for the retroreflected fields in the region $h > x_3 > 0$ the image source must be considered as a sink, with waves propagating towards the image source. Thus, the Green function must involve the incoming-wave representation. This can be understood from the ray diagram 6.4, which depicts the wave vector $\mathbf{k}'' = -\mathbf{k}^i$ for real $\mathbf{k}'$. The analysis can be extended by requiring that the location of the original point source is in complex space [39], whence the rays become Gaussian beams [9].

6.6 Problems

6.1 Show that, for the isotropic SQL boundary conditions to be valid for the sum of incident and reflected waves, the wave vectors must be defined by (6.20), whence the boundary condition (6.14) is split in two as (6.21), (6.22).

6.2 Derive the relation (6.42) between the polarization vectors of the mirror reflected wave and the incident wave.

6.3 Verify that the reflected fields (6.27) -(6.31) together with the incident field satisfy the condition (6.14) of the SQL boundary.

6.4 Show that a simple-isotropic impedance boundary satisfying $\mathbf{E}_t = Z_s \mathbf{n}\times\mathbf{H}_t$ is an SQL boundary only when it is either the PEC or the PMC boundary.

6.5 Find the fields reflected from the isotropic SQL boundary defined by the condition (6.15). Check the special cases of PEC and PMC boundaries.

6.6 Find the polarization vectors for the fields $\mathbf{E}'$, $\mathbf{E}''$ and $\mathbf{E}' + \mathbf{E}''$ reflected from the isotropic SQL boundary defined by (6.14) when the field $\mathbf{E}^i$ is normally incident and has right-hand circular polarization $\mathbf{E}^i = \mathbf{e}_+ E^i$.

Chapter 7

Scattering by Objects Defined by Boundary Conditions

Objects with a surface described by a given boundary condition interact with incident electromagnetic radiation in a manner that is determined by the condition. This chapter shows examples how the boundary condition affects the scattering properties of a particle. Most of the examples are computed for a spherical scattering object. Even though the spherical geometry is simple, it is a natural and accurate model for particles appearing in several domains in nature and engineering applications. As an example of area in which knowledge of the scattering properties of particles is essential is electromagnetic modeling of materials and the computation of their effective properties. The microstructure of a medium (being often a mixture of inclusions in homogeneous background matrix) and the scattering properties of the constituents determine the macroscopic constitutive parameters [82, 93].

7.1 Cross Sections and Efficiencies

The scattering constellation is shown in Figure 7.1. A plane wave is incident from the left, the scatterer is a sphere with radius a, and a given homogeneous boundary condition is assumed to hold on its surface.

When a propagating plane wave encounters and hits an object, the fields become scattered, absorbed, and diffracted. For a sphere, the scattering perturbation can be calculated analytically using the Lorenz–Mie analysis [77, 81]. In addition to scatter radiation, a dissipative object also absorbs energy. The extinction due to the scatterer is the sum of these two effects. The capacity of the sphere to distort the incoming radiation is measured by its scattering, absorption, and extinction cross sections and efficiencies .

The efficiencies depend on the type of the boundary condition and the size

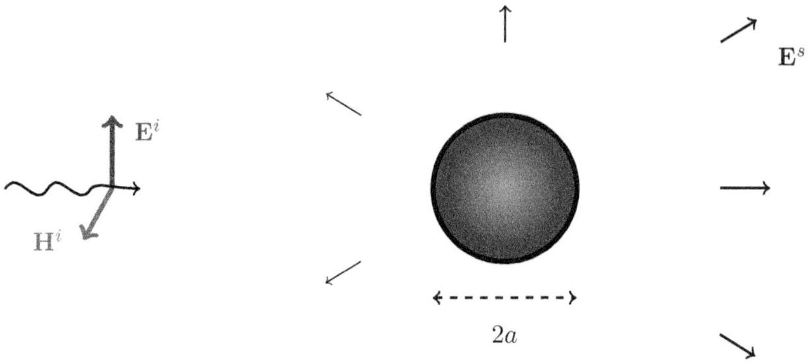

Figure 7.1: Plane-wave scattering by a BC sphere.

of the scatterer. The size parameter is defined by

$$x = ka, \tag{7.1}$$

with k being the free-space wave number and a the radius of the sphere.

The (dimensionless) scattering efficiency is the total scattering cross section normalized by the geometrical cross section of the sphere, and the three efficiencies can be calculated from the series [3]

$$Q_{\rm sca} = \frac{2}{x^2}\sum_{n=1}^{\infty}(2n+1)\left(|a_n|^2+|b_n|^2\right), \tag{7.2}$$

$$Q_{\rm ext} = \frac{2}{x^2}\sum_{n=1}^{\infty}(2n+1){\rm Re}\,\{a_n+b_n\}, \tag{7.3}$$

$$Q_{\rm abs} = Q_{\rm ext} - Q_{\rm sca}, \tag{7.4}$$

and here the Mie coefficients a_n (electric multipoles) and b_n (magnetic multipoles) depend on the character of the boundary condition on the sphere surface.[1] The expansions (7.2)–(7.4) are infinite series. For small spheres, they converge rapidly but as the electrical size increases, more and more terms are needed to maintain accuracy. The Wiscombe criterion [115], is commonly used for the necessary number of terms.

In the following sections, the scattering and absorption characteristics of scatterers with certain boundary conditions are analyzed.

[1] In electromagnetic scattering literature, these two series of coefficients are also refered to as TM_r modes (a_n) and TE_r modes (b_n), where the transversality of the fields is with respect to the radial coordinate r. For example, an electric dipole radiates a magnetic field that has no radial component whereas the radiated electric field has both the tangential and radial parts.

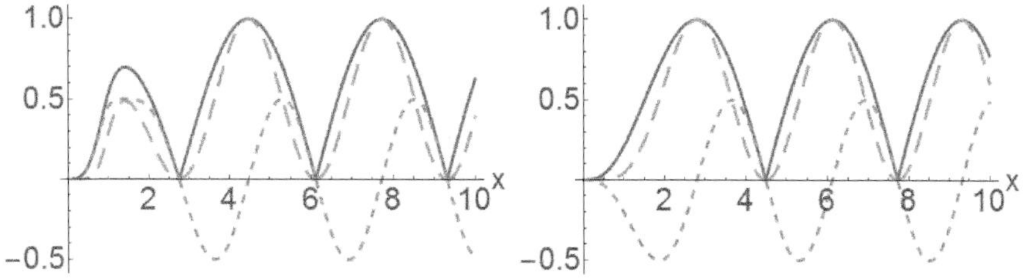

Figure 7.2: The electric (left) and magnetic (right) dipole Mie-coefficients a_1 and b_1 for the PEC sphere as function of the size parameter x. Solid blue: absolute value, dashed orange: real part, dotted green: imaginary part.

7.2 PEC, PMC, and PEMC Objects

For a perfect electric conductor sphere, one can show that the Mie coefficients depend on the size parameter x according to

$$a_{n,\mathrm{PEC}} = \frac{(xj_n(x))'}{\left(xh_n^{(2)}(x)\right)'} \quad \text{and} \quad b_{n,\mathrm{PEC}} = \frac{j_n(x)}{h_n^{(2)}(x)} \tag{7.5}$$

where j_n denotes the spherical Bessel function and $h_n^{(2)}$ the spherical Hankel function of the second kind.[2] Due to the lossless character of the PEC boundary condition, the absorption cross section is zero, and consequently the scattering and extinction efficiencies are the same.

The behavior of the fundamental Mie coefficients a_1 and b_1 is visualized in Figure 7.2. These complex-valued functions of x display interesting properties. For example, note the connections $\mathrm{Re}\{a_1\} = |a_1|^2$ and $\mathrm{Re}\{b_1\} = |b_1|^2$. The imaginary part fluctuates between $-1/2$ and $+1/2$.

In the small-sphere limit, the dipoles dominate. The effect of higher multipole orders takes effect once x increases, as is seen in Figure 7.3 which shows the amplitude of the first three electric and magnetic coefficients ($n = 1, 2, 3$).

Figure 7.4 displays the scattering efficiency of a PEC sphere as function of its size. As can be seen, a sphere small compared to the wavelength is a very weak scatterer. At low frequencies, the scattered power is proportional to the fourth power of x. This low-frequency (or small-scatterer) domain is called the Rayleigh scattering regime. When x increases, there are a couple of weak oscillations in the scattering curve, and finally, for high frequencies the efficiency approaches the value 2 in this case of lossless scatterer. The scattering cross section being twice the geometrical cross section of the scatterer is in agreement with the so-called extinction paradox [3, 4, 30].

[2] The choice of the Hankel function depends on the time-harmonic notation; with $\exp(j\omega t)$, the second kind is needed.

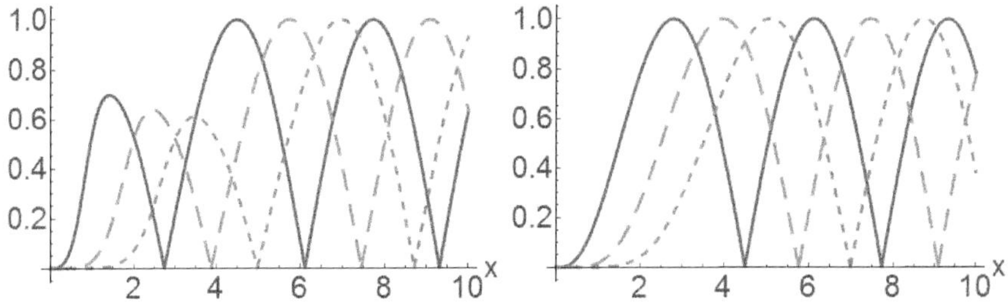

Figure 7.3: The absolute value of the electric a_n (left) and magnetic b_n (right) dipole Mie-coefficients for the PEC sphere as function of the size parameter x. Solid blue: $n = 1$, dashed orange: $n = 2$, dotted green: $n = 3$.

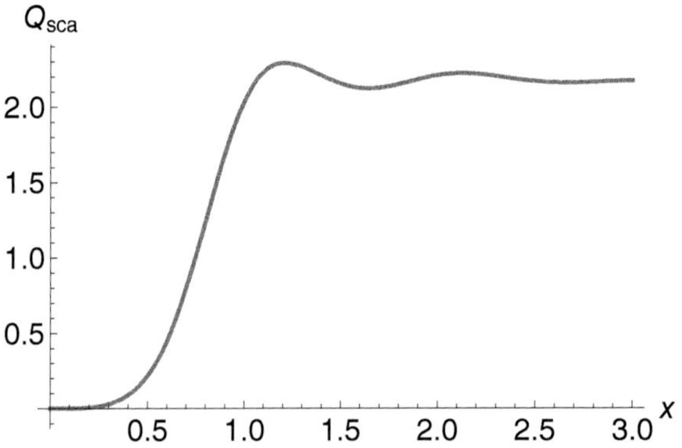

Figure 7.4: The scattering efficiency of a sphere with PEC boundary condition as function of the size parameter x.

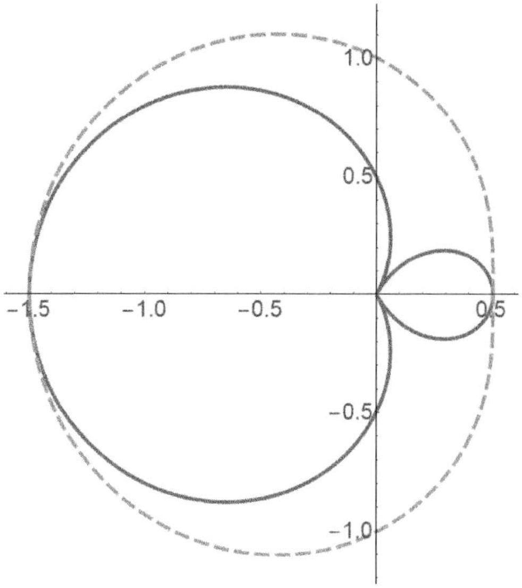

Figure 7.5: The scattering pattern of a small PEC sphere with size parameter $x = 0.1$. The amplitude of the scattered electric field is shown in different directions. The plane wave is incident from the left (note the strong backscattering). Blue solid: E-plane, orange dashed: H-plane. (The scale is linear.)

Scattering patterns

The bistatic radar cross sections of a PEC sphere for an electrically small and an electrically large sphere are shown in Figures 7.5 and 7.6. The scattering is not rotationally symmetric: in the E-plane (the plane of the incident wave vector and the incident electric field) it has more lobes than in the H plane. Furthermore it can be seen that a small PEC sphere is a strong backscatterer: the backscattered field amplitude is three times that of the forward scattered field (9 dB difference). However, when the size of the sphere increases (Figure 7.6), the scattering pattern changes drastically, and forward scattering becomes dominant. In this case (where the size parameter of the sphere is $x = 5$) the forward/backward field ratio is already 4.9, corresponding to 13.8 dB.

In addition to the total efficiencies given by Equations (7.2)–(7.4), two other important parameters are the efficiencies for backscattering $Q_{\rm b}$ and forward scattering $Q_{\rm f}$:

$$Q_{\rm b} = \frac{1}{x^2}\left|\sum_{n=1}^{\infty}(2n+1)(-1)^n\,(a_n - b_n)\right|^2, \tag{7.6}$$

$$Q_{\rm f} = \frac{1}{x^2}\left|\sum_{n=1}^{\infty}(2n+1)\,(a_n + b_n)\right|^2. \tag{7.7}$$

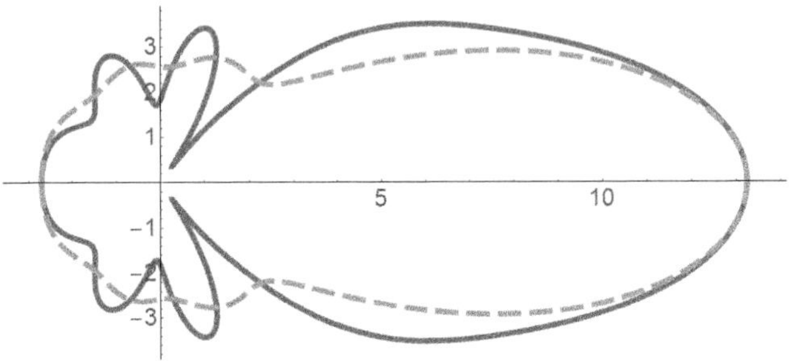

Figure 7.6: As in Figure 7.5, for a larger PEC sphere ($x = 5$). The amplitude scale is normalized differently in the two figures.

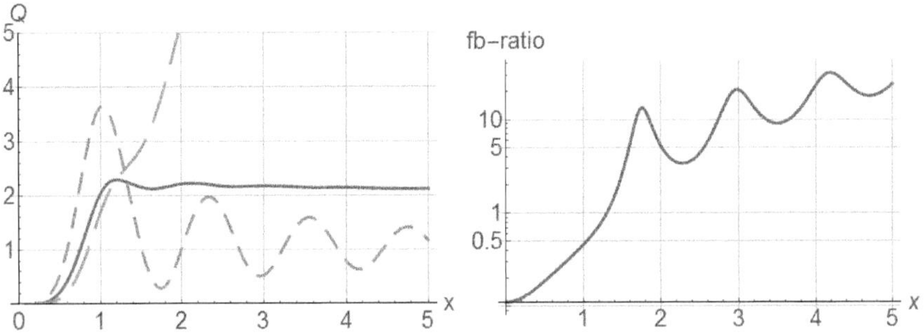

Figure 7.7: Total scattering Q_{sca} (solid blue), forward scattering (dashed orange), and backscattering (short-dashed green) efficiencies for a PEC sphere. Right panel: the ratio between the forward and backward scattering cross sections as the size of the sphere changes.

The drastic changes in the front-to-back ratio of PEC-sphere scattering are displayed in Figure 7.7. Indeed, backscattering dominates for small, and forward scattering for large spheres. At high frequencies, backscattering efficiency starts to oscillate with increasing x but finally settles at value 1, half of the total scattering efficiency limit.

Modeling PEC sphere by penetrable object

At microwave frequencies, the complex permittivity

$$\epsilon = \epsilon' - j\,\frac{\sigma}{\omega} \tag{7.8}$$

includes in addition to the real permittivity ϵ', also the imaginary part that depends on the conductivity σ and angular frequency ω. From this it follows that a good conductor has a very large absolute value of permittivity, at least in frequencies below the optical domain. Hence it is common to model a perfectly

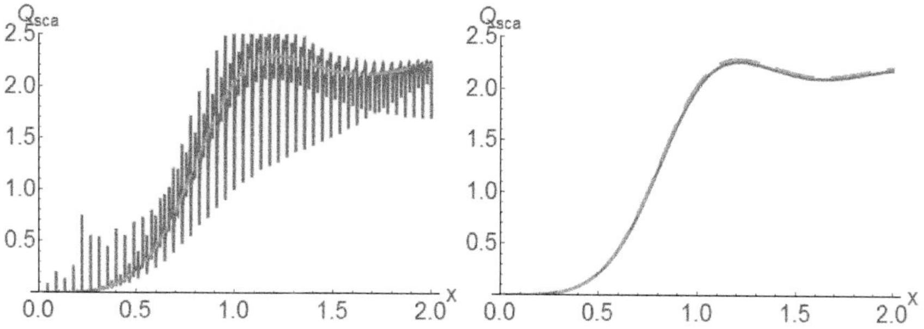

Figure 7.8: Comparison of the scattering efficiency of a PEC sphere (dashed orange line) and a high-permittivity sphere ($\epsilon_r = 5000(1 - j \cdot 0.001)$, solid blue line), left panel, and that by a low-impedance sphere ($\epsilon_r = 300, \mu_r = 1/300$, right panel), as functions of size parameter x.

conducting material having infinite permittivity. Also from the point of view of electrostatics this makes sense by considering a point charge over a dielectric surface with permittivity ϵ_r. As is known from image theory [35], Sec. 4.4.3, the effect of the surface can be taken into account by removing the dielectric half space and adding an image charge into the mirror point with magnitude $-(\epsilon_r - 1)/(\epsilon_r + 1)$ times the primary charge. To make this match with the PEC boundary condition (electric field perpendicular to the surface), the image charge has to be the negative of the original one, leading to $\epsilon_r \to \infty$.

However, once we move into electrodynamics, also the magnetic field has to be taken into account. Applying Faraday law in the electromagnetic time-harmonic case ($\nabla \times \mathbf{E} = -j\omega\mathbf{B}$), one can conclude that also the magnetic flux density $\mathbf{B}$ has to vanish inside a medium where the electric field is identically zero. In addition we have the magnetic constitutive relation $\mathbf{B} = \mu\mathbf{H}$. While $\mathbf{B}$ is identically zero, at the same time no condition is set on the magnetic field strength. To ensure consistency, the permeability μ has to vanish [101]. In addition, because of the continuity of the normal component of the magnetic flux density over an interface, a corollary follows: the normal component of the magnetic flux density is zero on a PEC boundary ($\mathbf{n} \cdot \mathbf{B} = 0$).

To demonstrate that to model scattering by PEC particles more effectively by "zero-impedance" ($\mu = 0, \epsilon = \infty$) rather than a plain infinite-epsilon assumption, Figure 7.8 compares the performance of two models in predicting the scattering efficiency as a function x.

The PEC behavior is captured once the parameters approach extreme values in both penetrable cases. However, for the low-impedance approximation, the curve is well reproduced already for moderate material parameters $\epsilon_r = 1/\mu_r = 300$, while the high-permittivity approximations suffers from resonances, always thinner and denser when ϵ_r increases. Here the permittivity is $\epsilon_r = 5000(1 - j \cdot 0.01)$ where the small losses are added to dampen slightly the very sharp resonances.

Another view to the necessity of both the electric and magnetic responses comes from the message of Figure 7.5 where the scattering pattern of a small PEC sphere is shown in two perpendicular scattering planes. A small sphere responds like a dipole. However, the dipole radiation is rotationally symmetric which is not the case here. Indeed, the pattern is well approximated by perpendicular electric and magnetic dipoles where the electric dipole is twice in magnitude of the magnetic one.

PMC sphere

The duality between PEC and PMC boundary conditions can be applied to draw conclusions on perfect magnetic conductor scatterers from the observations for PEC scattering. To model electromagnetically a PMC object with a penetrable dielectric–magnetic object of the same geometry, its permeability μ should be large and permittivity ϵ small. This is consistent with the observation that for a PMC sphere, the role of the Mie coefficients are interchanged compared to the PEC case (7.5):

$$a_{n,\mathrm{PMC}} = b_{n,\mathrm{PEC}}, \quad b_{n,\mathrm{PMC}} = a_{n,\mathrm{PEC}}. \tag{7.9}$$

And obviously, in terms of the scattering characteristics, the PMC sphere has an equal efficiency than a PEC sphere of the same size:

$$Q_{\mathrm{PMC}}(x) = Q_{\mathrm{PEC}}(x). \tag{7.10}$$

Due to the lossless character of both spheres, their absorption efficiencies vanish, and this equality holds for all three efficiencies. However, in terms of the scattering patterns of PMC and PEC spheres, the E-plane and H-plane patterns are interchanged.

PEMC sphere

The non-reciprocal character of the PEMC boundary condition leads to a rotation in the reflected field polarization, as was analyzed in Section 3 of Chapter 2. Hence also in the scattering process by a PEMC sphere, cross-polarized effects appear. To illustrate this, let us consider an electrically small sphere with PEMC boundary condition.

When such a sphere is exposed to an electromagnetic wave, the incident electric and magnetic fields induce both electric and magnetic dipole moments, respectively. Hence the resultant dipole moments are no longer aligned with the exciting fields, and their strengths and directions are dictated by the PEMC parameter M, as shown in Figure 7.9.

The connection between the incident fields $(\mathbf{E}^i, \mathbf{H}^i)$ and the induced dipole moments $(\mathbf{p}_e, \mathbf{p}_m)$ can be written in the matrix form [96]:

$$\begin{pmatrix} \eta_o \mathbf{p}_e \\ \mathbf{p}_m \end{pmatrix} = \sqrt{\mu_o \epsilon_o}\, V \frac{3/2}{1+(M\eta_o)^2} \begin{pmatrix} 2(M\eta_o)^2 - 1 & 3\,M\eta_o \\ 3\,M\eta_o & 2-(M\eta_o)^2 \end{pmatrix} \begin{pmatrix} \mathbf{E}^i \\ \eta_o \mathbf{H}^i \end{pmatrix}. \tag{7.11}$$

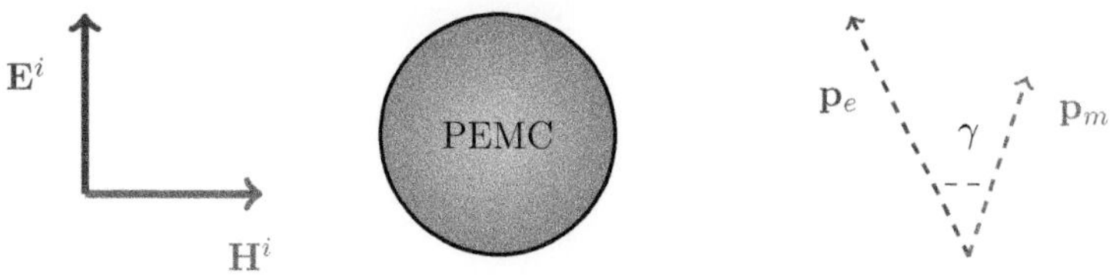

Figure 7.9: Incident plane wave induces in a small PEMC sphere electric and magnetic dipole moments whose directions and strength are dependent on the PEMC parameter M.

The angle γ between the dipole moments is hence given by

$$\cos\gamma = \frac{3\,M\eta_o}{\sqrt{[(4+(M\eta_o)^2]\,[4(M\eta_o)^2+1]}} \tag{7.12}$$

where V is the volume of the scatterer (the scattered field is proportional to the volume in the Rayleigh limit.)

The electric and magnetic dipole moments are perpendicular to each other only in the cases $M = 0$ (PMC) and $1/M = 0$ (PEC). The minimum angle $\gamma = \arccos(3/5) \approx 53°$ happens for $M\eta_o = \pm 1$, in which case the amplitudes of the dipole moments are equal. In the PEC case, the normalized electric dipole moment is twice that of the magnetic one, while in the PMC case $|\mathbf{p}_m|/\eta_o = 2|\mathbf{p}_e|$.

The field scattered by a small PEMC sphere can be computed from these dipole moments. Figure 7.10 shows the scattering patterns of some special cases: $M\eta_o = 0, 1, \infty$. The PEC and PMC cases are dual to each other, E- and H-planes show different patterns, and there is no cross-polarized scattering. However, once the PEMC parameter has the special value $M\eta_o = 1$, the co-polarized patterns become smaller although remaining finite, and the cross-polarized scattering dominates. In particular, backscattering is totally cross-polarized. Furthermore, the scattering patterns are rotationally symmetric, unlike in the PEC and PMC cases.

7.3 DB and D'B'-Boundary Objects

Another class of fundamental, parameter-free boundary conditions are the DB and D'B' conditions, analyzed in Chapter 4. Due to their lossless character,[3] spheres with DB or D'B' boundary have vanishing absorption efficiency, and consequently their scattering and extinction efficiencies are equal.

[3] DB and D'B' surfaces are mixed-impedance surfaces, acting like PEC or PMC boundaries depending on the polarization of the incident wave (see Table 4.1). They therefore inherit the losslessness property of PEC and PMC.

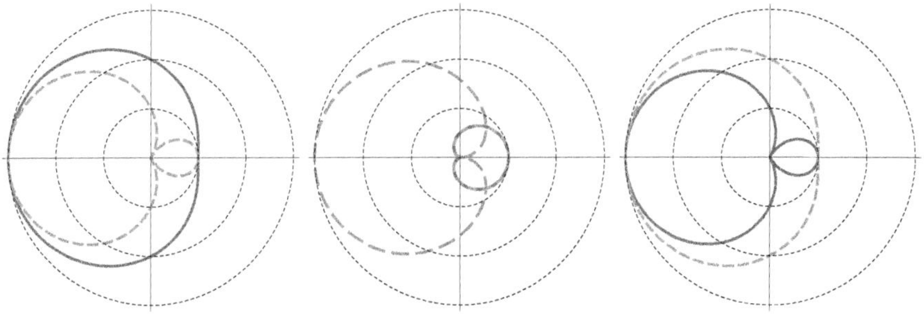

(a) $M = \infty$ (PEC) (b) $M\eta_o = 1$ (c) $M = 0$ (PMC)

Figure 7.10: The electric field amplitude scattering patterns of small PEMC spheres for three values of the PEMC parameter M. Co-polarized scattering in E-plane (solid blue), co-polarized in H-plane (dashed orange), cross-polarization (dashed green, the same in both planes). Incidence is from the left-hand side.

Scattering efficiencies

The efficiencies of DB and D'B' spheres can be shown to be [112] connected in a simple manner to the Mie coefficients of a PEC sphere by

$$Q_{\rm DB} = \frac{4}{x^2}\sum_{n=1}^{\infty}(2n+1)\left|b_{n,\rm PEC}\right|^2, \tag{7.13}$$

$$Q_{\rm D'B'} = \frac{4}{x^2}\sum_{n=1}^{\infty}(2n+1)\left|a_{n,\rm PEC}\right|^2. \tag{7.14}$$

Figure 7.11 shows the scattering efficiencies of PEC, DB, and D'B' spheres. In this lossless case, scattering equals extinction. In terms of scattering efficiency (which integrates the scattered power from all bistatic directions into a single number), the PEC and PMC spheres are equal. Their scattering patterns are the same, only 90° rotated from each other. In fact, also the scattering efficiency of a PEMC sphere is also equal to the PEC and PMC spheres. As can be seen, the PEC sphere scattering falls "between" the two other ones: for small scatterers, D'B' dominates whereas in the large-scatterer domain, DB is the strongest. Indeed, the scattering efficiency of a PEC sphere is the average of those of DB and D'B' spheres:

$$Q_{\rm PEC} = \frac{1}{2}\left(Q_{\rm DB} + Q_{\rm D'B'}\right). \tag{7.15}$$

It follows that all three spheres scatter equally for the size $x = 2.0333$ whence $Q_{\rm sca} = 2.2158$.

Due to the symmetry of the DB and D'B' boundaries with respect to the electric and magnetic fields, the E-plane and H-plane scattering patterns are the same in both cases.

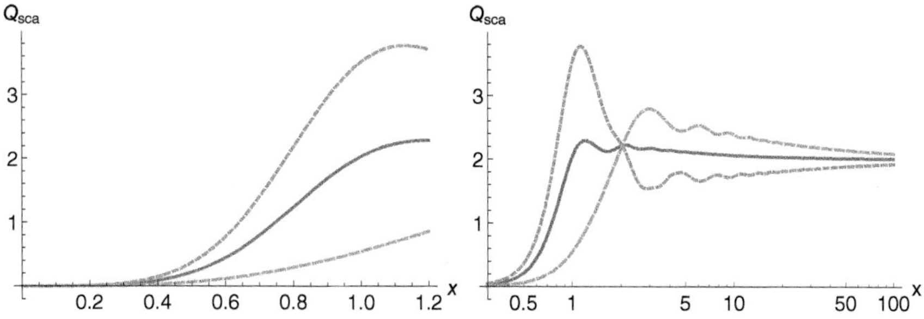

Figure 7.11: Scattering efficiencies of PEC (solid blue), DB (dashed orange), and D'B' (dotted green) spheres as functions of the size parameter.

For low frequencies, the lowest multipoles determine scattering. In this Rayleigh scattering regime, we can neglect all higher terms except the dipoles ($n = 1$). In electrostatics, a dielectric sphere is described by its normalized polarizability α which is a simple function of the relative permittivity ϵ_r:

$$\alpha = 3\frac{\epsilon_r - 1}{\epsilon_r + 2} \tag{7.16}$$

and the connection to the scattering efficiency is [98, 113]

$$Q_{\text{sca}} = \frac{8}{27}|\alpha|^2 x^4. \tag{7.17}$$

Likewise, for a magnetic sphere ($\mu_r \neq 1$), the magnetic dipole dominates its response. And for a PEC sphere, both dipoles are needed. This leads to the following low-frequency dependencies on x of the PEC, DB, and D'B' spheres:

$$Q_{\text{PEC}} = \frac{10}{3}x^4, \quad Q_{\text{DB}} = \frac{4}{3}x^4, \quad Q_{\text{D'B'}} = \frac{16}{3}x^4. \tag{7.18}$$

The accuracy of the Rayleigh approximation is surprisingly high even if the sphere has a moderate size. For a PEC sphere, the approximation $10x^4/3$ is accurate to 10% up to size parameter 0.8, as shown in Figure 7.12. The Rayleigh approximation contains both the electric and magnetic dipole contributions owing to their comparable magnitudes as was discussed in Section 7.2.

Scattering patterns

Figure 7.13 displays the scattering pattern of DB and D'B' spheres. As can be seen, for size parameter $x = 1$, the D'B' sphere is a clearly stronger scatterer, while in the case of a larger sphere ($x = 3$), the situation is the opposite, in agreement with the efficiency curves of Figure 7.11.

Another noteworthy property in the scattering patterns is the vanishing backscattering. It has been known for a long time [110] that axisymmetric

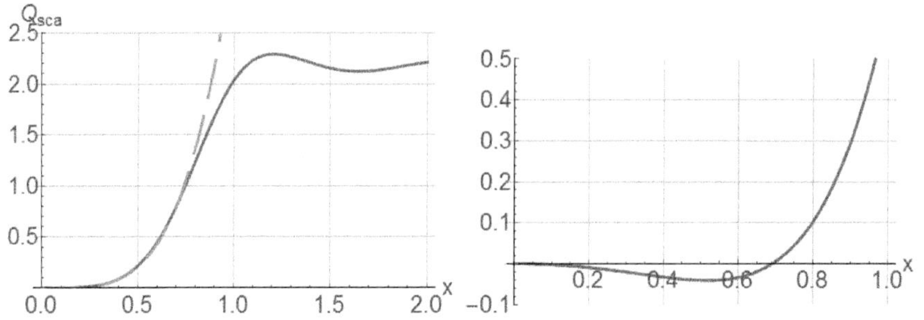

Figure 7.12: The scattering efficiency of a PEC sphere: full Mie result (solid blue) and the low-frequency Rayleigh approximation (dashed orange). Right panel: relative difference between the two efficiencies.

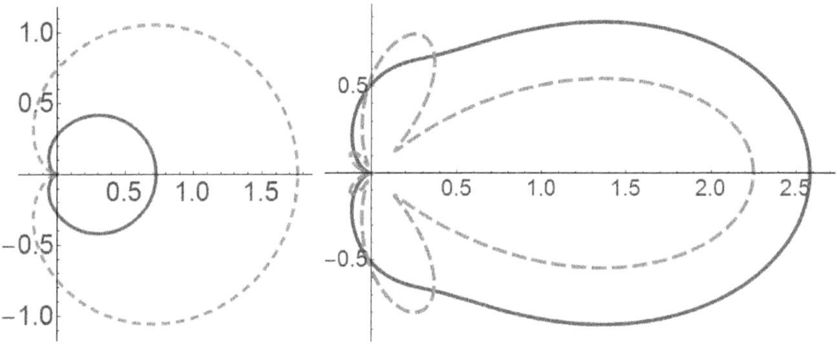

Figure 7.13: Scattering patterns (electric field amplitude, linear scale) of DB (solid blue), and D'B' (dashed orange) for small ($x = 1$, left) and larger ($x = 3$, right) spheres. Note the backscattering zero for all cases.

objects with equal permittivity and permeability are invisible for monostatic radar. However, the DB and D'B' spheres possess the same zero-backscattering property due to the fact that they are of self-dual character [62].

Note, however, that the property of zero backscattering requires also a symmetry of the object with respect to the excitation such that it looks similar to the electric and the magnetic field of the incident wave. This can be demonstrated by analyzing the scattering from a cube which is a less symmetric form than the sphere. Figure 7.14 plots the bistatic scattering cross section from a DB-boundary cube for two excitations: firstly, the wave is incident normal to a face and polarized parallel to the edge, and secondly, the cube is rotated by 45°: the wave is incident along the face diagonal, hitting the edge in an angle of 45°, and polarized parallel to the edge. The edge length of the cube is one-half of a wavelength, $L = \lambda/2$.

We can observe from the figure that in the first case of the geometrically

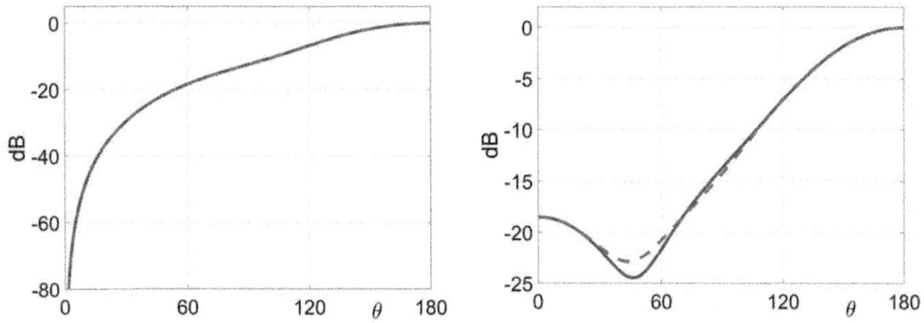

Figure 7.14: Bistatic scattering functions by a cube with DB boundary condition. Left: symmetric excitation normal to the face of the cube; Right: the cube is rotated by 45°. Solid blue: E-plane, dashed orange: H-plane. The size of the cube is $kL = \pi$. The scattering angle is 0° for backscattering and 180° for forward scattering. (Computations by Dr. Pasi Ylä-Oijala (Aalto University).)

symmetric excitation, backscattering vanishes, and the E-plane and H-plane scattering patterns are equal. However, when the excitation symmetry is broken in the second case, the scattering patterns are slightly different and backscattering has a finite value (in this case the front-to-back ratio is $-18.5\,\text{dB}$). The computations leading to the results in Figure 7.14 have been performed by a surface-integral-equation method discretized using the method of moments with Galerkin's testing and first order linear basis functions [62, 105].

7.4 Impedance-Boundary Objects

The analysis of scattering and extinction by spheres with impedance boundary condition leads to very interesting physical phenomena. Let us consider properties of spheres where the impedance dyadic (3.41) is simply-isotropic, equivalent to a scalar Z_s. In this case, a lossless scatterer requires $Z_s = R_s + jX_s$ to be purely imaginary, in other words reactive, whereas for cases with non-vanishing real part of Z_s, the sphere is either lossy ($R_s > 0$) or active ($R_s < 0$).

For a homogeneous and isotropic impedance boundary condition (IBC) sphere with impedance Z_s, the Mie coefficients a_n and b_n read [102]

$$a_{n,\text{IBC}} = \frac{x\,j_{n-1}(x) - n\,j_n(x) - j(Z_s/\eta_o)\,x\,j_n(x)}{x\,h_{n-1}^{(2)}(x) - n\,h_n^{(2)}(x) - j(Z_s/\eta_o)\,x\,h_n^{(2)}(x)}, \tag{7.19}$$

$$b_{n,\text{IBC}} = \frac{x\,j_{n-1}(x) - n\,j_n(x) - (j\,\eta_o/Z_s)\,x\,j_n(x)}{x\,h_{n-1}^{(2)}(x) - n\,h_n^{(2)}(x) - (j\,\eta_o/Z_s)\,x\,h_n^{(2)}(x)}. \tag{7.20}$$

from which forms the PEC coefficients (7.5) follow directly. Furthermore, the interchange of the roles of the electric and magnetic coefficients (Eqn. (7.9)) for

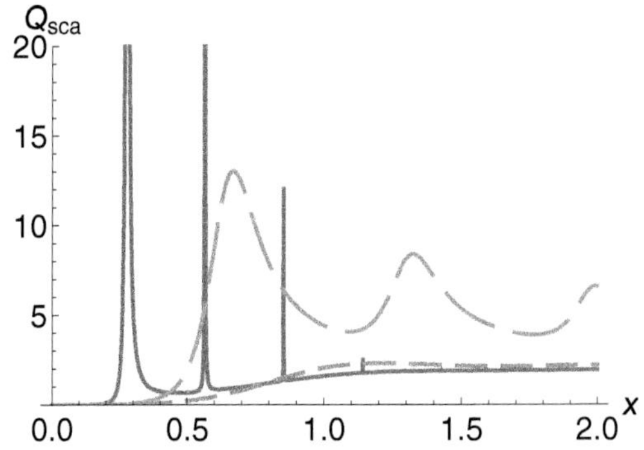

Figure 7.15: The scattering efficiency of lossless impedance spheres as function of the size parameter x. The surface impedance is $Z_s = -j\,0.3\eta_o$ (solid blue), $-j\,\eta_o$ (dashed orange), and 0 (PEC, short-dashed green).

a PMC sphere ($1/Z_s = 0$) is transparent in (7.19)–(7.20).

Lossless IBC spheres

For lossless impedance spheres, again scattering equals extinction. However, unlike in the case of PEC, DB, and D'B' spheres, this function depends not only on the size parameter x but also on Z_s, and the dependence is far from trivial. Figure 7.15 displays $Q_{\rm sca}$ as function of x for three values for the surface reactance. For $Z_s = 0$, we recover the rather smooth PEC scattering curve but for impedances with imaginary part deviating from zero, a strong resonance structure appears. The scattering peaks become stronger and sharper for small (negative) imaginary values of Z_s, in other words capacitive reactances. As can approximately be seen in the figure, these resonances appear for sizes $x = -nX_s/\eta_o$ where X_s is the imaginary part of Z_s, and n is integer. Indeed, the peaks are due to the magnetic multipolar resonances of order n. ,

In addition to these magnetic resonance peaks for small spheres with small capacitive reactance values, another series of resonances (due to the electric multipoles) can be observed for inductive lossless spheres. However, these take place for large values of the reactance X_s. In fact, the following equality holds between the two domains (the proof is left as exercise):

$$Q_{\rm sca}(x, X_s/\eta_o) = Q_{\rm sca}(x, -\eta_o/X_s), \tag{7.21}$$

showing that for small spheres, the inductive resonances require large X_s values: $nX_s = 1/x$.

This effect can be observed in the contour plot 7.16, showing two ranges of strong scattering domains on both the negative and positive X_s values. The gentle PEC scattering curve ($X_s = 0$) is a vertical line in the plot, located in the valley between the two high-amplitude ranges.

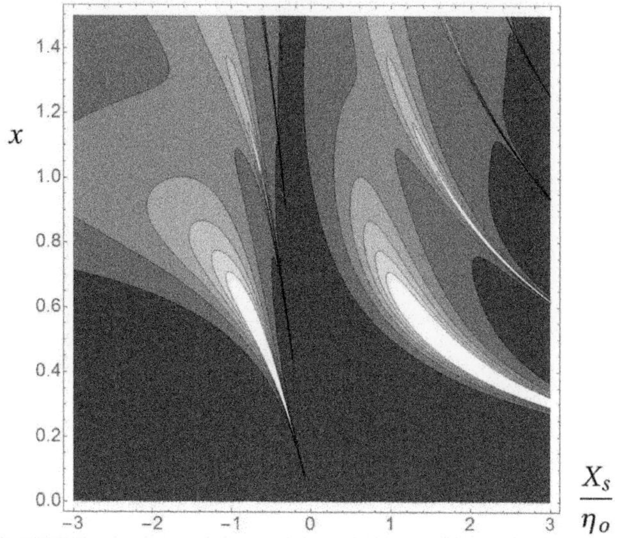

Figure 7.16: The scattering efficiency of lossless impedance spheres as function of the size parameter x and the normalized surface reactance X_s/η_o. The blue-to-white scale corresponds to values from small to large, showing the sharpening of the resonances for small spheres.

In the computations for the illustration in Figure 7.16, X_s has been assumed independent of the size parameter x. It is worth emphasizing that the variation of the efficiency is not explicitly in terms of the frequency. The size parameter x is essentially the product of frequency and absolute size of the sphere. Hence it can describe scattering functions of the sphere for varying frequency or a single-frequency response of a distribution of spheres of varying size. If the reactance X_s is a function of frequency or the absolute size of the sphere, the resonance spectra in the picture will be distorted.

Scattering patterns

The behavior of the scattering pattern of lossless impedance spheres is particularly interesting. A small PEC sphere is a fair backscatterer as can be seen in Figure 7.5. However, small IBC spheres have a very different character: forward scattering dominates. This can be seen in Figure 7.17 which show the front-to-back ratio (forward scattering efficiency divided by the backscattering efficiency, cf. Equations (7.6) and (7.7)) for spheres with certain inductive surface impedances. The ratio increases without limit when x decreases. For electrically large spheres the forward scattering increases strongly as is also the case for the PEC sphere. However, as seen in Figure 7.17, the increase of the front-to-back ratio of the PEC sphere with large x is not as oscillating as for

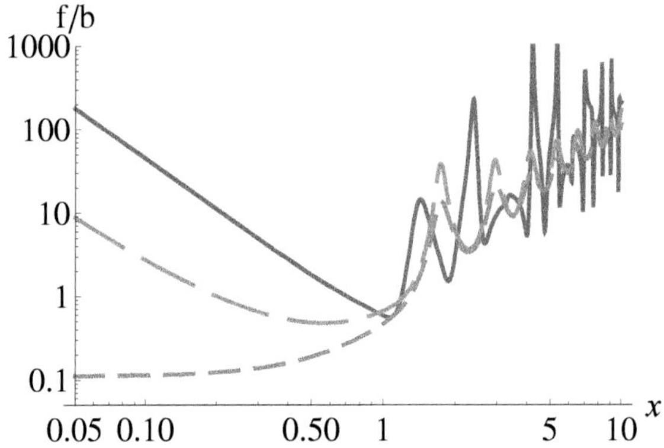

Figure 7.17: The forward-to-backward scattering efficiency ratio of lossless impedance spheres as function of the size parameter x. Solid blue line: $Z_s = j\,\eta_o$, dashed orange: $Z_s = 0.1\,j\,\eta_o$, short-dashed green: PEC.

the lossless IBC spheres.

However, as seen in the scattering efficiency plots, certain reactance values lead to a very narrow resonances for small spheres. This phenomenon is also observed in the scattering pattern asymmetry, and it happens for either small negative reactance values (capacitive impedance), or large positive reactances in the inductive range.

Therefore, despite the general trend of dominating forward scattering, there are combinations of size parameter and reactance for which the sphere scatters mostly into the back direction, for example the case in Figure 7.18 at around $x = 0.5$, $X_s = -0.1\eta_o$. Furthermore, due to the fact that $a_{n,\mathrm{IBC}}(X_s/\eta_o) = b_{n,\mathrm{IBC}}(-\eta_o/X_s)$, the same behavior of low forward scattering for spheres of certain size repeats itself for inductive surface-impedance spheres with $X_s = +10\eta_o$.

A look at the Taylor expansions of the Mie coefficients explains the dominance of the forward scattering for small lossless IBC spheres. For small x, we can expand the lowest multipole expressions (the dipole terms $n = 1$ from (7.19) and (7.20)) for $Z_s = j\,X_s$ as

$$a_{1,\mathrm{IBC}} = \frac{2jx^3}{3} + j\,x^4\frac{X_s}{\eta_o} + O\left(x^{9/2}\right), \qquad (7.22)$$

$$b_{1,\mathrm{IBC}} = \frac{2jx^3}{3} - j\,x^4\frac{\eta_o}{X_s} + O\left(x^{9/2}\right). \qquad (7.23)$$

For higher-order coefficients, the small-size expansions begin with higher powers of x: the first significant term of the expansion of an nth order Mie coefficient is proportional to x^{2n+1}. Hence the limiting behavior of the forward and backward scattering comes for the first-order coefficients. Using (7.6) and

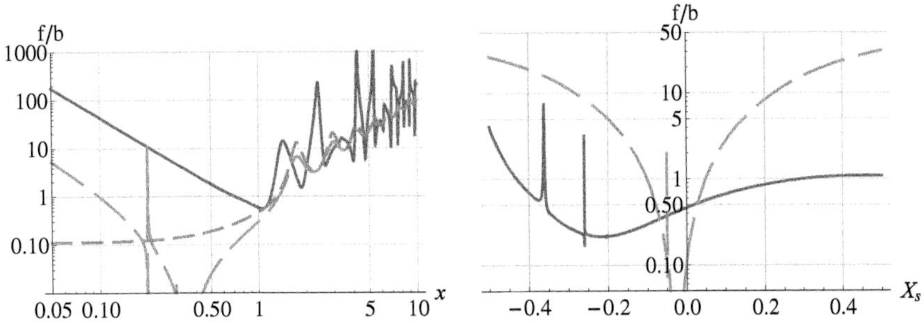

Figure 7.18: Left: front-to-back ratio of efficiencies for lossless IBC spheres as function of the size parameter x (solid blue line: $Z_s = -j\,\eta_o$ (the same as for $Z_s = +j\,\eta_o$), dashed orange: $Z_s = -0.1\,j\,\eta_o$, short-dashed green: PEC). Right: the same as function of the normalized reactance X_s/η_o (solid blue line: $x = 1$, dashed orange: $x = 0.1$).

(7.7), the forward-scattering and backscattering efficiencies for $x \ll 1$ read

$$Q_\mathrm{f} \approx 16x^4, \quad Q_\mathrm{b} \approx 9\left(X_s + \frac{1}{X_s}\right)^2 x^6. \tag{7.24}$$

The difference in the x dependencies is due to the canceling of the x^3 term in the Mie coefficients for backscattering. Therefore the front-to-back ratio increases without limit as the size decreases, as seen in Figures 7.17 and 7.18. The expansion (7.24) fails for the PEC and PMC limiting cases for which it is known that backscattering is 9.5 dB stronger than forward scattering for subwavelength spheres, as seen in Figure 7.5.

Lossy IBC spheres

For lossy spheres, the description becomes more complicated because extinction does not equal scattering. Absorption efficiency covers the difference between the two (cf. Eq. (7.4)). Figure 7.19 shows the three efficiencies of a lossy impedance sphere with surface impedance $Z_s = 10\eta_o$. For small spheres, absorption dominates over scattering but once the size parameter increases over the value $x = 1$, scattering counts mostly in extinction. The extinction paradox applies: $Q_\mathrm{ext} \to 2$ for large spheres. However, this convergence is much slower than for a PEC sphere. For instance, for $x = 1000$, the efficiencies are $Q_\mathrm{sca} = 1.649$, $Q_\mathrm{abs} = 0.368$, and $Q_\mathrm{ext} = 2.017$.

A particularly interesting case is when the surface impedance equals that of free space. Figure 7.20 shows the radiation pattern of a sphere with a surface impedance $Z_s = \eta_o$. Increasing size leads to increased scattering but the radiation pattern is rather smooth. Due to the impedance matching the sphere is a zero-backscattering object (ZBO). Furthermore, the scattering pattern is equal

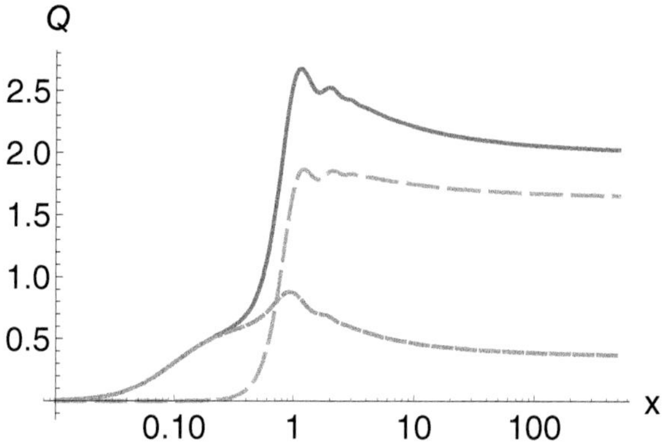

Figure 7.19: The extinction (solid blue), scattering (dashed orange), and absorption (short-dashed green) efficiencies of lossy impedance spheres ($Z_s = 10\eta_o$) as functions of the size parameter x.

in the E and H-planes. This property makes the impedance-matched scatterer similar to DB and D'B' objects (cf. Fig. 7.13).

The magnitude of the surface impedance Z_s does not affect very strongly the scattering efficiency like the size parameter does. This can be observed in the curves of Figure 7.21 where it is seen that impedance-matched spheres are slightly stronger scatterers in the case of small spheres, whereas the when the size parameter increases into values more than $x = 0.5$, scattering is higher for spheres where the impedance deviates from $Z_s = \eta_o$ and achieves maximum for the PEC and PMC cases.

The same figure also shows the effect of Z_s on absorption. As expected, impedance-matched spheres are lossier than those where Z_s is very different from η_o. However, impedance-matched spheres have maximum losses only for large spheres ($x > 0.36$). If the size parameter is smaller than this value, maximum absorption requires an impedance $Z_s = x$ or $Z_s = 1/x$.

Another view on the balance between losses and scattering is given by the albedo A which is the ratio between scattering and extinction. This quantity has a value between zero (for totally absorbing body) and unity (completely scattering object):

$$A = \frac{Q_{\text{sca}}}{Q_{\text{ext}}}. \tag{7.25}$$

Figure 7.22 displays how the albedo of a lossy IBC sphere varies when the size parameter changes. From a very low albedo in the Rayleigh regime it increases with size. Furthermore, as expected, an impedance-matched sphere ($Z_s = \eta_o$) has a lower albedo than one with impedance contrast. For increasing size of the sphere, the increase of the albedo starts to saturate.

As comparison, the albedo behavior of a lossy penetrable sphere is plotted in Figure 7.22. There it is seen that, in addition to the decrease of the overall

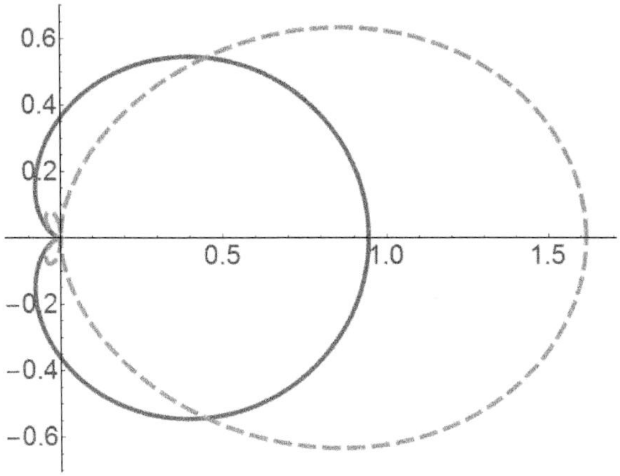

Figure 7.20: The scattering pattern (electric field amplitude, linear scale) of a lossy IBC sphere with surface impedance $Z_s = \eta_o$. Size parameter $x = 1$ (solid blue), $x = 2$ (dashed orange). Note vanishing backscattering.

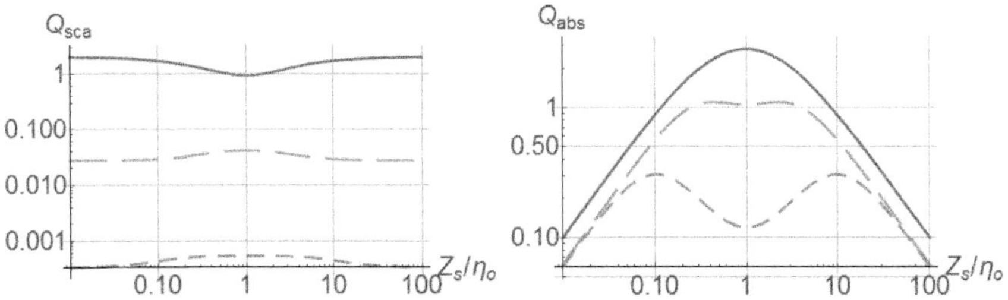

Figure 7.21: Scattering (left) and absorption (right) efficiencies of lossy IBC spheres as function of Z_s. Size parameter varies: $x = 1$ (solid blue), $x = 0.3$ (dashed orange), and $x = 0.1$ (short-dashed green).

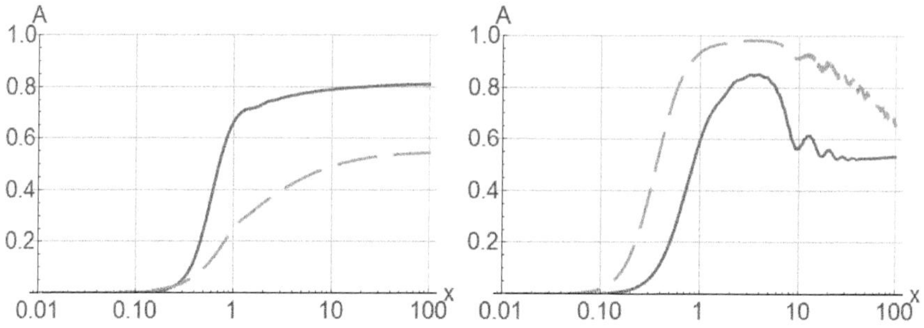

Figure 7.22: The albedo of a lossy IBC-boundary sphere (left) with $Z_s = 10\eta_o$ (solid blue) and $Z_s = \eta_o$ (dashed orange) as function of size parameter. For comparison, on the right: the albedo of a penetrable lossy sphere with relative permittivity $\epsilon_r = 2 - j\,0.1$ (solid blue) and $\epsilon_r = 2 - j\,0.01$ (dashed orange).

albedo with increasing imaginary part of the permittivity, the dependence on the size is not as monotonous as for the IBC sphere. Although the Rayleigh region is characterized by a very low albedo, its maximum is reached at values of the order of $x = 4$ after which it starts to decrease.

7.5 Problems

7.1 From the full Lorenz–Mie solution of the scattering efficiency $Q_{\text{sca,PEC}}$ for the PEC sphere (Eqs. (7.2) and (7.5)), derive the Rayleigh approximation (7.18), valid for electrically small spheres (cf. Fig. 7.12).

7.2 Show, using properties of the Mie coefficients a_n and b_n for PEC, PMC, DB, and D'B' spheres, that they are lossless scatterers. In other words, prove that their absorption efficiencies vanish; $Q_{\text{abs}} = 0$ for all four cases.

7.3 Using the Mie coefficients for an isotropic IBC sphere (7.19)–(7.20), prove the property (7.21) between capacitive and inductive lossless spheres.

7.4 Study the electric and magnetic dipole moments $\mathbf{p}_e$ and $\mathbf{p}_m$ induced in a PEMC sphere by a plane wave with incident fields $\mathbf{E}^i = E_o\mathbf{u}_x, \mathbf{H}^i = E_o/\eta_o\mathbf{u}_y$. Plot the angle between the dipole moments as function of the PEMC parameter M. Plot also the dependence of the amplitudes of the moments on M.

7.5 The scattering efficiencies of spheres with given boundary condition are functions of their electrical size. Find the maximum Q_{sca} for PEC, PMC, DB, and D'B' boundary spheres. What are the size parameters x for these maxima?

7.6 How well is scattering by a PEC sphere accounted for by the dipole approximation? In other words, compare Q_{sca} for a PEC sphere against the expression where only the first-order multipoles (electric and magnetic dipoles) are included (only the term $n = 1$ remains in (7.2)).

7.7 Show that a sphere of arbitrary size and with an impedance boundary condition $Z_s = \eta_o$ is a ZBO object (zero backscattering object).

7.8 As seen in Figures 7.17 and 7.18, spheres with a lossless impedance boundary condition tend to scatter mostly into the front direction. However, certain combinations for the size parameter and the reactance may lead to a very small forward scattering. Find conditions for x and X_s for this to happen.

Appendix A

Electromagnetic Formulas

Plane Waves

Incident and reflected time-harmonic plane waves are defined by

$$\mathbf{E}^i(\mathbf{r}) = \mathbf{E}^i \exp(-j\mathbf{k}^i \cdot \mathbf{r}), \quad \mathbf{H}^i(\mathbf{r}) = \mathbf{H}^i \exp(-j\mathbf{k}^i \cdot \mathbf{r}) \tag{A.1}$$

$$\mathbf{E}^r(\mathbf{r}) = \mathbf{E}^r \exp(-j\mathbf{k}^r \cdot \mathbf{r}), \quad \mathbf{H}^r(\mathbf{r}) = \mathbf{H}^r \exp(-j\mathbf{k}^r \cdot \mathbf{r}) \tag{A.2}$$

$$\mathbf{k}^i = -k_n\mathbf{n} + \mathbf{k}_t, \quad \mathbf{k}^r = k_n\mathbf{n} + \mathbf{k}_t, \tag{A.3}$$

$$\mathbf{k}^i \cdot \mathbf{k}^i = \mathbf{k}^r \cdot \mathbf{k}^r = k_o^2 = \omega^2\mu_o\epsilon_o \tag{A.4}$$

The plane-wave fields satisfy the Maxwell equations as

$$\mathbf{k} \times \mathbf{E} = \omega\mathbf{B} = k_o\eta_o\mathbf{H}, \tag{A.5}$$

$$\mathbf{k} \times \eta_o\mathbf{H} = -\omega\mathbf{D} = -k_o\mathbf{E}. \tag{A.6}$$

Any plane wave in isotropic medium can be decomposed in TE and TM parts as

$$\begin{pmatrix} \mathbf{E} \\ \mathbf{H} \end{pmatrix} = \begin{pmatrix} \mathbf{E}_{TE} \\ \mathbf{H}_{TE} \end{pmatrix} + \begin{pmatrix} \mathbf{E}_{TM} \\ \mathbf{H}_{TM} \end{pmatrix}, \tag{A.7}$$

satisfying

$$\mathbf{n} \cdot \mathbf{E}_{TE} = 0, \quad \mathbf{n} \cdot \mathbf{H}_{TM} = 0, \tag{A.8}$$

and

$$\mathbf{E}_{TE} = \frac{k_o}{k_t^2}(\mathbf{n} \times \mathbf{k})(\mathbf{n} \cdot \eta_o\mathbf{H}), \tag{A.9}$$

$$\mathbf{H}_{TE} = \frac{1}{k_t^2}(k_t^2\mathbf{n} - k_n\mathbf{k}_t)(\mathbf{n} \cdot \mathbf{H}), \tag{A.10}$$

$$\mathbf{E}_{TM} = \frac{1}{k_t^2}(k_t^2\mathbf{n} - k_n\mathbf{k}_t)(\mathbf{n} \cdot \mathbf{E}), \tag{A.11}$$

$$\eta_o\mathbf{H}_{TM} = -\frac{k_o}{k_t^2}(\mathbf{n} \times \mathbf{k})(\mathbf{n} \cdot \mathbf{E}). \tag{A.12}$$

Relations of tangential plane-wave field components

$$\eta_o\mathbf{H}_t^i = -\overline{\overline{\mathsf{J}}}_t\cdot\mathbf{E}_t^i, \qquad \mathbf{E}_t^i = \overline{\overline{\mathsf{J}}}_t\cdot\eta_o\mathbf{H}_t^i \tag{A.13}$$

$$\eta_o\mathbf{H}_t^r = \overline{\overline{\mathsf{J}}}_t\cdot\mathbf{E}_t^r, \qquad \mathbf{E}_t^r = -\overline{\overline{\mathsf{J}}}_t\cdot\eta_o\mathbf{H}_t^r \tag{A.14}$$

$$\mathbf{E}^r = \overline{\overline{\mathsf{R}}}\cdot\mathbf{E}^i \tag{A.15}$$

$$\mathbf{H}^i = -\overline{\overline{\mathsf{J}}}_t\cdot\overline{\overline{\mathsf{R}}}\cdot\overline{\overline{\mathsf{J}}}_t\cdot\mathbf{H}^i \tag{A.16}$$

Properties of the Dyadic $\overline{\overline{\mathsf{J}}}_t$

$$\overline{\overline{\mathsf{J}}}_t = \frac{1}{k_ok_n}((\mathbf{n}\times\mathbf{k})\mathbf{k}_t + k_n^2\mathbf{n}\times\overline{\overline{\mathsf{I}}}) \tag{A.17}$$

$$= \frac{1}{k_ok_nk_t^2}(k_o^2(\mathbf{n}\times\mathbf{k})\mathbf{k}_t - k_n^2\mathbf{k}_t(\mathbf{n}\times\mathbf{k})) \tag{A.18}$$

$$= \frac{1}{2k_ok_n}((\mathbf{n}\times\mathbf{k})\mathbf{k}_t + \mathbf{k}_t(\mathbf{n}\times\mathbf{k}) + (k_o^2+k_n^2)\mathbf{n}\times\overline{\overline{\mathsf{I}}}) \tag{A.19}$$

$$\overline{\overline{\mathsf{J}}}_t^2 = -\overline{\overline{\mathsf{I}}}_t, \qquad \overline{\overline{\mathsf{J}}}_t{}^{\times}_{\times}\mathbf{nn} = -\overline{\overline{\mathsf{J}}}_t^T, \qquad \overline{\overline{\mathsf{J}}}_t{}^{\times}_{\times}\overline{\overline{\mathsf{I}}}_t = 0 \tag{A.20}$$

$$\mathrm{tr}\overline{\overline{\mathsf{J}}}_t = 0, \qquad \mathrm{tr}(\mathbf{n}\times\overline{\overline{\mathsf{J}}}_t) = \mathrm{tr}(\overline{\overline{\mathsf{J}}}_t\times\mathbf{n}) = -k_o/k_n \tag{A.21}$$

$$\overline{\overline{\mathsf{J}}}_t^{(2)} = \mathbf{nn}, \qquad \det{}_t\overline{\overline{\mathsf{J}}}_t = \mathrm{tr}\overline{\overline{\mathsf{J}}}_t^{(2)} = -1 \tag{A.22}$$

$$\mathbf{n}\times\overline{\overline{\mathsf{J}}}_t = -\frac{1}{k_ok_n}(\mathbf{k}_t\mathbf{k}_t + k_n^2\overline{\overline{\mathsf{I}}}_t), \tag{A.23}$$

$$= -\frac{k_n}{k_o}\frac{(\mathbf{n}\times\mathbf{k})(\mathbf{n}\times\mathbf{k})}{\mathbf{k}_t^2} - \frac{k_o}{k_n}\frac{\mathbf{k}_t\mathbf{k}_t}{\mathbf{k}_t^2} \tag{A.24}$$

$$\overline{\overline{\mathsf{J}}}_t\times\mathbf{n} = -\frac{1}{k_ok_n}((\mathbf{n}\times\mathbf{k})(\mathbf{n}\times\mathbf{k}) + k_n^2\overline{\overline{\mathsf{I}}}_t) \tag{A.25}$$

$$= -\frac{k_o}{k_n}\frac{(\mathbf{n}\times\mathbf{k})(\mathbf{n}\times\mathbf{k})}{\mathbf{k}_t^2} - \frac{k_n}{k_o}\frac{\mathbf{k}_t\mathbf{k}_t}{\mathbf{k}_t^2} \tag{A.26}$$

$$\mathbf{k}_t\cdot\overline{\overline{\mathsf{J}}}_t = -(k_n/k_o)\mathbf{n}\times\mathbf{k}_t, \qquad (\mathbf{n}\times\mathbf{k})\cdot\overline{\overline{\mathsf{J}}}_t = (k_o/k_n)\mathbf{k}_t \tag{A.27}$$

$$\overline{\overline{\mathsf{J}}}_t\cdot\mathbf{k}_t = (k_o/k_n)\mathbf{n}\times\mathbf{k}, \qquad \overline{\overline{\mathsf{J}}}_t\cdot(\mathbf{n}\times\mathbf{k}) = -(k_n/k_o)\mathbf{k}_t \tag{A.28}$$

$$\overline{\overline{\mathsf{J}}}_t \pm j\overline{\overline{\mathsf{I}}}_t = \frac{1}{k_ok_nk_t^2}(k_o\mathbf{n}\times\mathbf{k} \pm jk_n\mathbf{k}_t)(k_o\mathbf{k}_t \pm jk_n\mathbf{n}\times\mathbf{k}) \tag{A.29}$$

$$\overline{\overline{\mathsf{J}}}_t\cdot(k_o\mathbf{n}\times\mathbf{k} \pm jk_n\mathbf{k}_t) = \pm j(k_o\mathbf{n}\times\mathbf{k} \pm jk_n\mathbf{k}_t) \tag{A.30}$$

Polarization vector

Polarizations properties of a time-harmonic vector $\mathbf{A}(t)$, depending on a complex vector $\mathbf{a} = \mathbf{a}_{re} + j\mathbf{a}_{im}$ as

$$\mathbf{A}(t) = \Re\{\mathbf{a}e^{j\omega t}\} = \mathbf{a}_{re}\cos\omega t - \mathbf{a}_{im}\sin\omega t, \tag{A.31}$$

can be studied in terms of the polarization vector $\mathbf{p}(\mathbf{a})$ defined by [35]

$$\mathbf{p}(\mathbf{a}) = \frac{\mathbf{a}\times\mathbf{a}^*}{j\mathbf{a}\cdot\mathbf{a}^*} = -2\frac{\mathbf{a}_{re}\times\mathbf{a}_{im}}{\mathbf{a}_{re}^2+\mathbf{a}_{im}^2} \tag{A.32}$$

$$= 2\frac{\mathbf{A}(0)\times\mathbf{A}(\pi/2\omega)}{\mathbf{A}^2(0)+\mathbf{A}^2(\pi/2\omega)}. \tag{A.33}$$

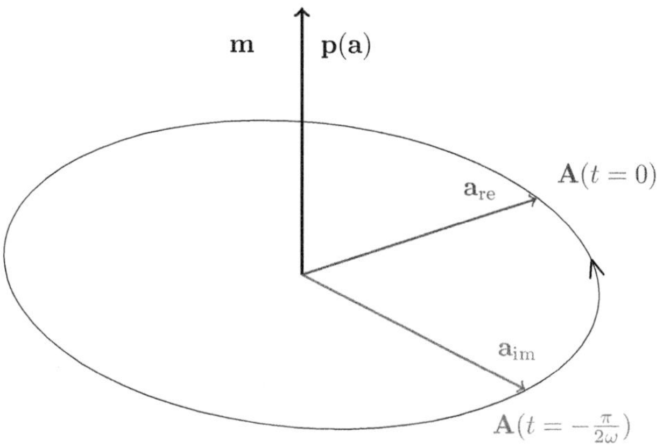

Figure A.1: The polarization vector $\mathbf{p}(\mathbf{a})$ is a real vector, normal to the polarization ellipse of the time-harmonic vector $\mathbf{A}(t)$ pointing into the direction of right-handed rotation.

It has the following properties

- $\mathbf{p}(\mathbf{a}) = 0$ corresponds to linearly polarized vectors $\mathbf{a}$ satisfying $\mathbf{a}\times\mathbf{a}^* = 0$.

- $\mathbf{p}^*(\mathbf{a}) = \mathbf{p}(\mathbf{a})$, whence $\mathbf{p}(\mathbf{a})$ is a real vector. It is orthogonal to the plane of the ellipse of $\mathbf{A}(t)$ and points to the direction for which $\mathbf{A}(t)$ has right-hand sense of rotation.

- $\mathbf{p}(\mathbf{a}^*) = -\mathbf{p}(\mathbf{a})$, or change in sense of rotation changes the direction of $\mathbf{p}(\mathbf{a})$.

- $\mathbf{p}(\lambda\mathbf{a}) = \mathbf{p}(\mathbf{a})$, $\lambda \neq 0$, or magnitude and phase of $\mathbf{A}(t)$ have no effect on the polarization vector.

- $|\mathbf{p}(\mathbf{a})| = 2e/(e^2+1) \le 1$ where e is the ellipticity (axial ratio) of the vector $\mathbf{A}(t)$. $|\mathbf{p}(\mathbf{a})| = 1$ corresponds to $e = 1$, i.e., any circularly polarized vector $\mathbf{A}(t)$.

- For $\mathbf{p}(\mathbf{a}) \neq 0$, with the unit vector $\mathbf{m} = \mathbf{p}(\mathbf{a})/|\mathbf{p}(\mathbf{a})|$ and $\mathbf{a}' = \mathbf{a}\cos\theta + \mathbf{m} \times \mathbf{a}\sin\theta$, we have $\mathbf{p}(\mathbf{a}') = \mathbf{p}(\mathbf{a})$ for any real θ. Since $\mathbf{a}'$ corresponds to the vector $\mathbf{A}(t)$ rotated by the angle θ in its plane, the polarization vector is independent of the orientation of the polarization ellipse in its plane.

GBC Boundaries

The most general linear and local boundary conditions can be represented in the form

$$a_{1n}\mathbf{n}\cdot\mathbf{E} + \mathbf{a}_{1t}\cdot\mathbf{E} + \mathbf{b}_{1t}\cdot\eta_o\mathbf{H}_t = 0, \quad (A.34)$$

$$b_{2n}\mathbf{n}\cdot\eta_o\mathbf{H} + \mathbf{a}_{2t}\cdot\mathbf{E} + \mathbf{b}_{2t}\cdot\eta_o\mathbf{H} = 0, \quad (A.35)$$

with scalar parameters a_{1n}, b_{2n} and vector parameters $\mathbf{a}_{1t}, \mathbf{a}_{2t}, \mathbf{b}_{1t}$ and $\mathbf{b}_{2t}$ defining the boundary. Various special cases considered under different labels can be summarized in the following table:

GBC	a_{1n}/b_{2n}	$\mathbf{a}_{1t}/\mathbf{a}_{2t}$	$\mathbf{b}_{1t}/\mathbf{b}_{2t}$
Impedance	$0/0$	$\mathbf{a}_{1t}/\mathbf{a}_{2t}$	$\mathbf{b}_{1t}/\mathbf{b}_{2t}$
DB	a_{1n}/b_{2n}	$0/0$	$0/0$
SH	$0/0$	$\mathbf{a}_{1t}/0$	$0/\mathbf{a}_{1t}$
SHDB	a_{1n}/a_{1n}	$\mathbf{a}_{1t}/0$	$0/-\mathbf{a}_{1t}$
GSH	$0/0$	$\mathbf{a}_{1t}/0$	$0/\mathbf{b}_{2t}$
GSHDB	a_{1n}/b_{2n}	$\mathbf{a}_{1t}/0$	$0/\mathbf{b}_{2t}$
E	$a_{1n}/0$	$\mathbf{a}_{1t}/\mathbf{a}_{2t}$	$0/0$
H	$0/b_{2n}$	$0/0$	$\mathbf{b}_{1t}/\mathbf{b}_{2t}$
EH	a_{1n}/b_{2n}	$0/\mathbf{a}_{2t}$	$\mathbf{b}_{1t}/0$

Problems

A.1 Derive the property $\mathbf{p}(\mathbf{a}') = \mathbf{p}(\mathbf{a})$ of the polarization vector $\mathbf{a}$ when $\mathbf{m} = \mathbf{p}(\mathbf{a})/|\mathbf{p}(\mathbf{a})|$ and $\mathbf{a}' = \mathbf{a}\cos\theta + \mathbf{m}\times\mathbf{a}\sin\theta$ for real θ.

A.2 Find the polarization vector of a vector $\mathbf{a}_t$ of the form

$$\mathbf{a}_t = \mathbf{b}_t + j\mathbf{n}\times\mathbf{b}_t,$$

where $\mathbf{b}_t$ is a complex vector satisfying $\mathbf{n}\cdot\mathbf{b}_t = 0$.

Appendix B

Dyadics

Dyadics are linear mappings between vectors. They were introduced by J.W. Gibbs in a non-published but widely circulated pamphlet in 1884 [14]. A textbook [13], compiled by E.B. Wilson in 1901 from professor Gibbs' lectures, contained an extended exposition of dyadic algebra. The present collection of identities has been reproduced with some additions from [35]. See also [109].

Definitions

$$\overline{\overline{\mathsf{A}}} = \sum_{i=1}^{3} \mathbf{a}_i\mathbf{c}_i, \quad \overline{\overline{\mathsf{A}}}^T = \sum_{i=1}^{3} \mathbf{c}_i\mathbf{a}_i \tag{B.1}$$

$$\overline{\overline{\mathsf{I}}} = \sum_{i=1}^{3} \mathbf{e}_i\mathbf{e}_i, \quad \mathbf{e}_i \cdot \mathbf{e}_j = \delta_{ij} \tag{B.2}$$

$$\overline{\overline{\mathsf{A}}}^2 = \overline{\overline{\mathsf{A}}} \cdot \overline{\overline{\mathsf{A}}}, \tag{B.3}$$

$$\overline{\overline{\mathsf{A}}}^{-2} = (\overline{\overline{\mathsf{A}}}^{-1})^2 = (\overline{\overline{\mathsf{A}}}^2)^{-1} \tag{B.4}$$

$$\overline{\overline{\mathsf{A}}} : \overline{\overline{\mathsf{B}}} = \sum_{i=1}^{3} \mathbf{a}_i\mathbf{c}_j : \sum_{j=1}^{3} \mathbf{b}_j\mathbf{d}_j = \sum_{i,j=1}^{3} (\mathbf{a}_i \cdot \mathbf{b}_j)(\mathbf{c}_i \cdot \mathbf{d}_j) \tag{B.5}$$

$$\overline{\overline{\mathsf{A}}} {}^\times_\times \overline{\overline{\mathsf{B}}} = \sum_{i=1}^{3} \mathbf{a}_i\mathbf{c}_j {}^\times_\times \sum_{j=1}^{3} \mathbf{b}_j\mathbf{d}_j = \sum_{i,j=1}^{3} (\mathbf{a}_i \times \mathbf{b}_j)(\mathbf{c}_i \times \mathbf{d}_j) \tag{B.6}$$

$$\overline{\overline{\mathsf{A}}}^{(2)} = \frac{1}{2}\overline{\overline{\mathsf{A}}} {}^\times_\times \overline{\overline{\mathsf{A}}} \tag{B.7}$$

$$= (\mathbf{a}_1 \times \mathbf{a}_2)(\mathbf{c}_1 \times \mathbf{c}_2) + (\mathbf{a}_2 \times \mathbf{a}_3)(\mathbf{c}_2 \times \mathbf{c}_3) + (\mathbf{a}_3 \times \mathbf{a}_1)(\mathbf{c}_3 \times \mathbf{c}_1) \tag{B.8}$$

$$\overline{\overline{\mathsf{A}}}^{(-2)} = (\overline{\overline{\mathsf{A}}}^{-1})^{(2)} = (\overline{\overline{\mathsf{A}}}^{(2)})^{-1} \tag{B.9}$$

$$\mathrm{tr}\overline{\overline{\mathsf{A}}} = \overline{\overline{\mathsf{A}}} : \overline{\overline{\mathsf{I}}} = \sum_{i=1}^{3} \mathbf{a}_i \cdot \overline{\overline{\mathsf{I}}} \cdot \mathbf{c}_i = \sum_{i=1}^{3} \mathbf{a}_i \cdot \mathbf{c}_i \tag{B.10}$$

$$\mathrm{tr}\overline{\overline{\mathsf{I}}} = 3 \tag{B.11}$$

$$\mathrm{spm}\overline{\overline{\mathsf{A}}} = \mathrm{tr}\overline{\overline{\mathsf{A}}}^{(2)} = \frac{1}{2}\overline{\overline{\mathsf{A}}}{}^{\times}_{\times}\overline{\overline{\mathsf{A}}} : \overline{\overline{\mathsf{I}}} \tag{B.12}$$

$$\mathrm{spm}\overline{\overline{\mathsf{I}}} = 1 \tag{B.13}$$

$$\mathrm{det}\overline{\overline{\mathsf{A}}} = \frac{1}{6}\overline{\overline{\mathsf{A}}}{}^{\times}_{\times}\overline{\overline{\mathsf{A}}} : \overline{\overline{\mathsf{A}}} = (\mathbf{a}_1 \times \mathbf{a}_2 \cdot \mathbf{a}_3)(\mathbf{c}_1 \times \mathbf{c}_2 \cdot \mathbf{c}_3) \tag{B.14}$$

$$\mathrm{det}\overline{\overline{\mathsf{I}}} = 1 \tag{B.15}$$

$$\mathrm{det}\overline{\overline{\mathsf{A}}} \neq 0 \;\leftrightarrow\; \overline{\overline{\mathsf{A}}} = \mathbf{a}_1\mathbf{c}_1 + \mathbf{a}_2\mathbf{c}_2 + \mathbf{a}_3\mathbf{c}_3 \tag{B.16}$$

$$\mathrm{det}\overline{\overline{\mathsf{A}}} = 0 \;\leftrightarrow\; \overline{\overline{\mathsf{A}}} = \mathbf{a}_1\mathbf{c}_1 + \mathbf{a}_2\mathbf{c}_2 \tag{B.17}$$

$$\overline{\overline{\mathsf{A}}}^{(2)} = 0 \;\leftrightarrow\; \overline{\overline{\mathsf{A}}} = \mathbf{a}_1\mathbf{c}_1 \tag{B.18}$$

$$\overline{\overline{\mathsf{A}}}^{T} = -\overline{\overline{\mathsf{A}}} \;\leftrightarrow\; \overline{\overline{\mathsf{A}}} = \mathbf{a} \times \overline{\overline{\mathsf{I}}} = \overline{\overline{\mathsf{I}}} \times \mathbf{a} \quad \overline{\overline{\mathsf{A}}} \text{ antisymmetric} \tag{B.19}$$

Identities

$$\overline{\overline{\mathsf{A}}} : \overline{\overline{\mathsf{B}}} = \overline{\overline{\mathsf{B}}} : \overline{\overline{\mathsf{A}}} \tag{B.20}$$

$$\overline{\overline{\mathsf{A}}}{}^{\times}_{\times}\overline{\overline{\mathsf{B}}} = \overline{\overline{\mathsf{B}}}{}^{\times}_{\times}\overline{\overline{\mathsf{A}}} \tag{B.21}$$

$$\begin{aligned} &= [(\overline{\overline{\mathsf{A}}} : \overline{\overline{\mathsf{I}}})(\overline{\overline{\mathsf{B}}} : \overline{\overline{\mathsf{I}}}) - \overline{\overline{\mathsf{A}}} : \overline{\overline{\mathsf{B}}}^{T}]\overline{\overline{\mathsf{I}}} - (\overline{\overline{\mathsf{A}}} : \overline{\overline{\mathsf{I}}})\overline{\overline{\mathsf{B}}}^{T} \\ &\quad -(\overline{\overline{\mathsf{B}}} : \overline{\overline{\mathsf{I}}})\overline{\overline{\mathsf{A}}}^{T} + [\overline{\overline{\mathsf{A}}} \cdot \overline{\overline{\mathsf{B}}} + \overline{\overline{\mathsf{B}}} \cdot \overline{\overline{\mathsf{A}}}]^{T} \end{aligned} \tag{B.22}$$

$$\overline{\overline{\mathsf{A}}}{}^{\times}_{\times}\overline{\overline{\mathsf{I}}} = (\overline{\overline{\mathsf{A}}} : \overline{\overline{\mathsf{I}}})\overline{\overline{\mathsf{I}}} - \overline{\overline{\mathsf{A}}}^{T} \tag{B.23}$$

$$\frac{1}{2}\overline{\overline{\mathsf{I}}}{}^{\times}_{\times}\overline{\overline{\mathsf{I}}} = \overline{\overline{\mathsf{I}}}^{(2)} = \overline{\overline{\mathsf{I}}} \tag{B.24}$$

$$\overline{\overline{\mathsf{A}}}^{(2)} = \overline{\overline{\mathsf{A}}}^{2T} - (\mathrm{tr}\overline{\overline{\mathsf{A}}})\overline{\overline{\mathsf{A}}}^{T} + (\mathrm{spm}\overline{\overline{\mathsf{A}}})\overline{\overline{\mathsf{I}}} \tag{B.25}$$

$$(\mathbf{a} \times \overline{\overline{\mathsf{I}}}){}^{\times}_{\times}\overline{\overline{\mathsf{I}}} = \mathbf{a} \times \overline{\overline{\mathsf{I}}} \tag{B.26}$$

$$(\mathbf{a} \times \overline{\overline{\mathsf{I}}}){}^{\times}_{\times}(\mathbf{b} \times \overline{\overline{\mathsf{I}}}) = \mathbf{ab} + \mathbf{ba} \tag{B.27}$$

$$\overline{\overline{\mathsf{S}}}{}^{\times}_{\times}(\mathbf{a} \times \overline{\overline{\mathsf{I}}}) = (\overline{\overline{\mathsf{S}}} \cdot \mathbf{a}) \times \overline{\overline{\mathsf{I}}} \quad (\overline{\overline{\mathsf{S}}}^{T} = \overline{\overline{\mathsf{S}}}) \tag{B.28}$$

$$(\overline{\overline{\mathsf{A}}} \times \mathbf{a}){}^{\times}_{\times}(\overline{\overline{\mathsf{B}}} \times \mathbf{a}) = (\overline{\overline{\mathsf{A}}}{}^{\times}_{\times}\overline{\overline{\mathsf{B}}}) \cdot \mathbf{aa} \tag{B.29}$$

$$(\mathbf{a} \times \overline{\overline{\mathsf{A}}}){}^{\times}_{\times}(\mathbf{a} \times \overline{\overline{\mathsf{B}}}) = \mathbf{aa} \cdot (\overline{\overline{\mathsf{A}}}{}^{\times}_{\times}\overline{\overline{\mathsf{B}}}) \tag{B.30}$$

$$\frac{1}{2}(\mathbf{a} \times \overline{\overline{\mathsf{I}}}){}^{\times}_{\times}(\mathbf{a} \times \overline{\overline{\mathsf{I}}}) = (\mathbf{a} \times \overline{\overline{\mathsf{I}}})^{(2)} = \mathbf{aa} \tag{B.31}$$

$$\overline{\overline{\mathsf{A}}} \cdot (\mathbf{a} \times \overline{\overline{\mathsf{A}}}^{T}) = (\overline{\overline{\mathsf{A}}}^{(2)} \cdot \mathbf{a}) \times \overline{\overline{\mathsf{I}}} \tag{B.32}$$

$$\overline{\overline{\mathsf{A}}}{}^{\times}_{\times}(\overline{\overline{\mathsf{B}}}{}^{\times}_{\times}\overline{\overline{\mathsf{C}}}) = (\overline{\overline{\mathsf{A}}} : \overline{\overline{\mathsf{C}}})\overline{\overline{\mathsf{B}}} + (\overline{\overline{\mathsf{A}}} : \overline{\overline{\mathsf{B}}})\overline{\overline{\mathsf{C}}} - \overline{\overline{\mathsf{B}}} \cdot \overline{\overline{\mathsf{A}}}^{T} \cdot \overline{\overline{\mathsf{C}}} - \overline{\overline{\mathsf{C}}} \cdot \overline{\overline{\mathsf{A}}}^{T} \cdot \overline{\overline{\mathsf{B}}} \tag{B.33}$$

$$(\overline{\overline{\mathsf{A}}}{}^{\times}_{\times}\overline{\overline{\mathsf{B}}}) : \overline{\overline{\mathsf{C}}} = (\overline{\overline{\mathsf{A}}}{}^{\times}_{\times}\overline{\overline{\mathsf{C}}}) : \overline{\overline{\mathsf{B}}} = \ldots \quad \text{(all permutations)} \tag{B.34}$$

$$\overline{\overline{\mathsf{I}}}{}^{\times}_{\times}(\overline{\overline{\mathsf{A}}}{}^{\times}_{\times}\overline{\overline{\mathsf{B}}}) = (\overline{\overline{\mathsf{A}}}:\overline{\overline{\mathsf{I}}})\overline{\overline{\mathsf{B}}} + (\overline{\overline{\mathsf{B}}}:\overline{\overline{\mathsf{I}}})\overline{\overline{\mathsf{A}}} - (\overline{\overline{\mathsf{A}}}\cdot\overline{\overline{\mathsf{B}}} + \overline{\overline{\mathsf{B}}}\cdot\overline{\overline{\mathsf{A}}}) \tag{B.35}$$

$$\overline{\overline{\mathsf{I}}}{}^{\times}_{\times}(\overline{\overline{\mathsf{I}}}{}^{\times}_{\times}\overline{\overline{\mathsf{A}}}) = \overline{\overline{\mathsf{A}}} + (\overline{\overline{\mathsf{A}}}:\overline{\overline{\mathsf{I}}})\overline{\overline{\mathsf{I}}} \tag{B.36}$$

$$\overline{\overline{\mathsf{I}}}{}^{\times}_{\times}(\overline{\overline{\mathsf{I}}}{}^{\times}_{\times}\overline{\overline{\mathsf{I}}}) = 4\overline{\overline{\mathsf{I}}} \tag{B.37}$$

$$(\overline{\overline{\mathsf{A}}}{}^{\times}_{\times}\overline{\overline{\mathsf{A}}}){}^{\times}_{\times}(\overline{\overline{\mathsf{A}}}{}^{\times}_{\times}\overline{\overline{\mathsf{A}}}) = 8(\overline{\overline{\mathsf{A}}}{}^{(2)})^{(2)} = 8\mathrm{det}\overline{\overline{\mathsf{A}}}\,\overline{\overline{\mathsf{A}}} \tag{B.38}$$

$$(\overline{\overline{\mathsf{A}}}{}^{(2)})^{(2)} = \overline{\overline{\mathsf{A}}}\mathrm{det}\overline{\overline{\mathsf{A}}} \tag{B.39}$$

$$\mathrm{det}(\overline{\overline{\mathsf{A}}}{}^{\times}_{\times}\overline{\overline{\mathsf{A}}}) = 8\mathrm{det}^2\overline{\overline{\mathsf{A}}} \tag{B.40}$$

$$\mathrm{det}(\overline{\overline{\mathsf{A}}}\cdot\overline{\overline{\mathsf{B}}}) = \mathrm{det}\overline{\overline{\mathsf{A}}}\ \mathrm{det}\overline{\overline{\mathsf{B}}} \tag{B.41}$$

$$\mathrm{det}(\mathbf{a}\times\overline{\overline{\mathsf{I}}}) = 0 \tag{B.42}$$

$$(\overline{\overline{\mathsf{A}}}{}^{\times}_{\times}\overline{\overline{\mathsf{B}}})\cdot(\overline{\overline{\mathsf{C}}}{}^{\times}_{\times}\overline{\overline{\mathsf{D}}}) = (\overline{\overline{\mathsf{A}}}\cdot\overline{\overline{\mathsf{C}}}){}^{\times}_{\times}(\overline{\overline{\mathsf{B}}}\cdot\overline{\overline{\mathsf{D}}}) + (\overline{\overline{\mathsf{A}}}\cdot\overline{\overline{\mathsf{D}}}){}^{\times}_{\times}(\overline{\overline{\mathsf{B}}}\cdot\overline{\overline{\mathsf{C}}}) \tag{B.43}$$

$$(\overline{\overline{\mathsf{A}}}{}^{\times}_{\times}\overline{\overline{\mathsf{A}}})\cdot(\overline{\overline{\mathsf{B}}}{}^{\times}_{\times}\overline{\overline{\mathsf{B}}}) = 2(\overline{\overline{\mathsf{A}}}\cdot\overline{\overline{\mathsf{B}}}){}^{\times}_{\times}(\overline{\overline{\mathsf{A}}}\cdot\overline{\overline{\mathsf{B}}}) \tag{B.44}$$

$$(\overline{\overline{\mathsf{A}}}{}^{\times}_{\times}\overline{\overline{\mathsf{B}}})^2 = (\overline{\overline{\mathsf{A}}}{}^2){}^{\times}_{\times}(\overline{\overline{\mathsf{B}}}{}^2) + (\overline{\overline{\mathsf{A}}}\cdot\overline{\overline{\mathsf{B}}}){}^{\times}_{\times}(\overline{\overline{\mathsf{A}}}\cdot\overline{\overline{\mathsf{B}}}) \tag{B.45}$$

$$(\overline{\overline{\mathsf{A}}}{}^{\times}_{\times}\overline{\overline{\mathsf{A}}})^2 = 2(\overline{\overline{\mathsf{A}}}{}^2){}^{\times}_{\times}(\overline{\overline{\mathsf{A}}}{}^2) \tag{B.46}$$

$$(\overline{\overline{\mathsf{A}}}{}^{\times}_{\times}\overline{\overline{\mathsf{I}}})^2 = (\overline{\overline{\mathsf{A}}}{}^2){}^{\times}_{\times}\overline{\overline{\mathsf{I}}} + \overline{\overline{\mathsf{A}}}{}^{\times}_{\times}\overline{\overline{\mathsf{A}}} \tag{B.47}$$

$$(\overline{\overline{\mathsf{A}}}{}^{\times}_{\times}\overline{\overline{\mathsf{A}}})^T\cdot\overline{\overline{\mathsf{A}}} = \overline{\overline{\mathsf{A}}}\cdot(\overline{\overline{\mathsf{A}}}{}^{\times}_{\times}\overline{\overline{\mathsf{A}}})^T = \frac{1}{3}(\overline{\overline{\mathsf{A}}}{}^{\times}_{\times}\overline{\overline{\mathsf{A}}}:\overline{\overline{\mathsf{A}}})\overline{\overline{\mathsf{I}}} \tag{B.48}$$

$$\overline{\overline{\mathsf{A}}}{}^{(2)T}\cdot\overline{\overline{\mathsf{A}}} = \overline{\overline{\mathsf{A}}}\cdot\overline{\overline{\mathsf{A}}}{}^{(2)T} = \mathrm{det}\overline{\overline{\mathsf{A}}}\,\overline{\overline{\mathsf{I}}} \tag{B.49}$$

$$\overline{\overline{\mathsf{A}}}{}^{-1} = \frac{\overline{\overline{\mathsf{A}}}{}^{(2)T}}{\mathrm{det}\overline{\overline{\mathsf{A}}}} \quad (\mathrm{det}\overline{\overline{\mathsf{A}}} \neq 0) \tag{B.50}$$

Two-Dimensional Dyadics

$$\mathbf{n}\cdot\overline{\overline{\mathsf{A}}}_t = \overline{\overline{\mathsf{A}}}_t\cdot\mathbf{n} = 0 \tag{B.51}$$

$$\overline{\overline{\mathsf{A}}}_t:\overline{\overline{\mathsf{B}}}_t = \overline{\overline{\mathsf{B}}}_t:\overline{\overline{\mathsf{A}}}_t = \mathrm{tr}(\overline{\overline{\mathsf{A}}}_t\cdot\overline{\overline{\mathsf{B}}}_t{}^T) \tag{B.52}$$

$$\overline{\overline{\mathsf{A}}}_t{}^{\times}_{\times}\overline{\overline{\mathsf{B}}}_t = \mathbf{nn}\ \mathrm{tr}(\overline{\overline{\mathsf{A}}}_t{}^{\times}_{\times}\overline{\overline{\mathsf{B}}}_t) \tag{B.53}$$

$$\mathrm{det}_t\overline{\overline{\mathsf{A}}}_t = \mathrm{tr}\overline{\overline{\mathsf{A}}}_t{}^{(2)} = \frac{1}{2}[(\mathrm{tr}\overline{\overline{\mathsf{A}}}_t)^2 - \mathrm{tr}\overline{\overline{\mathsf{A}}}_t{}^2] \tag{B.54}$$

$$\mathrm{det}_t(\overline{\overline{\mathsf{A}}}_t\cdot\overline{\overline{\mathsf{B}}}_t) = \mathrm{tr}(\overline{\overline{\mathsf{A}}}_t{}^{(2)}\cdot\overline{\overline{\mathsf{B}}}_t{}^{(2)}) = \overline{\overline{\mathsf{A}}}_t{}^{(2)}:\overline{\overline{\mathsf{B}}}_t{}^{(2)T} \tag{B.55}$$

$$\mathrm{det}(\overline{\overline{\mathsf{A}}}_t{}^{\times}_{\times}\overline{\overline{\mathsf{A}}}_t) = 8\mathrm{tr}(\overline{\overline{\mathsf{A}}}_t{}^{(2)}) = 8(\mathrm{det}\overline{\overline{\mathsf{A}}}_t)^2 \tag{B.56}$$

$$\mathrm{det}_t(\overline{\overline{\mathsf{A}}}_t\cdot\overline{\overline{\mathsf{B}}}_t) = \mathrm{det}_t\overline{\overline{\mathsf{A}}}_t\ \mathrm{det}_t\overline{\overline{\mathsf{B}}}_t \tag{B.57}$$

$$\mathrm{det}_t(\overline{\overline{\mathsf{A}}}_t\cdot\overline{\overline{\mathsf{B}}}_t + \alpha\overline{\overline{\mathsf{I}}}_t) = \mathrm{det}_t(\overline{\overline{\mathsf{B}}}_t\cdot\overline{\overline{\mathsf{A}}}_t + \alpha\overline{\overline{\mathsf{I}}}_t) \tag{B.58}$$

$$\mathbf{a}_t\times(\overline{\overline{\mathsf{A}}}_t{}^{\times}_{\times}\overline{\overline{\mathsf{B}}}_t) = \overline{\overline{\mathsf{B}}}_t\times(\mathbf{a}_t\cdot\overline{\overline{\mathsf{A}}}_t) + \overline{\overline{\mathsf{A}}}_t\times(\mathbf{a}_t\cdot\overline{\overline{\mathsf{B}}}_t) \tag{B.59}$$

$$(\overline{\overline{\mathsf{A}}}_t {}^\times_\times \overline{\overline{\mathsf{B}}}_t) \times \mathbf{a}_t = (\overline{\overline{\mathsf{A}}}_t \cdot \mathbf{a}_t) \times \overline{\overline{\mathsf{B}}}_t + (\overline{\overline{\mathsf{B}}}_t \cdot \mathbf{a}_t) \times \overline{\overline{\mathsf{A}}}_t \quad \text{(B.60)}$$

$$(\overline{\overline{\mathsf{A}}}_t \cdot \mathbf{a}_t) \times (\overline{\overline{\mathsf{A}}}_t \cdot \mathbf{b}_t) = \overline{\overline{\mathsf{A}}}_t{}^{(2)} \cdot (\mathbf{a}_t \times \mathbf{b}_t) = \mathbf{n}\ \mathrm{det}_t\overline{\overline{\mathsf{A}}}\ (\mathbf{n} \cdot \mathbf{a}_t \times \mathbf{b}_t) \quad \text{(B.61)}$$

$$(\mathbf{a}_t \cdot \overline{\overline{\mathsf{A}}}_t) \times (\mathbf{b}_t \cdot \overline{\overline{\mathsf{A}}}_t) = (\mathbf{a}_t \times \mathbf{b}_t) \cdot \overline{\overline{\mathsf{A}}}_t{}^{(2)} = \mathbf{n}\ \mathrm{det}_t\overline{\overline{\mathsf{A}}}\ (\mathbf{n} \cdot \mathbf{a}_t \times \mathbf{b}_t) \quad \text{(B.62)}$$

$$\overline{\overline{\mathsf{A}}}_t{}^{-1} = \frac{\overline{\overline{\mathsf{A}}}_t{}^T {}^\times_\times \mathbf{nn}}{\mathrm{det}_t\overline{\overline{\mathsf{A}}}_t} \quad \text{(B.63)}$$

$$\overline{\overline{\mathsf{A}}}_t{}^T \cdot \mathbf{n} \times \overline{\overline{\mathsf{A}}}_t = \mathrm{det}_t\overline{\overline{\mathsf{A}}}_t\ \mathbf{n} \times \overline{\overline{\mathsf{I}}} = \overline{\overline{\mathsf{A}}}_t \cdot \mathbf{n} \times \overline{\overline{\mathsf{A}}}_t^T \quad \text{(B.64)}$$

$$\mathbf{nn} {}^\times_\times \overline{\overline{\mathsf{A}}}_t = \mathrm{tr}\overline{\overline{\mathsf{A}}}_t\ \overline{\overline{\mathsf{I}}}_t - \overline{\overline{\mathsf{A}}}_t^T \quad \text{(B.65)}$$

Expansion of Unit Dyadic

The unit dyadic $\overline{\overline{\mathsf{I}}}$ can be expanded in terms of a given vector basis $\mathbf{a}_1, \mathbf{a}_2, \mathbf{a}_3$ satisfying $\Delta = \mathbf{a}_1 \cdot (\mathbf{a}_2 \times \mathbf{a}_3) \neq 0$ as

$$\overline{\overline{\mathsf{I}}} = \frac{1}{\Delta}(\mathbf{a}_1(\mathbf{a}_2 \times \mathbf{a}_3) + \mathbf{a}_2(\mathbf{a}_3 \times \mathbf{a}_1) + \mathbf{a}_3(\mathbf{a}_1 \times \mathbf{a}_2)) \quad \text{(B.66)}$$

$$= \frac{1}{\Delta}((\mathbf{a}_2 \times \mathbf{a}_3)\mathbf{a}_1 + (\mathbf{a}_3 \times \mathbf{a}_1)\mathbf{a}_2 + (\mathbf{a}_1 \times \mathbf{a}_2)\mathbf{a}_3). \quad \text{(B.67)}$$

For $\mathbf{a}_3 = \mathbf{n}$, $\mathbf{a}_1 = \mathbf{a}_{1t}$, $\mathbf{a}_2 = \mathbf{a}_{2t}$ and $\Delta = \mathbf{n} \cdot (\mathbf{a}_{1t} \times \mathbf{a}_{2t})$, the 2D unit dyadic has the expansion

$$\overline{\overline{\mathsf{I}}}_t = \frac{1}{\Delta}(\mathbf{a}_{1t}(\mathbf{a}_{2t} \times \mathbf{n}) + \mathbf{a}_{2t}(\mathbf{n} \times \mathbf{a}_{1t})) \quad \text{(B.68)}$$

$$= \frac{1}{\Delta}((\mathbf{a}_{2t} \times \mathbf{n})\mathbf{a}_{1t} + (\mathbf{n} \times \mathbf{a}_{1t})\mathbf{a}_{2t}). \quad \text{(B.69)}$$

Problems

B.1 Prove the identity

$$\mathbf{a} \times \overline{\overline{\mathsf{I}}} = \overline{\overline{\mathsf{I}}} \times \mathbf{a},$$

valid for any vector $\mathbf{a}$.

B.2 Starting from the identity

$$\overline{\overline{\mathsf{A}}}^{(2)} = \overline{\overline{\mathsf{A}}}^{2T} - \mathrm{tr}\overline{\overline{\mathsf{A}}}\ \overline{\overline{\mathsf{A}}}^T + \mathrm{tr}\overline{\overline{\mathsf{A}}}^{(2)}\ \overline{\overline{\mathsf{I}}}$$

derive the corresponding identity for $\overline{\overline{\mathsf{A}}} {}^\times_\times \overline{\overline{\mathsf{B}}}$.

B.3 Expanding the two-dimensional dyadic $\overline{\overline{\mathsf{A}}}_t$ in the form

$$\overline{\overline{\mathsf{A}}}_t = A_{11}\mathbf{e}_1\mathbf{e}_1 + A_{12}\mathbf{e}_1\mathbf{e}_2 + A_{21}\mathbf{e}_2\mathbf{e}_1 + A_{22}\mathbf{e}_2\mathbf{e}_2$$

in the orthonormal base $(\mathbf{e}_1, \mathbf{e}_2, \mathbf{e}_3 = \mathbf{n})$, prove the rule

$$\overline{\overline{\mathsf{A}}}_t {}^\times_\times \mathbf{nn} + \overline{\overline{\mathsf{A}}}_t^T = (\mathrm{tr}\overline{\overline{\mathsf{A}}}_t)\overline{\overline{\mathsf{I}}}_t.$$

B.4 Prove the identity

$$\overline{\overline{\mathsf{A}}}\cdot(\mathbf{a}\times\overline{\overline{\mathsf{A}}}^T) = (\overline{\overline{\mathsf{A}}}^{(2)}\cdot\mathbf{a})\times\overline{\overline{\mathsf{I}}}$$

B.5 Prove the identity

$$\overline{\overline{\mathsf{A}}}^{(2)} = \overline{\overline{\mathsf{A}}}^{2T} - \mathrm{tr}\overline{\overline{\mathsf{A}}}\,\overline{\overline{\mathsf{A}}}^T + \mathrm{spm}\overline{\overline{\mathsf{A}}}\,\overline{\overline{\mathsf{I}}}$$

and apply it to derive the Cayley-Hamilton equation,

$$\overline{\overline{\mathsf{A}}}^3 - \mathrm{tr}\overline{\overline{\mathsf{A}}}\,\overline{\overline{\mathsf{A}}}^2 + \mathrm{spm}\overline{\overline{\mathsf{A}}}\,\overline{\overline{\mathsf{A}}} - \det\overline{\overline{\mathsf{A}}}\,\overline{\overline{\mathsf{I}}} = 0,$$

valid for any dyadic $\overline{\overline{\mathsf{A}}}$.

B.6 Applying the identity of the previous problem, derive the Cayley-Hamilton equation

$$\overline{\overline{\mathsf{A}}}_t^2 - \mathrm{tr}\overline{\overline{\mathsf{A}}}_t\,\overline{\overline{\mathsf{A}}}_t + \mathrm{spm}\overline{\overline{\mathsf{A}}}_t\,\overline{\overline{\mathsf{I}}}_t = 0,$$

valid for any 2D dyadic $\overline{\overline{\mathsf{A}}}_t$.

B.7 Find the inverse of the dyadic $\overline{\overline{\mathsf{A}}} = \alpha\overline{\overline{\mathsf{I}}} + \mathbf{a}\times\overline{\overline{\mathsf{I}}}$ by solving $\mathbf{x}$ from the equation $\overline{\overline{\mathsf{A}}}\cdot\mathbf{x} = \mathbf{y}$. Verify the result by comparing with $\overline{\overline{\mathsf{A}}}^{-1} = \overline{\overline{\mathsf{A}}}^{(2)T}/\det\overline{\overline{\mathsf{A}}}$.

Appendix C

Four-Dimensional Formalism

Equations involving electromagnetic fields and sources can be expressed in elegant and concise form by applying four-dimensional formalism in the coordinate-independent form originally suggested for electromagnetics by G. A. Deschamps [10]. In this chapter, a brief introduction to the formalism is given together with its application to boundary conditions. For a more complete discussion, one may consult [44, 72].

Notation

The formalism is based on two linear spaces: $\mathbb{E}_1$ containing vectors $\mathbf{a}, \mathbf{b} \cdots$ and $\mathbb{F}_1$ containing one-forms (dual vectors) $\boldsymbol{\alpha}, \boldsymbol{\beta} \cdots$. The electric and magnetic fields $\mathbf{E}, \mathbf{H}$ are examples of one-forms. Using the anticommutative exterior product $\wedge$ other linear spaces are formed: $\mathbb{E}_2$ containing bivectors $\sum \mathbf{a}_i \wedge \mathbf{b}_i$ and $\mathbb{F}_2$ containing two-forms (dual bivectors) $\sum \boldsymbol{\alpha}_i \wedge \boldsymbol{\beta}_i$. More complicated products like $\mathbf{a}\wedge\mathbf{b}\wedge\mathbf{c}$ and $\boldsymbol{\alpha}\wedge\boldsymbol{\beta}\wedge\boldsymbol{\gamma}$ will give rise to trivectors and quadrivectors or three-forms and four-forms. Because we consider spaces of vectors of dimension 4, there are no five-forms or quintivectors.

Dyadic algebra introduced by J.W. Gibbs as a coordinate-free representation of linear mappings in the three-dimensional vector space [13, 35] can be generalized to four dimensions as shown in [44, 72]. Dyadics mapping vectors to vectors form a linear space denoted by $\mathbb{E}_1\mathbb{F}_1$. Other spaces are denoted similarly, e.g., those mapping two-forms to two-forms belong to the space $\mathbb{F}_2\mathbb{E}_2$.

Products

Different products of multivectors and multiforms (dual multivectors) are listed below.

- The wedge product of vectors is associative, $(\mathbf{a} \wedge \mathbf{b}) \wedge \mathbf{c} = \mathbf{a} \wedge (\mathbf{b} \wedge \mathbf{c})$, and anticommutative, $\mathbf{a} \wedge \mathbf{b} = -\mathbf{b} \wedge \mathbf{a}$. For a vector $\mathbf{a}$ and a bivector

$\mathbf{B} = \mathbf{b}_1 \wedge \mathbf{b}_2$ the rule is commutative,

$$\mathbf{a} \wedge \mathbf{B} = \mathbf{a} \wedge \mathbf{b}_1 \wedge \mathbf{b}_2 = -\mathbf{b}_1 \wedge \mathbf{a} \wedge \mathbf{b}_2 = \mathbf{b}_1 \wedge \mathbf{b}_2 \wedge \mathbf{a} = \mathbf{B} \wedge \mathbf{a}, \tag{C.1}$$

and similarly for the wedge product of a one-form and a two-form. More general rules can be found by following the same pattern.

Bivectors form a space of 6 dimensions and trivectors a space of 4 dimensions while quadrivectors are multiples of any other quadrivector whence they are similar to scalars.

- The bar product between a vector $\mathbf{a}$ and a one-form $\boldsymbol{\alpha}$ is denoted by $\mathbf{a}|\boldsymbol{\alpha} = \boldsymbol{\alpha}|\mathbf{a}$ and the result is a scalar. For a given basis of vectors $\{\mathbf{e}_i\}$ one can define the reciprocal basis of one-forms $\{\boldsymbol{\varepsilon}_j\}$ satisfying

$$\mathbf{e}_i|\boldsymbol{\varepsilon}_j = \boldsymbol{\varepsilon}_j|\mathbf{e}_i = \delta_{ij}. \tag{C.2}$$

Expanding $\mathbf{a} = \sum a_i \mathbf{e}_i$, $\boldsymbol{\alpha} = \sum \alpha_j \boldsymbol{\varepsilon}_j$, we have

$$\mathbf{a}|\boldsymbol{\alpha} = \sum_1^4 a_i \alpha_i. \tag{C.3}$$

Here we agree that $\mathbf{e}_1, \mathbf{e}_2, \mathbf{e}_3$ denote spatial basis vectors and $\mathbf{e}_4$ is the temporal basis vector. Similarly, $\boldsymbol{\varepsilon}_1, \boldsymbol{\varepsilon}_2, \boldsymbol{\varepsilon}_3$ correspond to spatial and $\boldsymbol{\varepsilon}_4$ to temporal one-forms of the reciprocal basis.

- A given basis of vectors and the reciprocal basis of one-forms generate the set of basis bivectors and two-forms as $\mathbf{e}_{ij} = \mathbf{e}_i \wedge \mathbf{e}_j$, $\boldsymbol{\varepsilon}_{ij} = \boldsymbol{\varepsilon}_i \wedge \boldsymbol{\varepsilon}_j$. This can be continued to trivectors $\mathbf{e}_{ijk}$ and three-forms $\boldsymbol{\varepsilon}_{123}$. There is only one basis quadrivector $\mathbf{e}_{1234}$ and basis four-form $\boldsymbol{\varepsilon}_{1234}$.

- The contraction product $\rfloor$ between a basis vector $\mathbf{a}$ and two-form $\boldsymbol{\beta} \wedge \boldsymbol{\gamma}$ is defined by the rule

$$\mathbf{a}\rfloor(\boldsymbol{\beta} \wedge \boldsymbol{\gamma}) = \boldsymbol{\beta}(\mathbf{a}|\boldsymbol{\gamma}) - \boldsymbol{\gamma}(\mathbf{a}|\boldsymbol{\beta}), \tag{C.4}$$

and the result is a one-form. (C.4) can be identified as the 4D version of the Gibbsian bac-cb rule for 3D vectors. The same rule can be written in another form as

$$(\boldsymbol{\beta} \wedge \boldsymbol{\gamma})\lfloor\mathbf{a} = \boldsymbol{\gamma}(\mathbf{a}|\boldsymbol{\beta}) - \boldsymbol{\beta}(\mathbf{a}|\boldsymbol{\gamma}). \tag{C.5}$$

Comparing to (C.4) leads to the more general rule

$$\mathbf{a}\rfloor\boldsymbol{\Phi} = -\boldsymbol{\Phi}\lfloor\mathbf{a}, \tag{C.6}$$

satisfied by any vector $\mathbf{a}$ and two-form $\boldsymbol{\Phi}$.

Dyadics

- The dyadic product of a vector $\mathbf{a} \in \mathbb{E}_1$ and a dual vector $\boldsymbol{\alpha} \in \mathbb{F}_1$ is presented by the classical Gibbsian 'no sign' notation as $\mathbf{a}\boldsymbol{\alpha}$ and the result is in the space of dyadics denoted by $\mathbb{E}_1\mathbb{F}_1$. It defines a mapping of a vector $\mathbf{b}$ to a multiple of $\mathbf{a}$ as

$$(\mathbf{a}\boldsymbol{\alpha})|\mathbf{b} = \mathbf{a}(\boldsymbol{\alpha}|\mathbf{b}), \tag{C.7}$$

as a simple associative rule. More generally, any linear mapping from vector to another vector can be expressed in terms of a dyadic $\overline{\overline{\mathsf{A}}} \in \mathbb{E}_1\mathbb{F}_1$

$$\overline{\overline{\mathsf{A}}} = \sum \mathbf{a}_i \boldsymbol{\alpha}_i. \tag{C.8}$$

Correspondingly, $\overline{\overline{\mathsf{A}}}^T$, the transpose of $\overline{\overline{\mathsf{A}}}$ is in the space $\mathbb{F}_1\mathbb{E}_1$

$$\overline{\overline{\mathsf{A}}}^T = \sum \boldsymbol{\alpha}_i \mathbf{a}_i \tag{C.9}$$

and it maps one-forms to one-forms. In the same way we can define dyadic spaces $\mathbb{E}_1\mathbb{E}_1, \mathbb{F}_1\mathbb{F}_1$ and, more generally, spaces like $\mathbb{E}_p\mathbb{F}_p$, $\mathbb{F}_p\mathbb{E}_p$,, which define mappings between multivectors and/or multiforms of the same dimension.

- The bar product $\overline{\overline{\mathsf{A}}}_1|\overline{\overline{\mathsf{A}}}_2$ between two dyadics $\overline{\overline{\mathsf{A}}}_1, \overline{\overline{\mathsf{A}}}_2 \in \mathbb{E}_p\mathbb{F}_p$ or $\in \mathbb{F}_p\mathbb{E}_p$ gives a dyadic in the same space. In analogy to Gibbsian double-dot product [13, 35], double-bar product can be defined as $(\mathbf{a}\boldsymbol{\alpha})||(\boldsymbol{\beta}\mathbf{b}) = (\mathbf{a}|\boldsymbol{\beta})(\boldsymbol{\alpha}|\mathbf{b})$ and, more generally, for dyadics $\overline{\overline{\mathsf{A}}}_i \in \mathbb{E}_1\mathbb{F}_1$ as $\overline{\overline{\mathsf{A}}}_1||\overline{\overline{\mathsf{A}}}_2^T$, whose result is a scalar. The double-bar product $\overline{\overline{\mathsf{A}}}||\overline{\overline{\mathsf{B}}}$ can also be defined for dyadics $\overline{\overline{\mathsf{A}}} \in \mathbb{E}_p\mathbb{E}_p$, $\overline{\overline{\mathsf{B}}} \in \mathbb{F}_p\mathbb{F}_p$. Similarly, the incomplete double duality products $\overline{\overline{\mathsf{A}}}\lfloor\lfloor\overline{\overline{\mathsf{B}}}$ and $\overline{\overline{\mathsf{B}}}\rfloor\rfloor\overline{\overline{\mathsf{A}}}$ can be defined between dyadics in certain spaces. For example, for $\overline{\overline{\mathsf{A}}} \in \mathbb{E}_p\mathbb{E}_p$ and $\overline{\overline{\mathsf{B}}} \in \mathbb{F}_q\mathbb{F}_q$ with $p > q$, the resulting dyadic lies in the space $\mathbb{E}_{p-q}\mathbb{E}_{p-q}$.

- The unit dyadic $\overline{\overline{\mathsf{I}}} \in \mathbb{E}_1\mathbb{F}_1$ maps any vector to itself, $\overline{\overline{\mathsf{I}}}|\mathbf{a} = \mathbf{a}$, and its transpose maps any one-form to itself, $\overline{\overline{\mathsf{I}}}^T|\boldsymbol{\alpha} = \boldsymbol{\alpha}$. Its expansion

$$\overline{\overline{\mathsf{I}}} = \sum \mathbf{e}_i \boldsymbol{\varepsilon}_i \tag{C.10}$$

is independent of the chosen vector and the reciprocal one-form bases.

- A symmetric dyadic $\overline{\overline{\mathsf{G}}} = \sum \mathbf{a}_i\mathbf{a}_i \in \mathbb{E}_1\mathbb{E}_1$ defines a dot product between two one-forms: $\boldsymbol{\alpha} \cdot \boldsymbol{\beta} = \boldsymbol{\beta} \cdot \boldsymbol{\alpha} = \boldsymbol{\alpha}|\overline{\overline{\mathsf{G}}}|\boldsymbol{\beta}$.

- In spaces $\mathbb{E}_p\mathbb{E}_p$ and $\mathbb{F}_p\mathbb{F}_p$ the transpose of any dyadic is in the same space. In these cases symmetric dyadics can be defined to satisfy $\overline{\overline{\mathsf{A}}}^T = \overline{\overline{\mathsf{A}}}$ and antisymmetric dyadics $\overline{\overline{\mathsf{A}}}^T = -\overline{\overline{\mathsf{A}}}$. The most general antisymmetric dyadic $\overline{\overline{\mathsf{A}}} \in \mathbb{E}_1\mathbb{E}_1$ can be expressed in terms of a bivector $\mathbf{A}$ as

$$\overline{\overline{\mathsf{A}}} = \overline{\mathsf{I}}\rfloor\mathbf{A}. \tag{C.11}$$

In fact, applying the bac-cab rule (C.4), for each term in $\mathbf{A} = \sum \mathbf{a}_j \wedge \mathbf{b}_j$, we can write the antisymmetric expression

$$\overline{\mathsf{I}}\rfloor(\mathbf{a}_j \wedge \mathbf{b}_j) = \sum \mathbf{e}_i\boldsymbol{\varepsilon}_i\rfloor(\mathbf{a}_j \wedge \mathbf{b}_j) = \sum \mathbf{e}_i\boldsymbol{\varepsilon}_i|(\mathbf{b}_j\mathbf{a}_j - \mathbf{a}_j\mathbf{b}_j) = \mathbf{b}_j\mathbf{a}_j - \mathbf{a}_j\mathbf{b}_j, \tag{C.12}$$

the right side of which can be interpreted as belonging to an expansion of an antisymmetric dyadic.

Products of dyadics

- Denoting the three-dimensional (Euclidean) spatial basis vectors by $\mathbf{e}_1, \mathbf{e}_2, \mathbf{e}_3$ and by $\mathbf{e}_4 = \mathbf{e}_\tau$ the temporal basis vector, the reciprocal basis one-forms $\boldsymbol{\varepsilon}_1, \boldsymbol{\varepsilon}_2, \boldsymbol{\varepsilon}_3, \boldsymbol{\varepsilon}_4 = \mathbf{d}\tau$ satisfy $\mathbf{e}_i|\boldsymbol{\varepsilon}_j = \delta_{ij}$.

- The sign $|$ denotes the scalar product between a p-vector and a p-form while $\rfloor$ or $\lfloor$ denotes the contraction between a p-vector and a q-form [10]. Double products $\,^\wedge_\wedge, ||, \rfloor\rfloor, \lfloor\lfloor$ follow laws similar to those defined in the Gibbsian dyadic algebra.

- The double-wedge product of two dyadics $\overline{\overline{\mathsf{A}}}, \overline{\overline{\mathsf{B}}} \in \mathbb{E}_1\mathbb{F}_1$ is the counterpart of the double-cross product between two Gibbsian dyadics [13, 35] and it is defined as

$$(\mathbf{a}\boldsymbol{\alpha})\,^\wedge_\wedge(\mathbf{b}\boldsymbol{\beta}) = (\mathbf{a} \wedge \mathbf{b})(\boldsymbol{\alpha} \wedge \boldsymbol{\beta}), \tag{C.13}$$

and similarly for sums of such products:

$$\overline{\overline{\mathsf{A}}}\,^\wedge_\wedge\overline{\overline{\mathsf{B}}} = \sum \mathbf{a}_i\boldsymbol{\alpha}_i\,^\wedge_\wedge \sum \mathbf{b}_j\boldsymbol{\beta}_j = \sum(\mathbf{a}_i \wedge \mathbf{b}_j)(\boldsymbol{\alpha}_i \wedge \boldsymbol{\beta}_j). \tag{C.14}$$

The result is in the dyadic space $\mathbb{E}_2\mathbb{F}_2$. Similarly we can define products $\overline{\overline{\mathsf{A}}}\,^\wedge_\wedge\overline{\overline{\mathsf{B}}}\,^\wedge_\wedge\overline{\overline{\mathsf{C}}}$ etc.

- The double-wedge square of a dyadic $\overline{\overline{\mathsf{A}}} = \sum \mathbf{a}_i\boldsymbol{\alpha}_i \in \mathbb{E}_1\mathbb{F}_1$ is

$$\begin{aligned} \overline{\overline{\mathsf{A}}}^{(2)} = \frac{1}{2}\overline{\overline{\mathsf{A}}}\,^\wedge_\wedge\overline{\overline{\mathsf{A}}} &= \mathbf{a}_{12}\boldsymbol{\alpha}_{12} + \mathbf{a}_{23}\boldsymbol{\alpha}_{23} + \mathbf{a}_{31}\boldsymbol{\alpha}_{31} \\ &\quad + \mathbf{a}_{14}\boldsymbol{\alpha}_{14} + \mathbf{a}_{24}\boldsymbol{\alpha}_{24} + \mathbf{a}_{34}\boldsymbol{\alpha}_{34}. \end{aligned} \tag{C.15}$$

Similarly we can define $\overline{\overline{\mathsf{A}}}^{(3)}$ and $\overline{\overline{\mathsf{A}}}^{(4)}$ as

$$\overline{\overline{\mathsf{A}}}^{(3)} = \frac{1}{3!}\overline{\overline{\mathsf{A}}}\,^\wedge_\wedge\overline{\overline{\mathsf{A}}}\,^\wedge_\wedge\overline{\overline{\mathsf{A}}} = \mathbf{a}_{123}\boldsymbol{\alpha}_{123} + \mathbf{a}_{234}\boldsymbol{\alpha}_{234} + \cdots, \tag{C.16}$$

$$\overline{\overline{\mathsf{A}}}^{(4)} = \frac{1}{4!}\overline{\overline{\mathsf{A}}}\,^\wedge_\wedge\overline{\overline{\mathsf{A}}}\,^\wedge_\wedge\overline{\overline{\mathsf{A}}}\,^\wedge_\wedge\overline{\overline{\mathsf{A}}} = \mathbf{a}_{1234}\boldsymbol{\alpha}_{1234} = \mathbf{a}_N\boldsymbol{\alpha}_N. \tag{C.17}$$

- The unit dyadic mapping any bivector to itself is

$$\overline{\overline{\mathsf{I}}}^{(2)} = \frac{1}{2}\overline{\overline{\mathsf{I}}}{}^{\wedge}_{\wedge}\overline{\overline{\mathsf{I}}} = \mathbf{e}_1\boldsymbol{\varepsilon}_1 + \mathbf{e}_2\boldsymbol{\varepsilon}_2 + \mathbf{e}_3\boldsymbol{\varepsilon}_3 + \mathbf{e}_4\boldsymbol{\varepsilon}_4. \tag{C.18}$$

The unit dyadic mapping a two-form to itself is the transpose $\overline{\overline{\mathsf{I}}}^{(2)T}$. For any $\overline{\overline{\mathsf{A}}} \in \mathbb{E}_1\mathbb{F}_1$ we can write $\overline{\overline{\mathsf{A}}}^{(4)} = \mathrm{det}\overline{\overline{\mathsf{A}}}\,\overline{\overline{\mathsf{I}}}^{(4)} = \mathrm{det}\overline{\overline{\mathsf{A}}}\,\mathbf{e}_N\boldsymbol{\varepsilon}_N$ where the scalar $\mathrm{det}\overline{\overline{\mathsf{A}}}$ is called the determinant of $\overline{\overline{\mathsf{A}}}$. It satisfies

$$\mathrm{det}\overline{\overline{\mathsf{A}}} = \mathrm{tr}\overline{\overline{\mathsf{A}}}^{(4)} = \overline{\overline{\mathsf{A}}}^{(4)}||\overline{\overline{\mathsf{I}}}^{(4)T} = \mathbf{a}_N|\boldsymbol{\alpha}_N. \tag{C.19}$$

- If vectors are mapped through the dyadic $\overline{\overline{\mathsf{A}}}$, bivectors are mapped through $\overline{\overline{\mathsf{A}}}^{(2)}$ as

$$(\overline{\overline{\mathsf{A}}}|\mathbf{a}_1)\wedge(\overline{\overline{\mathsf{A}}}|\mathbf{a}_2) = \overline{\overline{\mathsf{A}}}^{(2)}|(\mathbf{a}_1\wedge\mathbf{a}_2), \tag{C.20}$$

and, trivectors as

$$(\overline{\overline{\mathsf{A}}}|\mathbf{a}_1)\wedge(\overline{\overline{\mathsf{A}}}|\mathbf{a}_2)\wedge(\overline{\overline{\mathsf{A}}}|\mathbf{a}_3) = \overline{\overline{\mathsf{A}}}^{(3)}|(\mathbf{a}_1\wedge\mathbf{a}_2\wedge\mathbf{a}_3). \tag{C.21}$$

- The inverse of a dyadic can be formed much in the same manner as in the Gibbsian three-dimensional case [13, 35]. For example, the inverse of a dyadic $\overline{\overline{\mathsf{A}}} \in \mathbb{E}_1\mathbb{F}_1$ has the form [44]

$$\overline{\overline{\mathsf{A}}}^{-1} = \frac{\overline{\overline{\mathsf{I}}}^{(4)}\lfloor\lfloor\overline{\overline{\mathsf{A}}}^{(3)T}}{\mathrm{det}\overline{\overline{\mathsf{A}}}} \in \mathbb{E}_1\mathbb{F}_1, \tag{C.22}$$

and it satisfies

$$\overline{\overline{\mathsf{A}}}^{-1}|\overline{\overline{\mathsf{A}}} = \overline{\overline{\mathsf{A}}}|\overline{\overline{\mathsf{A}}}^{-1} = \overline{\overline{\mathsf{I}}}. \tag{C.23}$$

Identities

An identity is an equation which is valid for any values of its arguments. Certain number of identities is essential in any analysis using differential forms because they reduce the need to expand expressions in terms of basis vectors and their basic relations. Identities can be derived by expanding multivectors, multiforms and dyadics in terms of basis vectors and basis one-forms. Here we just give a few examples taken from [44].

As an identity involving dyadics let us take the following one,

$$(\overline{\overline{\mathsf{A}}}{}^{\wedge}_{\wedge}\overline{\overline{\mathsf{B}}})\lfloor\lfloor\overline{\overline{\mathsf{C}}}^T = (\overline{\overline{\mathsf{A}}}||\overline{\overline{\mathsf{C}}}^T)\overline{\overline{\mathsf{B}}} + (\overline{\overline{\mathsf{B}}}||\overline{\overline{\mathsf{C}}}^T)\overline{\overline{\mathsf{A}}} - \overline{\overline{\mathsf{A}}}|\overline{\overline{\mathsf{C}}}|\overline{\overline{\mathsf{B}}} - \overline{\overline{\mathsf{B}}}|\overline{\overline{\mathsf{C}}}|\overline{\overline{\mathsf{A}}}, \tag{C.24}$$

valid for three dyadics $\overline{\overline{\mathsf{A}}}, \overline{\overline{\mathsf{B}}}, \overline{\overline{\mathsf{C}}} \in \mathbb{F}_1\mathbb{E}_1$. The result is another dyadic in the same space $\mathbb{F}_1\mathbb{E}_1$. This identity is the counterpart of one for Gibbsian dyadics [35]

$$(\overline{\overline{\mathsf{A}}}{}^{\times}_{\times}\overline{\overline{\mathsf{B}}}){}^{\times}_{\times}\overline{\overline{\mathsf{C}}} = (\overline{\overline{\mathsf{A}}}:\overline{\overline{\mathsf{C}}})\overline{\overline{\mathsf{B}}} + (\overline{\overline{\mathsf{B}}}:\overline{\overline{\mathsf{C}}})\overline{\overline{\mathsf{A}}} - \overline{\overline{\mathsf{A}}}\cdot\overline{\overline{\mathsf{C}}}^T\cdot\overline{\overline{\mathsf{B}}} - \overline{\overline{\mathsf{B}}}\cdot\overline{\overline{\mathsf{C}}}^T\cdot\overline{\overline{\mathsf{A}}}. \tag{C.25}$$

A useful special case of (C.24) is obtained as the special case $\overline{\overline{\mathsf{B}}} = \overline{\overline{\mathsf{A}}}$:

$$\overline{\overline{\mathsf{A}}}^{(2)}\lfloor\lfloor\overline{\overline{\mathsf{C}}}^T = (\overline{\overline{\mathsf{A}}} : \overline{\overline{\mathsf{C}}})\overline{\overline{\mathsf{A}}} - \overline{\overline{\mathsf{A}}}|\overline{\overline{\mathsf{C}}}|\overline{\overline{\mathsf{A}}}. \quad \text{(C.26)}$$

The same identity is also valid for dyadics in the metric spaces $\overline{\overline{\mathsf{A}}} \in \mathbb{E}_1\mathbb{E}_1$ and $\overline{\overline{\mathsf{C}}} \in \mathbb{F}_1\mathbb{F}_1$or $\overline{\overline{\mathsf{A}}} \in \mathbb{F}_1\mathbb{F}_1$ and $\overline{\overline{\mathsf{C}}} \in \mathbb{E}_1\mathbb{E}_1$.

Electromagnetic Fields

The basic electromagnetic fields are represented by two quantities, the electromagnetic two-forms $\mathbf{\Phi} \in \mathbb{F}_2$ and $\mathbf{\Psi} \in \mathbb{F}_2$, each of which involve six scalar components. They can be decomposed as [1]

$$\mathbf{\Phi} = c\mathbf{B} + \mathbf{E} \wedge \boldsymbol{\varepsilon}_4, \quad \text{(C.27)}$$

$$\mathbf{\Psi} = c\mathbf{D} - \mathbf{H} \wedge \boldsymbol{\varepsilon}_4, \quad \text{(C.28)}$$

in terms of two spatial two-forms **B** and **D** and two spatial one-forms **E** and **H**. The one-form fields can be expanded as

$$\mathbf{E} = E_1\boldsymbol{\varepsilon}_1 + E_2\boldsymbol{\varepsilon}_2 + E_3\boldsymbol{\varepsilon}_3, \quad \text{(C.29)}$$

$$\mathbf{H} = H_1\boldsymbol{\varepsilon}_1 + H_2\boldsymbol{\varepsilon}_2 + H_3\boldsymbol{\varepsilon}_3, \quad \text{(C.30)}$$

and the two-form fields as

$$\mathbf{B} = B_{12}\boldsymbol{\varepsilon}_{12} + B_{23}\boldsymbol{\varepsilon}_{23} + B_{31}\boldsymbol{\varepsilon}_{31}, \quad \text{(C.31)}$$

$$\mathbf{D} = D_{12}\boldsymbol{\varepsilon}_{12} + D_{23}\boldsymbol{\varepsilon}_{23} + D_{31}\boldsymbol{\varepsilon}_{31}. \quad \text{(C.32)}$$

The one-form differential operator **d** has the expansion

$$\mathbf{d} = \boldsymbol{\varepsilon}_1\partial_{x_1} + \boldsymbol{\varepsilon}_2\partial_{x_2} + \boldsymbol{\varepsilon}_3\partial_{x_3} + \boldsymbol{\varepsilon}_4\partial_{x_4}, \quad \text{(C.33)}$$

where the x_i are components of the position vector

$$\mathbf{x} = \mathbf{e}_1x_1 + \mathbf{e}_2x_2 + \mathbf{e}_3x_3 + \mathbf{e}_4x_4, \quad \text{(C.34)}$$

of which $x_4 = ct$ is the temporal component.

In this formalism, the Maxwell equations take the simple form

$$\mathbf{d} \wedge \mathbf{\Phi} = \boldsymbol{\gamma}_m \quad \text{(C.35)}$$

$$\mathbf{d} \wedge \mathbf{\Psi} = \boldsymbol{\gamma}_e, \quad \text{(C.36)}$$

where $\boldsymbol{\gamma}_m$ and $\boldsymbol{\gamma}_e$ represent magnetic and electric charge three-forms. They can expressed in terms of spatial magnetic and electric charge three-forms, and electric and magnetic current two-forms, as

$$\boldsymbol{\gamma}_m = c\boldsymbol{\rho}_m - \mathbf{J}_m \wedge \boldsymbol{\varepsilon}_4, \quad \text{(C.37)}$$

$$\boldsymbol{\gamma}_e = c\boldsymbol{\rho}_e - \mathbf{J}_e \wedge \boldsymbol{\varepsilon}_4. \quad \text{(C.38)}$$

[1]Notation applied here differs from that of [10, 44] and [72] by the factor $c = 1/\sqrt{\mu_o\epsilon_o}$.

When expanded in their spatial and temporal parts, the Maxwell equations (C.35) and (C.36) take the form

$$\mathbf{d}_s \wedge \mathbf{B} = \rho_m, \tag{C.39}$$
$$\mathbf{d}_s \wedge \mathbf{E} + \partial_\tau \mathbf{B} = -\mathbf{J}_m, \tag{C.40}$$

and

$$\mathbf{d}_s \wedge \mathbf{D} = \rho_e, \tag{C.41}$$
$$-\mathbf{d}_s \wedge \mathbf{H} + \partial_\tau \mathbf{D} = -\mathbf{J}_e, \tag{C.42}$$

Boundary Conditions

Defining a boundary surface in terms of a spatial one-form ε_3 as $\varepsilon_3|\mathbf{x} = x_3 = 0$, the GBC boundary conditions can be expressed as two linear relations between the electromagnetic two-forms as

$$\varepsilon_3 \wedge (\boldsymbol{\alpha}_1 \wedge \boldsymbol{\Phi} + \boldsymbol{\beta}_1 \wedge \eta_o \boldsymbol{\Psi}) = 0, \tag{C.43}$$
$$\varepsilon_3 \wedge (\boldsymbol{\alpha}_2 \wedge \boldsymbol{\Phi} + \boldsymbol{\beta}_2 \wedge \eta_o \boldsymbol{\Psi}) = 0, \tag{C.44}$$

where $\boldsymbol{\alpha}_1 - \boldsymbol{\beta}_2$ are four dimensionless one-forms defining the nature of the boundary. The conditions (C.43) and (C.44) are four-forms which are equivalent to two scalar conditions because any four-form can be expressed as a scalar multiple of the basis four-form $\varepsilon_N = \varepsilon_{1234}$.

For a one-form defined by its spatial and temporal components as

$$\boldsymbol{\alpha} = \boldsymbol{\alpha}_s + \alpha_4 \varepsilon_4, \tag{C.45}$$

we can expand

$$\varepsilon_3 \wedge \boldsymbol{\alpha} \wedge \boldsymbol{\Phi} = \varepsilon_3 \wedge (\boldsymbol{\alpha}_s + \alpha_4 \varepsilon_4) \wedge (c\mathbf{B} + \mathbf{E} \wedge \varepsilon_4) \tag{C.46}$$
$$= \varepsilon_3 \wedge \boldsymbol{\alpha}_s \wedge c\mathbf{B} + \varepsilon_3 \wedge (\alpha_4 c\mathbf{B} + \boldsymbol{\alpha}_s \wedge \mathbf{E}) \wedge \varepsilon_4 \tag{C.47}$$
$$= \varepsilon_3 \wedge (\alpha_4 c\mathbf{B} + \boldsymbol{\alpha}_s \wedge \mathbf{E}) \wedge \varepsilon_4. \tag{C.48}$$

The term $\varepsilon_3 \wedge \boldsymbol{\alpha}_s \wedge \mathbf{B}$ is a spatial four-form, which vanishes identically. Applying similar expansions to (C.43) and (C.44), they are reduced to the spatial three-form conditions

$$\varepsilon_3 \wedge (\alpha_{14} c\mathbf{B} + \boldsymbol{\alpha}_{1s} \wedge \mathbf{E} + \beta_{14} c\eta_o \mathbf{D} - \boldsymbol{\beta}_{1s} \wedge \eta_o \mathbf{H}) = 0, \tag{C.49}$$
$$\varepsilon_3 \wedge (\alpha_{24} c\mathbf{B} + \boldsymbol{\alpha}_{2s} \wedge \mathbf{E} + \beta_{24} c\eta_o \mathbf{D} - \boldsymbol{\beta}_{2s} \wedge \eta_o \mathbf{H}) = 0, \tag{C.50}$$

where $\alpha_{14} - \beta_{24}$ are four scalars and $\boldsymbol{\alpha}_{1s} - \boldsymbol{\beta}_{2s}$ are four dimensionless spatial one-forms. (C.49) and (C.50) correspond to the GBC conditions (5.6) and (5.7) in the three-dimensional vector representation of electromagnetic fields.

Special Cases

Different special cases of the GBC boundary conditions (C.49) and (C.50) can be defined as follows:

- DB Boundary, $\varepsilon_3 \wedge \mathbf{D} = 0, \quad \varepsilon_3 \wedge \mathbf{B} = 0,$

$$\boldsymbol{\alpha}_{1s} = \beta_{14} = \boldsymbol{\beta}_{1s} = \alpha_{24} = \boldsymbol{\alpha}_{2s} = \boldsymbol{\beta}_{2s} = 0, \tag{C.51}$$

- SH Boundary, $\varepsilon_3 \wedge \boldsymbol{\alpha}_s \wedge \mathbf{E} = 0, \quad \varepsilon_3 \wedge \boldsymbol{\alpha}_s \wedge \mathbf{H} = 0,$

$$\alpha_{14} = \beta_{14} = \boldsymbol{\beta}_{1s} = \alpha_{24} = \boldsymbol{\alpha}_{2s} = \beta_{24} = 0, \tag{C.52}$$

$$\boldsymbol{\alpha}_{1s} = \boldsymbol{\beta}_{2s} = \boldsymbol{\alpha}_s, \tag{C.53}$$

- SHDB Boundary, $\varepsilon_3 \wedge (\alpha c\mathbf{B} + \boldsymbol{\alpha}_s \wedge \mathbf{E}) = 0, \quad \varepsilon_3 \wedge (\alpha c\mathbf{D} - \boldsymbol{\alpha}_s \wedge \mathbf{H}) = 0$

$$\beta_{14} = \boldsymbol{\beta}_{1s} = \alpha_{24} = \boldsymbol{\alpha}_{2s} = 0, \tag{C.54}$$

$$\alpha_{14} = \beta_{24} = \alpha, \quad \boldsymbol{\alpha}_{1s} = \boldsymbol{\beta}_{2s} = \boldsymbol{\alpha}_s, \tag{C.55}$$

- GSH Boundary, $\varepsilon_3 \wedge \boldsymbol{\alpha}_s \wedge \mathbf{E} = 0, \quad \varepsilon_3 \wedge \boldsymbol{\beta}_s \wedge \mathbf{H} = 0,$

$$\alpha_{14} = \beta_{14} = \boldsymbol{\beta}_{1s} = \alpha_{24} = \boldsymbol{\alpha}_{2s} = \beta_{24} = 0, \tag{C.56}$$

$$\boldsymbol{\alpha}_{1s} = \boldsymbol{\alpha}_s, \quad \boldsymbol{\beta}_{2s} = \boldsymbol{\beta}_s, \tag{C.57}$$

- GSHDB Boundary, $\varepsilon_3 \wedge (\alpha c\mathbf{B} + \boldsymbol{\alpha}_s \wedge \mathbf{E}) = 0, \quad \varepsilon_3 \wedge (\beta c\mathbf{D} + \boldsymbol{\beta}_s \wedge \mathbf{H}) = 0$

$$\alpha_{14} = \alpha, \quad \boldsymbol{\alpha}_{1s} = \boldsymbol{\alpha}_s, \quad \beta_{14} = \boldsymbol{\beta}_{1s} = 0 \tag{C.58}$$

$$\alpha_{24} = \boldsymbol{\alpha}_{2s} = 0, \quad \beta_{24} = \beta, \quad \boldsymbol{\beta}_{2s} = -\boldsymbol{\beta}_s, \tag{C.59}$$

Appendix D

Solutions to Problems

1.1 Starting from (1.16), we set

$$M_d\eta_o = \frac{C + AM\eta_o}{D + BM\eta_o} = M\eta_o,$$

which leads to the equation

$$B(M\eta_o)^2 + (D - A)M\eta_o - C = 0.$$

This has the solutions

$$\begin{aligned} M\eta_o &= \frac{1}{2B}(A - D \pm \sqrt{(A-D)^2 + 4BC}) \\ &= \frac{1}{2B}(A - D \pm \sqrt{(A+D)^2 - 4}), \end{aligned}$$

where we have applied $AD - BC = 1$. For the transformation (1.19), we obtain

$$M\eta_o = \pm\frac{1}{2\sin\varphi}\sqrt{4\cos^2\varphi - 4} = \pm j.$$

1.2 Invoking (1.13), (1.3) and (1.12), we can proceed

$$\begin{aligned} \begin{pmatrix} \eta_o\mathbf{D}_d \\ \mathbf{B}_d \end{pmatrix} &= \begin{pmatrix} D & -C \\ -B & A \end{pmatrix}\begin{pmatrix} \eta_o\mathbf{D} \\ \mathbf{B} \end{pmatrix} \\ &= \begin{pmatrix} D & -C \\ -B & A \end{pmatrix}\begin{pmatrix} \overline{\overline{\epsilon}}\eta_o & \overline{\overline{\xi}} \\ \overline{\overline{\zeta}} & \overline{\overline{\mu}}/\eta_o \end{pmatrix}\cdot\begin{pmatrix} \mathbf{E} \\ \eta_o\mathbf{H} \end{pmatrix} \\ &= \begin{pmatrix} D & -C \\ -B & A \end{pmatrix}\begin{pmatrix} \overline{\overline{\epsilon}}\eta_o & \overline{\overline{\xi}} \\ \overline{\overline{\zeta}} & \overline{\overline{\mu}}/\eta_o \end{pmatrix}\cdot\begin{pmatrix} D & -B \\ -C & A \end{pmatrix}\begin{pmatrix} \mathbf{E}_d \\ \eta_o\mathbf{H}_d \end{pmatrix}, \end{aligned}$$

and compare to

$$\begin{pmatrix} \eta_o\mathbf{D}_d \\ \mathbf{B}_d \end{pmatrix} = \begin{pmatrix} \overline{\overline{\epsilon}}_d\eta_o & \overline{\overline{\xi}}_d \\ \overline{\overline{\zeta}}_d & \overline{\overline{\mu}}_d/\eta_o \end{pmatrix}\cdot\begin{pmatrix} \mathbf{E}_d \\ \eta_o\mathbf{H}_d \end{pmatrix},$$

from which, through matrix multiplications, the rule (1.17) can be identified. Obviously, $\overline{\overline{\xi}} - \overline{\overline{\zeta}}$ is invariant in the duality transformation.

1.3 From (1.17) for $\overline{\overline{\epsilon}}_d = \overline{\overline{\epsilon}} = \epsilon_o\overline{\overline{\mathsf{I}}}$, $\overline{\overline{\xi}}_d = \overline{\overline{\zeta}}_d = \overline{\overline{\xi}} = \overline{\overline{\zeta}} = 0$ and $\overline{\overline{\mu}}_d = \overline{\overline{\mu}} = \mu_o\overline{\overline{\mathsf{I}}}$ we obtain

$$\begin{aligned} \overline{\overline{\epsilon}}_d &= (D^2 + C^2)\epsilon_o\overline{\overline{\mathsf{I}}} = \epsilon_o\overline{\overline{\mathsf{I}}}, \\ \overline{\overline{\xi}}_d = \overline{\overline{\zeta}}_d &= -(BD\eta_o\epsilon_o + AC\mu_o/\eta_o)\overline{\overline{\mathsf{I}}} = 0, \\ \overline{\overline{\mu}}_d &= (B^2 + A^2)\mu_o\overline{\overline{\mathsf{I}}} = \mu_o\overline{\overline{\mathsf{I}}}. \end{aligned}$$

These lead us to a set of three conditions for the transformation parameters,

$$D^2 + C^2 = 1, \qquad B^2 + A^2 = 1, \qquad BD + AC = 0,$$

together with the original condition $AD - BC = 1$

From $(1 - B^2)(1 - C^2) = (AD)^2 = (1 + BC)^2 = 1 + 2BC + (BC)^2$ we obtain $(B + C)^2 = 0$, whence $C = -B$.

From $BD + AC = B(D - A) = 0$ we obtain $D = A$ or $B = C = 0$.

For $B = C = 0$, from $D^2 = A^2 = 1, AD = 1$ we obtain $A = D = \pm 1$, two trivial transformations which can be ignored.

For $D = A$, $B = -C$, the only restriction between A and B is $A^2 + B^2 = 1$. We can set $A = \cos\varphi = D$ and $B = \sin\varphi = -C$, whence (1.17) can be expressed in the form (1.20).

1.4 The property $\mathcal{Q}^T(\varphi) = \mathcal{Q}(-\varphi)$ can be clearly seen from the expression of the $\mathcal{Q}$ matrix. One can also easily check through matrix multiplication that the product $\mathcal{Q}(-\varphi)\mathcal{Q}(\varphi)$ yields the unit 4×4 matrix, whence $\mathcal{Q}(-\varphi) = \mathcal{Q}^{-1}(\varphi)$.

1.5 Setting $M_1\eta_o = \tan\vartheta_1$ and $M_2\eta_o = \tan\vartheta_2$, from the condition (1.26) we obtain the transformed admittances

$$M_{id}\eta_o = \tan(\vartheta_1 + \varphi), \qquad M_{2d}\eta_o = \tan(\vartheta_2 + \varphi),$$

where φ defines a duality transformation which leaves the simple-isotropic medium invariant. Requiring

$$M_{1d} = -M_{2d} = M_d$$

leads to

$$\vartheta_1 + \varphi = -(\vartheta_2 + \varphi),$$

whence the transformation parameter must be chosen as

$$\varphi = -\frac{1}{2}(\vartheta_1 + \vartheta_2).$$

Thus, we have

$$M_d\eta_o = \tan(\vartheta_1+\varphi) = \tan\frac{\vartheta_1-\vartheta_2}{2} = \frac{\tan\vartheta_1/2 - \tan\vartheta_2/2}{1+\tan\vartheta_1/2\tan\vartheta_2/2}$$

Expanding

$$\begin{aligned}\tan\vartheta/2 &= \frac{\sin\vartheta/2}{\cos\vartheta/2} = \sqrt{\frac{1-\cos\vartheta}{1+\cos\vartheta}} = \frac{\sin\vartheta}{1+\cos\vartheta} \\ &= \frac{\tan\vartheta}{1+\sqrt{1+\tan^2\vartheta}} = \frac{M\eta_o}{1+\sqrt{1+M^2\eta_o^2}}\end{aligned}$$

for ϑ_1 and ϑ_2, we obtain the relation $M_d(M_1, M_2)$.

1.6 Expressing (1.43) in the form

$$\begin{aligned}\overline{\overline{\mathsf{J}}}_t &= \frac{1}{k_ok_nk_t^2}\mathbf{n}\times(k_o^2\mathbf{k}_t\mathbf{k}_t + k_n^2(\mathbf{n}\times\mathbf{k})(\mathbf{n}\times\mathbf{k})) \\ &= \frac{1}{k_ok_nk_t^2}(k_o^2(\mathbf{n}\times\mathbf{k})\mathbf{k}_t - k_n^2\mathbf{k}_t(\mathbf{n}\times\mathbf{k})),\end{aligned}$$

we can expand

$$\begin{aligned}\overline{\overline{\mathsf{J}}}_t^2 &= \frac{1}{k_o^2k_n^2k_t^4}(-k_o^2k_n^2(\mathbf{n}\times\mathbf{k})(\mathbf{k}\times\mathbf{n})k_t^2 - k_o^2k_n^2\mathbf{k}_t\mathbf{k}_tk_t^2) \\ &= -\frac{1}{k_t^2}((\mathbf{n}\times\mathbf{k})(\mathbf{n}\times\mathbf{k}) + \mathbf{k}_t\mathbf{k}_t) = -\overline{\overline{\mathsf{I}}}_t,\end{aligned}$$

and

$$\begin{aligned}\overline{\overline{\mathsf{J}}}_t^{(2)} &= \frac{1}{k_o^2k_n^2k_t^4}(k_o^2(\mathbf{n}\times\mathbf{k})\mathbf{k}_t - k_n^2\mathbf{k}_t(\mathbf{n}\times\mathbf{k}))^{(2)} \\ &= -\frac{1}{k_t^4}((\mathbf{n}\times\mathbf{k})\times\mathbf{k}_t)(\mathbf{k}_t\times(\mathbf{n}\times\mathbf{k})) \\ &= \mathbf{nn}.\end{aligned}$$

1.7 Expanding the vector product $(\mathbf{n}\times\mathbf{k})\times((\mathbf{n}\times\mathbf{k})\times\mathbf{E})$ in two ways and taking into account $\mathbf{k}\times\mathbf{E} = k_o\eta_o\mathbf{H}$ and $\mathbf{k}\cdot\mathbf{E} = 0$, we first obtain

$$\begin{aligned}(\mathbf{n}\times\mathbf{k})\times((\mathbf{n}\times\mathbf{k})\times\mathbf{E}) &= \mathbf{n}\times\mathbf{k}(\mathbf{n}\times\mathbf{k}\cdot\mathbf{E}) - (\mathbf{n}\times\mathbf{k})^2\mathbf{E} \\ &= \mathbf{n}\times\mathbf{k}(\mathbf{n}\cdot k_o\eta_o\mathbf{H}) - k_t^2\mathbf{E},\end{aligned}$$

and, second,

$$\begin{aligned}(\mathbf{n}\times\mathbf{k})\times((\mathbf{n}\times\mathbf{k})\times\mathbf{E}) &= (\mathbf{n}\times\mathbf{k})\times\mathbf{k}(\mathbf{n}\cdot\mathbf{E}) \\ &= (\mathbf{k}k_n - \mathbf{n}k_o^2)(\mathbf{n}\cdot\mathbf{E}) \\ &= (\mathbf{k}_tk_n - \mathbf{n}k_t^2)(\mathbf{n}\cdot\mathbf{E}).\end{aligned}$$

Equating these two expressions we obtain

$$k_o\mathbf{n}\times\mathbf{k}(\mathbf{n}\cdot\eta_o\mathbf{H}) - k_t^2\mathbf{E} = (k_n\mathbf{k} - k_o^2\mathbf{n})(\mathbf{n}\cdot\mathbf{E}),$$

which can be rewritten in the form (1.52). The rule (1.53) can be found similarly.

1.8 From (1.20), we straightforwardly obtain

$$(\overline{\overline{\xi}}_d - \overline{\overline{\zeta}}_d)/\sqrt{\mu_o\epsilon_o} = (\overline{\overline{\xi}} - \overline{\overline{\zeta}})/\sqrt{\mu_o\epsilon_o},$$

whence $\overline{\overline{\xi}} - \overline{\overline{\zeta}}$ is invariant in the duality transformation. Similarly, we obtain

$$\overline{\overline{\epsilon}}_d/\epsilon_o + \overline{\overline{\mu}}_d/\mu_o = \overline{\overline{\epsilon}}/\epsilon_o + \overline{\overline{\mu}}/\mu_o,$$

whence the sum of relative permittivity and permeability dyadics is invariant. For the remaining medium dyadics, (1.20) is reduced to

$$\begin{pmatrix} \overline{\overline{\epsilon}}_d/\epsilon_o - \overline{\overline{\mu}}_d/\mu_o \\ (\overline{\overline{\xi}}_d + \overline{\overline{\zeta}}_d)/\sqrt{\mu_o\epsilon_o} \end{pmatrix} = \begin{pmatrix} \cos 2\varphi & \sin 2\varphi \\ -\sin 2\varphi & \cos 2\varphi \end{pmatrix} \begin{pmatrix} \overline{\overline{\epsilon}}/\epsilon_o - \overline{\overline{\mu}}/\mu_o \\ (\overline{\overline{\xi}} + \overline{\overline{\zeta}})/\sqrt{\mu_o\epsilon_o} \end{pmatrix},$$

which involves a 2×2 rotation matrix. The full 4×4 matrix $\mathcal{Q}'(\varphi)$ can be easily written in terms of these results.

2.1 The Maxwell equations for a plane wave in the medium defined by (2.4) can be expressed as

$$\begin{aligned} \mathbf{k}\times\mathbf{E} &= \omega\mathbf{B} = \omega q(\mathbf{E} + (1/M)\mathbf{H}), \\ \mathbf{k}\times\mathbf{H} &= -\omega\mathbf{D} = -\omega q(M\mathbf{E} + \mathbf{H}), \end{aligned}$$

or, as

$$\begin{aligned} (\mathbf{k}\times\overline{\overline{\mathsf{I}}} - \omega q\overline{\overline{\mathsf{I}}})\cdot\mathbf{E} &= (\omega q/M)\mathbf{H}, \\ (\mathbf{k}\times\overline{\overline{\mathsf{I}}} + \omega q\overline{\overline{\mathsf{I}}}).\mathbf{H} &= -\omega q M\mathbf{E}. \end{aligned}$$

Eliminating **H**, these lead to the condition

$$(\mathbf{k}\times\overline{\overline{\mathsf{I}}})^2\cdot\mathbf{E} = 0,$$

whence the dispersion equation becomes

$$\det(\mathbf{k}\times\overline{\overline{\mathsf{I}}})^2 = (\det(\mathbf{k}\times\overline{\overline{\mathsf{I}}}))^2 = 0.$$

Because the determinant of a 3D antisymmetric dyadic is zero, the equation is identically satisfies for any **k** vector. Since this is valid for any q, there is no dispersion equation for a plane wave in a PEMC medium.

2.2 Expanding the dyadic $\overline{\overline{\mathsf{J}}}_t$ as

$$\begin{aligned}\overline{\overline{\mathsf{J}}}_t &= \frac{1}{k_o k_n}\mathbf{n}\times(\mathbf{k}_t\mathbf{k}_t + k_n^2\overline{\overline{\mathsf{I}}}_t) \\ &= \frac{1}{k_o k_n k_t^2}\mathbf{n}\times(k_t^2\mathbf{k}_t\mathbf{k}_t + k_n^2(\mathbf{k}_t\mathbf{k}_t + (\mathbf{n}\times\mathbf{k}_t)(\mathbf{n}\times\mathbf{k}_t)) \\ &= \frac{k_o}{k_n k_t^2}(\mathbf{n}\times\mathbf{k}_t)\mathbf{k}_t - \frac{k_n}{k_o k_t^2}\mathbf{k}_t(\mathbf{n}\times\mathbf{k}_t),\end{aligned}$$

we obtain the following rules:

$$\begin{aligned}\overline{\overline{\mathsf{J}}}_t\cdot\mathbf{k}_t &= \frac{k_o}{k_n}(\mathbf{n}\times\mathbf{k}_t) \\ \overline{\overline{\mathsf{J}}}_t\cdot(\mathbf{n}\times\mathbf{k}_t) &= -\frac{k_n}{k_o}\mathbf{k}_t,\end{aligned}$$

in terms of which the relation (2.11) results:

$$\overline{\overline{\mathsf{J}}}_t\cdot(k_n\mathbf{k}_t \mp jk_o\mathbf{n}\times\mathbf{k}_t) = \pm j(k_n\mathbf{k}_t \mp jk_o\mathbf{n}\times\mathbf{k}_t).$$

2.3 Considering a boundary surface defined by $f(\mathbf{r}) = 0$, and expressing the normal unit vector by

$$\mathbf{n} = \lambda(\mathbf{r})\nabla f(\mathbf{r}),$$

with

$$\lambda(\mathbf{r}) = 1/\sqrt{\nabla f(\mathbf{r})\cdot\nabla f(\mathbf{r})},$$

which satisfies $\mathbf{n}\cdot\mathbf{n} = 1$, the boundary condition (2.6) yields

$$\nabla\cdot(\mathbf{n}\times(\mathbf{H}+M\mathbf{E})) = \nabla_t\cdot(\mathbf{n}\times(\mathbf{H}+M\mathbf{E})) = 0.$$

Expanding

$$\begin{aligned}\nabla\cdot(\mathbf{n}\times(\mathbf{H}+M\mathbf{E})) &= (\nabla\times\mathbf{n})\cdot(\mathbf{H}+M\mathbf{E}) - \mathbf{n}\cdot(\nabla\times(\mathbf{H}+M\mathbf{E})) \\ &= (\nabla\times\mathbf{n})\cdot(\mathbf{H}+M\mathbf{E}) - j\omega\mathbf{n}\cdot(\mathbf{D}-M\mathbf{B}) = 0,\end{aligned}$$

and applying

$$\nabla\times\mathbf{n} = \nabla\times(\lambda\nabla f) = \nabla\lambda\times\nabla f = \frac{\nabla\lambda}{\lambda}\times\mathbf{n},$$

we obtain

$$\begin{aligned}(\nabla\times\mathbf{n})\cdot(\mathbf{H}+M\mathbf{E}) &= \left(\frac{\nabla\lambda}{\lambda}\times\mathbf{n}\right)\cdot(\mathbf{H}+M\mathbf{E}) \\ &= \frac{\nabla\lambda}{\lambda}\cdot(\mathbf{n}\times(\mathbf{H}+M\mathbf{E})) = 0.\end{aligned}$$

Combining these, we finally have

$$\nabla\cdot(\mathbf{n}\times(\mathbf{H}+M\mathbf{E})) = -j\omega\mathbf{n}\cdot(\mathbf{D}-M\mathbf{B}) = 0,$$

whence (2.6) yields the condition (2.7).

2.4 Starting from the eigenvector expansion

$$\overline{\overline{\mathsf{J}}}_t = \mathbf{x}_{t+}\mathbf{a}_{t+} + \mathbf{x}_{t-}\mathbf{a}_{t-},$$

the vectors $\mathbf{a}_{t+}, \mathbf{a}_{t-}$ can be obtained in terms of eigenvectors from

$$\begin{aligned}\overline{\overline{\mathsf{J}}}_t \cdot \mathbf{x}_{t+} &= \mathbf{x}_{t+}(\mathbf{a}_{t+} \cdot \mathbf{x}_{t+}) + \mathbf{x}_{t-}(\mathbf{a}_{t-} \cdot \mathbf{x}_{t+}) = +j\mathbf{x}_{t+}, \\ \overline{\overline{\mathsf{J}}}_t \cdot \mathbf{x}_{t-} &= \mathbf{x}_{t+}(\mathbf{a}_{t+} \cdot \mathbf{x}_{t-}) + \mathbf{x}_{t-}(\mathbf{a}_{t-} \cdot \mathbf{x}_{t-}) = -j\mathbf{x}_{t-}.\end{aligned}$$

Multiplying these respectively by $(\mathbf{n} \times \mathbf{x}_{t+})\cdot$ and $(\mathbf{n} \times \mathbf{x}_{t-})\cdot$ yields

$$(\mathbf{n} \cdot \mathbf{x}_{t+} \times \mathbf{x}_{t-})(\mathbf{a}_{t-} \cdot \mathbf{x}_{t+}) = (\mathbf{n} \cdot \mathbf{x}_{t-} \times \mathbf{x}_{t+})(\mathbf{a}_{t+} \cdot \mathbf{x}_{t-}) = 0,$$

whence $\mathbf{a}_{t+} = A_+\mathbf{n} \times \mathbf{x}_{t-}, \quad \mathbf{a}_{t-} = A_-\mathbf{n} \times \mathbf{x}_{t+}$ and

$$\overline{\overline{\mathsf{J}}}_t = A_+\mathbf{x}_{t+}(\mathbf{n} \times \mathbf{x}_{t-}) + A_-\mathbf{x}_{t-}(\mathbf{n} \times \mathbf{x}_{t+}),$$

From $\overline{\overline{\mathsf{J}}}_t \cdot \mathbf{x}_{t\pm} = \pm j\mathbf{x}_{t\pm}$ we obtain $A_+ = A_- = -j/(\mathbf{n} \cdot \mathbf{x}_{t+} \times \mathbf{x}_{t-})$, which leads to the expansion (2.17).

2.5 Expanding

$$\begin{aligned}\overline{\overline{\mathsf{J}}}_t &= \frac{1}{k_o k_n}((\mathbf{n} \times \mathbf{k}_t)\mathbf{k}_t + k_n^2 \mathbf{n} \times \overline{\overline{\mathsf{I}}}_t) \\ &= \frac{1}{k_o k_n k_t^2}(k_t^2(\mathbf{n} \times \mathbf{k})\mathbf{k}_t + k_n^2(\mathbf{n} \times \mathbf{k}_t)\mathbf{k}_t - k_n^2\mathbf{k}_t(\mathbf{n} \times \mathbf{k}_t)) \\ &= \frac{1}{k_o k_n k_t^2}(k_o^2(\mathbf{n} \times \mathbf{k})\mathbf{k}_t - k_n^2\mathbf{k}_t(\mathbf{n} \times \mathbf{k}_t)),\end{aligned}$$

we obtain

$$\begin{aligned}\overline{\overline{\mathsf{J}}}_t \pm j\overline{\overline{\mathsf{I}}}_t &= \frac{1}{k_o k_n k_t^2}(k_o^2(\mathbf{n} \times \mathbf{k})\mathbf{k}_t - k_n^2\mathbf{k}_t(\mathbf{n} \times \mathbf{k}_t) \\ &\quad \pm j k_o k_n(\mathbf{k}_t\mathbf{k}_t + (\mathbf{n} \times \mathbf{k}_t)(\mathbf{n} \times \mathbf{k}_t)) \\ &= \frac{1}{k_o k_n k_t^2}(k_o(\mathbf{n} \times \mathbf{k}_t) \pm j k_n \mathbf{k}_t)(k_o\mathbf{k}_t \pm j k_n(\mathbf{n} \times \mathbf{k}_t)).\end{aligned}$$

Because the dyadic is of the form $\mathbf{a}_t\mathbf{b}_t$, its 2D determinant vanishes and, hence, it has no inverse. The same can be obtained more compactly through the eigenvector expansions (2.16) and (2.17) as

$$\begin{aligned}\overline{\overline{\mathsf{J}}}_t \pm j\overline{\overline{\mathsf{I}}}_t &= \frac{1}{X}((j \pm j)\mathbf{x}_{t+}\mathbf{x}_{t-} + (j \mp j)\mathbf{x}_{t-}\mathbf{x}_{t+}) \times \mathbf{n}, \\ &= \frac{2j}{\mathbf{n} \cdot \mathbf{x}_{t+} \times \mathbf{x}_{t-}}\mathbf{x}_{t\pm}(\mathbf{x}_{t\mp} \times \mathbf{n}).\end{aligned}$$

2.6 The electric co-polarizability α_{ee} is equal to the cross-polarizability α_{em} for the following two PEMC parameters:

$$M\eta_o = \left\{ \begin{array}{l} M_1\eta_o \\ M_2\eta_o \end{array} \right. = \frac{3}{4} \pm \frac{\sqrt{17}}{4} \approx \left\{ \begin{array}{r} 1.78 \\ -0.281. \end{array} \right.$$

For these values, the polarizability components read

$$\mathcal{A}_{\rm PEMC}(M_1\eta_0) = \frac{3}{8}\begin{pmatrix} 1+\sqrt{17} & 1+\sqrt{17} \\ 1+\sqrt{17} & 3-\sqrt{17} \end{pmatrix} \approx \begin{pmatrix} 1.92 & 1.92 \\ 1.92 & -0.421 \end{pmatrix}$$

and

$$\mathcal{A}_{\rm PEMC}(M_2\eta_0) = \frac{3}{8}\begin{pmatrix} 1-\sqrt{17} & 1-\sqrt{17} \\ 1-\sqrt{17} & 3+\sqrt{17} \end{pmatrix} \approx \begin{pmatrix} -1.17 & -1.17 \\ -1.17 & 2.67 \end{pmatrix}.$$

Likewise, the magnetic co-polarizability matches the cross-polarizability term ($\alpha_{mm} = \alpha_{me}$) provided that

$$M\eta_o = \begin{cases} M_3\eta_o \\ M_4\eta_o \end{cases} = -\frac{3}{2} \pm \frac{\sqrt{17}}{2} \approx \begin{cases} 0.562 \\ -3.56. \end{cases}$$

Now the polarizability components are

$$\mathcal{A}_{\rm PEMC}(M_3\eta_0) = \frac{3}{8}\begin{pmatrix} 3-\sqrt{17} & 1+\sqrt{17} \\ 1+\sqrt{17} & 1+\sqrt{17} \end{pmatrix} \approx \begin{pmatrix} -0.421 & 1.92 \\ 1.92 & 1.92 \end{pmatrix}$$

and

$$\mathcal{A}_{\rm PEMC}(M_4\eta_0) = \frac{3}{8}\begin{pmatrix} 3+\sqrt{17} & 1-\sqrt{17} \\ 1-\sqrt{17} & 1-\sqrt{17} \end{pmatrix} \approx \begin{pmatrix} 2.67 & -1.17 \\ -1.17 & -1.17 \end{pmatrix}.$$

This means that the magnitude of the cross-polarizability component is larger than either of the co-polarizabilities in the ranges

$$0.562 < M\eta_o < 1.78 \quad \text{and} \quad -1.78 < M\eta_o < -0.562.$$

2.7 From the result (2.116), the minimum can be found for $\mu_{\rm eff}$. It takes place at

$$p_{\rm min} = \frac{\sqrt{2}\,M\eta_o - 2}{\sqrt{2}\,M\eta_o + 1}$$

and when this value is substituted into (2.116), the minimum for the permeability comes as

$$\mu_{r,{\rm eff,min}} = \frac{2\sqrt{2}\,M\eta_o - 1}{1 + (M\eta_o)^2}$$

These functions are plotted in Figure D.1. The derivative of the effective permeability at $p = 0$ is

$$\left.\frac{\partial\mu_{r,{\rm eff}}}{\partial p}\right|_{p=0} = \frac{3}{2}\,\frac{2-(M\eta_o)^2}{1+(M\eta_o)^2}$$

which means that it only starts to decrease for $|M\eta_o| > \sqrt{2}$. Hence there is no minimum for $\mu_{r,{\rm eff}s}$ for PEMC parameter values smaller than this.

(Likewise, for the dual case of $|M\eta_o| < 1$ which is closer to the PMC situation, the same quantitative conclusions can be drawn for the minimum of the relative effective permittivity $\epsilon_{r,{\rm eff}}$, by replacing $M\eta_o$ with $(M\eta_o)^{-1}$ in the equations above.)

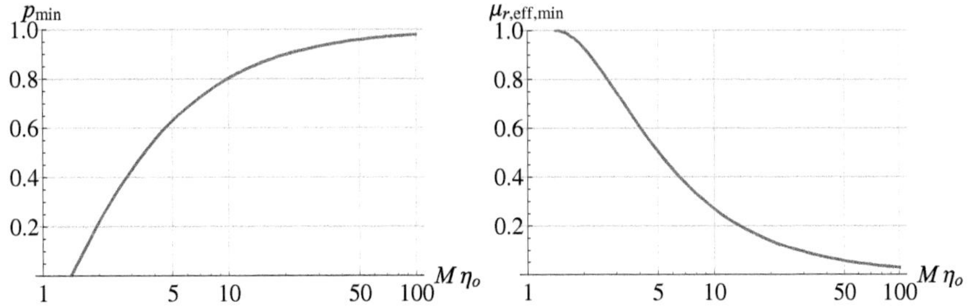

Figure D.1: The minimum of the effective relative permeability $\mu_{\rm eff}$ of a PEMC mixture in Figure 2.8 as function of the PEMC parameter $M\eta_o > 1$. Left: the volume fraction $p_{\rm min}$ for minimum, right: the minimum value of the permeability.

2.8 The polarizability matrix for a PEMC sphere

$$\mathcal{A}_{\rm PEMC} = \frac{1}{1+(M\eta_o)^2}\begin{pmatrix} \alpha_e(M\eta_o)^2+\alpha_m & (\alpha_e-\alpha_m)M\eta_o \\ (\alpha_e-\alpha_m)M\eta_o & \alpha_m(M\eta_o)^2+\alpha_e \end{pmatrix},$$

expanded for complex-valued PEMC parameter $M\eta_o = m + j\,n$, gives the components

$$\begin{aligned}
\alpha_{ee} &= \frac{3}{2}\cdot\frac{2(m^2+n^2)^2+m^2-(n^2+1)}{(1+m^2-n^2)^2+4m^2n^2} + j\,\frac{9mn}{(1+m^2-n^2)^2+4m^2n^2},\\
\alpha_{mm} &= \frac{3}{2}\cdot\frac{-(m^2+n^2)^2+m^2-(n^2-2)}{(1+m^2-n^2)^2+4m^2n^2} - j\,\frac{9mn}{(1+m^2-n^2)^2+4m^2n^2},\\
\alpha_{em} &= \frac{9}{2}\cdot\frac{m(1+m^2+n^2)}{(1+m^2-n^2)^2+4m^2n^2} + j\,\frac{9}{2}\cdot\frac{n(1-m^2-n^2)}{(1+m^2-n^2)^2+4m^2n^2},
\end{aligned}$$

showing that the co-polarizability components have equal and opposite imaginary parts. It is worth noting that despite the complex-valued M parameter, the determinant of the $\mathcal{A}_{\rm PEMC}$ matrix is constant:

$$\alpha_{ee}\alpha_{mm} - \alpha_{em}^2 = -\frac{9}{2}.$$

It can be also shown that the same antisymmetry of the imaginary parts as with the co-polarizability components also applies for the effective permittivity and effective permeability parameters for a mixture containing small PEMC spheres (Equations (2.114) and (2.116)):

$$\mathrm{Im}\{\epsilon_{r,\rm eff}\} = -\mathrm{Im}\{\mu_{r,\rm eff}\}$$

regardless of the volume fraction p or the PEMC parameter $M\eta_o = m+jn$.

2.9 The refractive index $n_{\rm eff} = \sqrt{\epsilon_{r,\rm eff}\mu_{r,\rm eff} - \chi^2_{r,\rm eff}}$ of the homogenized PEMC mixture is in fact the square root of the determinant of the material matrix in (2.113). It turns out to be

$$n_{\rm eff} = \sqrt{\frac{2+4p}{2+p}}$$

This simple function increasing from unity to $\sqrt{2}$ as p grows from 0 to 1 is independent of the PEMC parameter $M\eta_o$!

2.10 For $M \to \infty$, (2.71) and (2.72) become

$$\begin{aligned} \mathbf{E}(x_1) &= \frac{1}{\sqrt{1+(M\eta_o)^2}}(M\eta_o\mathbf{E}_d - \eta_o\mathbf{H}_d) \to \mathbf{E}_d \\ &= \mathbf{e}_2\eta_o ME \sin k_c x_1, \\ \eta_o\mathbf{H}(x_1) &= \frac{1}{\sqrt{1+(M\eta_o)^2}}(\mathbf{E}_d + M\eta_o^2\mathbf{H}_d) \to \eta_o\mathbf{H}_d \\ &= \frac{ME}{k_o}(-\mathbf{e}_1\beta\sin k_c x_1 + \mathbf{e}_3 jk_c\cos k_c x_1) \end{aligned}$$

For $M \to 0$, (2.71) and (2.72) become

$$\begin{aligned} \mathbf{E}(x_1) &= \frac{1}{\sqrt{1+(M\eta_o)^2}}(M\eta_o\mathbf{E}_d - \eta_o\mathbf{H}_d) \to -\eta_o\mathbf{H}_d \\ &= \frac{E}{k_o}(\mathbf{e}_1\beta\sin k_c x_1 - \mathbf{e}_3 jk_c\cos k_c x_1), \\ \eta_o\mathbf{H}(x_1) &= \frac{1}{\sqrt{1+(M\eta_o)^2}}(\mathbf{E}_d + M\eta_o^2\mathbf{H}_d) \to \mathbf{E}_d \\ &= \mathbf{e}_2 E\sin k_c x_1 \end{aligned}$$

2.11 Assuming a plane wave with wave vector $\mathbf{k}$, the relations $\mathbf{k}\times\mathbf{E} = k_o\eta_o\mathbf{H}$ and $\mathbf{k}\cdot\mathbf{E} = 0$ can be applied to the PEMC boundary condition as

$$\begin{aligned} k_o\eta_o(\mathbf{H}_t + M\mathbf{E}_t) &= (\mathbf{k}\times\mathbf{E})_t + k_o M\eta_o\mathbf{E}_t \\ &= ((\mathbf{n}\cdot\mathbf{k})\mathbf{n}\times\bar{\bar{\mathsf{I}}}_t + \frac{\mathbf{n}\times\mathbf{k}_t}{\mathbf{n}\cdot\mathbf{k}}\mathbf{k}_t + k_o M\eta_o\bar{\bar{\mathsf{I}}}_t)\cdot\mathbf{E}_t = 0, \end{aligned}$$

which makes sense only for $\mathbf{n}\cdot\mathbf{k} \neq 0$. Let us consider two cases separately. For the case $\mathbf{n}\cdot\mathbf{k} = 0$, the condition becomes

$$-\mathbf{n}\times\mathbf{k}E_n + k_o M\eta_o\mathbf{E}_t = 0,$$

whence $\mathbf{E}_t = A\mathbf{n}\times\mathbf{k}$ and the plane-wave fields become

$$\mathbf{E} = A(\mathbf{n}\times\mathbf{k} + k_o M\eta_o\mathbf{n}),$$

$$\mathbf{H} = \frac{1}{k_o\eta_o}\mathbf{k}_t \times \mathbf{E} = \frac{A}{k_o\eta_o}\mathbf{k}_t \times (\mathbf{n}\times\mathbf{k} + k_oM\eta_o\mathbf{n}) = \frac{Ak_o}{\eta_o}\mathbf{n} - MA\mathbf{n}\times\mathbf{k}.$$

These define a lateral wave satisfying the PEMC condition for any $\mathbf{k} = \mathbf{k}_t = k_o\mathbf{u}_t$.

For the case $\mathbf{n}\cdot\mathbf{k} \neq 0$, a matched wave requires vanishing of the 2D determinant of the dyadic multiplying $\mathbf{E}_t$, whence the condition for $\mathbf{k}$ becomes

$$\begin{aligned} 0 &= \mathrm{det}_t((\mathbf{n}\cdot\mathbf{k})\mathbf{n}\times\overline{\overline{\mathsf{I}}}_t + \frac{\mathbf{n}\times\mathbf{k}_t}{\mathbf{n}\cdot\mathbf{k}}\mathbf{k}_t + k_oM\eta_o\overline{\overline{\mathsf{I}}}_t) \\ &= (\mathbf{n}\cdot\mathbf{k})^2 + (k_oM\eta_o)^2 + \mathbf{k}_t\cdot\mathbf{k}_t = k_o^2(1+(M\eta_o)^2). \end{aligned}$$

Here we have applied the rules

$$\mathrm{det}_t\overline{\overline{\mathsf{A}}}_t = \mathrm{tr}\overline{\overline{\mathsf{A}}}_t^{(2)} = \mathbf{nn} : \overline{\overline{\mathsf{A}}}_t^{(2)}$$

$$\mathrm{tr}(\overline{\overline{\mathsf{A}}}_t {}^{\times}_{\times} \overline{\overline{\mathsf{I}}}_t) = \mathrm{tr}\overline{\overline{\mathsf{A}}}_t, \quad \mathrm{tr}(\overline{\overline{\mathsf{A}}}_t {}^{\times}_{\times} (\mathbf{n}\times\overline{\overline{\mathsf{I}}})) = \overline{\overline{\mathsf{A}}}_t : (\mathbf{n}\times\overline{\overline{\mathsf{I}}})$$

$$\mathrm{det}_t(\mathbf{n}\times\overline{\overline{\mathsf{I}}}) = 1, \quad \mathrm{det}_t\overline{\overline{\mathsf{I}}}_t = 1.$$

Thus, for a matched wave to exist in the case $\mathbf{n}\cdot\mathbf{k} \neq 0$, there is no condition for $\mathbf{k}$, but the PEMC admittance is limited to the special cases $M_\pm = \pm j/\eta_o$. To show that in this case the wave must be circularly polarized, we start from the PEMC condition

$$k_o\eta_o\mathbf{n}\times(\mathbf{H} + M_\pm\mathbf{E}) = \mathbf{n}\times(\mathbf{k}\times\mathbf{E}_\pm) \pm jk_o\mathbf{n}\times\mathbf{E}_\pm = 0,$$

from which we obtain

$$\begin{aligned} 0 &= (\mathbf{n}\times(\mathbf{k}\times\mathbf{E}_\pm))^2 + k_o^2(\mathbf{n}\times\mathbf{E}_\pm)^2 \\ &= (\mathbf{k}(\mathbf{n}\cdot\mathbf{E}_\pm) - (\mathbf{n}\cdot\mathbf{k})\mathbf{E}_\pm)^2 - k_o^2\mathbf{E}_{\pm t}^2 \\ &= k_o^2(\mathbf{n}\cdot\mathbf{E}_\pm)^2 + (\mathbf{n}\cdot\mathbf{k})^2\mathbf{E}_\pm^2 - k_o^2\mathbf{E}_{\pm t}^2 \\ &= (k_o^2 + (\mathbf{n}\cdot\mathbf{k})^2)\mathbf{E}_\pm^2. \end{aligned}$$

This yields $\mathbf{E}_\pm^2 = \mathbf{E}_\pm\cdot\mathbf{E}_\pm = 0$, whence $\mathbf{E}_\pm$ must be circularly polarized.

2.12 Applying the rules

$$\mathrm{tr}\overline{\overline{\mathsf{J}}}_t = 0, \quad \overline{\overline{\mathsf{J}}}_t {}^{\times}_{\times} \overline{\overline{\mathsf{I}}}_t = 0, \quad \overline{\overline{\mathsf{J}}}_t^{(2)} = \mathbf{nn}$$

from Appendix B, we can expand

$$\begin{aligned} \mathrm{tr}\overline{\overline{\mathsf{R}}}_t &= \cos 2\vartheta\,\mathrm{tr}\overline{\overline{\mathsf{I}}}_t = 2\cos 2\vartheta, \\ \overline{\overline{\mathsf{R}}}_t^{(2)} &= \cos^2 2\vartheta\,\overline{\overline{\mathsf{I}}}_t^{(2)} + \sin 2\vartheta\cos 2\vartheta\,\overline{\overline{\mathsf{J}}}_t {}^{\times}_{\times} \overline{\overline{\mathsf{I}}}_t + \sin^2 2\vartheta\,\overline{\overline{\mathsf{J}}}_t^{(2)} \\ &= (\cos^2 2\vartheta + \sin^2 2\vartheta)\mathbf{nn} = \mathbf{nn}, \end{aligned}$$

whence

$$\mathrm{det}_t\overline{\overline{\mathsf{R}}}_t = \mathrm{tr}\overline{\overline{\mathsf{R}}}_t^{(2)} = 1.$$

3.1 Because the 2D determinant has the properties

$$\mathrm{det}_t\overline{\overline{\mathsf{Z}}}_t = \mathrm{tr}\overline{\overline{\mathsf{Z}}}_t^{(2)} = \mathbf{nn} : \overline{\overline{\mathsf{Z}}}_t^{(2)} = \frac{1}{2}(\mathbf{nn}{}^{\times}_{\times}\overline{\overline{\mathsf{Z}}}_t) : \overline{\overline{\mathsf{Z}}}_t,$$

we can expand

$$\mathrm{tr}(Z_s\overline{\overline{\mathsf{I}}}_t + Z_n\mathbf{n}\times\overline{\overline{\mathsf{I}}} + \overline{\overline{\mathsf{Z}}}_{ta})^{(2)} = Z_s^2\mathrm{tr}\overline{\overline{\mathsf{I}}}_t^{(2)} + Z_n^2\mathrm{tr}(\mathbf{n}\times\overline{\overline{\mathsf{I}}})^{(2)} + \mathrm{tr}\overline{\overline{\mathsf{Z}}}_{ta}^{(2)}$$

$$+Z_sZ_n\mathrm{tr}(\overline{\overline{\mathsf{I}}}_t{}^{\times}_{\times}(\mathbf{n}\times\overline{\overline{\mathsf{I}}})) + Z_s\mathrm{tr}(\overline{\overline{\mathsf{I}}}_t{}^{\times}_{\times}\overline{\overline{\mathsf{Z}}}_{ta}) + Z_n\mathrm{tr}((\mathbf{n}\times\overline{\overline{\mathsf{I}}}){}^{\times}_{\times}\overline{\overline{\mathsf{Z}}}_{ta})$$

Substituting

$$\overline{\overline{\mathsf{I}}}_t{}^{(2)} = \mathbf{nn}, \quad \Rightarrow \quad \mathrm{tr}\overline{\overline{\mathsf{I}}}_t{}^{(2)} = 1$$

$$(\mathbf{n}\times\overline{\overline{\mathsf{I}}})^{(2)} = \mathbf{nn}, \quad \Rightarrow \quad \mathrm{tr}(\mathbf{n}\times\overline{\overline{\mathsf{I}}})^{(2)} = 1$$

$$\mathbf{nn}{}^{\times}_{\times}\overline{\overline{\mathsf{Z}}}_{ta} = -\overline{\overline{\mathsf{Z}}}_{ta}, \quad \Rightarrow \quad \mathrm{tr}\overline{\overline{\mathsf{Z}}}_{ta}^{(2)} = -\frac{1}{2}\overline{\overline{\mathsf{Z}}}_{ta} : \overline{\overline{\mathsf{Z}}}_{ta} = -\frac{1}{2}\mathrm{tr}\overline{\overline{\mathsf{Z}}}_{ta}^2$$

$$\mathrm{tr}(\overline{\overline{\mathsf{I}}}_t{}^{\times}_{\times}(\mathbf{n}\times\overline{\overline{\mathsf{I}}})) = (\mathbf{nn}{}^{\times}_{\times}\overline{\overline{\mathsf{I}}}_t) : (\mathbf{n}\times\overline{\overline{\mathsf{I}}}) = \mathrm{tr}(\mathbf{n}\times\overline{\overline{\mathsf{I}}}) = 0$$

$$\mathrm{tr}(\overline{\overline{\mathsf{I}}}_t{}^{\times}_{\times}\overline{\overline{\mathsf{Z}}}_{ta}) = (\mathbf{nn}{}^{\times}_{\times}\overline{\overline{\mathsf{I}}}_t) : \overline{\overline{\mathsf{Z}}}_{ta} = \mathrm{tr}\overline{\overline{\mathsf{Z}}}_{ta} = 0$$

$$\mathrm{tr}((\mathbf{n}\times\overline{\overline{\mathsf{I}}}){}^{\times}_{\times}\overline{\overline{\mathsf{Z}}}_{ta}) = (\mathbf{nn}{}^{\times}_{\times}(\mathbf{n}\times\overline{\overline{\mathsf{I}}})) : \overline{\overline{\mathsf{Z}}}_{ta} = (\mathbf{n}\times\overline{\overline{\mathsf{I}}}) : \overline{\overline{\mathsf{Z}}}_{ta} = 0$$

we obtain

$$\mathrm{det}_t(Z_s\overline{\overline{\mathsf{I}}}_t + Z_n\mathbf{n}\times\overline{\overline{\mathsf{I}}} + \overline{\overline{\mathsf{Z}}}_{ta}) = Z_s^2 + Z_n^2 - \frac{1}{2}\mathrm{tr}\overline{\overline{\mathsf{Z}}}_{ta}^2.$$

3.2 Starting from the PEMC impedance dyadic $\overline{\overline{\mathsf{Z}}}_t = (1/M)\mathbf{n}\times\overline{\overline{\mathsf{I}}}$, the reflection dyadic (3.48) can be expanded as

$$\begin{aligned}\overline{\overline{\mathsf{R}}}_t &= \overline{\overline{\mathsf{I}}}_t - 2M\eta_o(\overline{\overline{\mathsf{J}}}_t + M\eta_o\overline{\overline{\mathsf{I}}}_t)^{-1} \\ &= \overline{\overline{\mathsf{I}}}_t + \frac{2M\eta_o}{1+(M\eta_o)^2}(\overline{\overline{\mathsf{J}}}_t - M\eta_o\overline{\overline{\mathsf{I}}}_t) \\ &= \frac{1-(M\eta_o)^2}{1+(M\eta_o)^2}\overline{\overline{\mathsf{I}}}_t + \frac{2M\eta_o}{1+(M\eta_o)^2}\overline{\overline{\mathsf{J}}}_t.\end{aligned}$$

3.3 First expand (3.51),

$$\begin{aligned}\Delta &= \mathrm{det}_t(Z_s\mathbf{n}\times\overline{\overline{\mathsf{J}}}_t - \eta_o\overline{\overline{\mathsf{I}}}_t) = \mathrm{tr}(Z_s\mathbf{n}\times\overline{\overline{\mathsf{J}}}_t - \eta_o\overline{\overline{\mathsf{I}}}_t)^{(2)} \\ &= \mathbf{nn} : (Z_s^2(\mathbf{n}\times\overline{\overline{\mathsf{I}}})^{(2)}\cdot\overline{\overline{\mathsf{J}}}_t^{(2)} - Z_s\eta_o(\mathbf{n}\times\overline{\overline{\mathsf{J}}}_t){}^{\times}_{\times}\overline{\overline{\mathsf{I}}}_t + \eta_o^2\mathbf{nn}) \\ &= Z_s^2 + Z_s\eta_o\frac{1}{k_ok_n}(k_t^2 + 2k_n^2) + \eta_o^2 = \frac{1}{k_ok_n}(Z_sk_o + \eta_ok_n)(Z_sk_n + \eta_ok_o),\end{aligned}$$

and substitute in (3.50),

$$\begin{aligned}k_ok_n\Delta\overline{\overline{\mathsf{R}}}_t &= (k_ok_n\Delta - 2k_ok_n\eta_o^2)\overline{\overline{\mathsf{I}}}_t + 2k_ok_nZ_s\eta_o\overline{\overline{\mathsf{J}}}_t\times\mathbf{n} \\ &= (k_ok_n\Delta - 2k_ok_n\eta_o^2)\overline{\overline{\mathsf{I}}}_t - 2Z_s\eta_o((\mathbf{n}\times\mathbf{k}_t)(\mathbf{n}\times\mathbf{k}_t) + k_n^2\overline{\overline{\mathsf{I}}}_t) \\ &= (Z_sk_n - \eta_ok_o)(Z_sk_o + \eta_ok_n)\overline{\overline{\mathsf{I}}}_t - 2Z_s\eta_o(\mathbf{n}\times\mathbf{k}_t)(\mathbf{n}\times\mathbf{k}_t)\end{aligned}$$

Setting $\overline{\overline{\mathsf{I}}}_t = (\mathbf{k}_t\mathbf{k}_t + (\mathbf{n}\times\mathbf{k}_t)(\mathbf{n}\times\mathbf{k}_t)/k_t^2$ and $k_t^2 = k_o^2 - k_n^2$, we obtain (3.75) with (3.84) and (3.83) as

$$\begin{aligned} R_{TM} &= (Z_s k_n - \eta_o k_o)(Z_s k_o + \eta_o k_n)/k_o k_n \Delta, \\ R_{TE} &= ((Z_s k_n - \eta_o k_o)(Z_s k_o + \eta_o k_n) - 2Z_s\eta_o(k_o^2 - k_n^2))/k_o k_n \Delta. \end{aligned}$$

3.4 Ssubstituting $\mathbf{E}^i = \mathbf{E}_a^i = E_a\mathbf{a}_t \times \mathbf{k}^i$ in (3.210) yields

$$\mathbf{E}^r = \frac{E_a}{\Delta}(\mathbf{a}_t \times \mathbf{k}^r)(\mathbf{b}_t \times \mathbf{k}^i)\cdot(\mathbf{a}_t \times \mathbf{k}^i) = -E_a(\mathbf{a}_t \times \mathbf{k}^r) = \mathbf{E}_a^r,$$

and

$$\mathbf{E}_a^r + \mathbf{E}_a^i = E_a\mathbf{a}_t \times (\mathbf{k}^i - \mathbf{k}^r) = -2k_n E_a \mathbf{a}_t \times \mathbf{n},$$

which satisfies the boundary condition $\mathbf{a}_t\cdot(\mathbf{E}_a^r + \mathbf{E}_a^i) = 0$. On the other hand, from

$$\begin{aligned} k_o\eta_o\mathbf{H}_a^r &= \mathbf{k}^r \times \mathbf{E}_a^r = \frac{1}{\Delta}(\mathbf{k}^r \times (\mathbf{a}_t \times \mathbf{k}^r))(\mathbf{b}_t \cdot (\mathbf{k}^i \times \mathbf{E}_a^i) \\ &= \frac{1}{\Delta}(\mathbf{k}^r \times (\mathbf{a}_t \times \mathbf{k}^r))k_o\eta_o(\mathbf{b}_t\cdot\mathbf{H}_a^i), \end{aligned}$$

we obtain

$$\mathbf{b}_t\cdot\mathbf{H}_a^r = \frac{1}{\Delta}(\mathbf{b}_t \times \mathbf{k}^r)\cdot(\mathbf{a}_t \times \mathbf{k}^r))(\mathbf{b}_t\cdot\mathbf{H}_a^i) = -\mathbf{b}_t\cdot\mathbf{H}_a^i,$$

whence the second boundary condition $\mathbf{b}_t\cdot(\mathbf{H}_a^r + \mathbf{H}_a^i) = 0$ is satisfied.

Substituting $\mathbf{E}^i = \mathbf{E}_b^i = E_b((\mathbf{b}_t \times \mathbf{k}^i)\times\mathbf{k}^i)$ in (3.210) yields

$$\begin{aligned} \mathbf{E}^r &= \frac{-E_b}{\Delta}((\mathbf{b}_t \times \mathbf{k}^r)\times\mathbf{k}^r)(\mathbf{a}_t\cdot((\mathbf{b}_t \times \mathbf{k}^i)\times\mathbf{k}^i)) \\ &= -E_b((\mathbf{b}_t \times \mathbf{k}^r)\times\mathbf{k}^r) = \mathbf{E}_b^r. \end{aligned}$$

Expressing

$$k_o\eta_o\mathbf{H}_b^i = E_b k_o^2(\mathbf{b}_t \times \mathbf{k}^i), \qquad k_o\eta_o\mathbf{H}_b^r = -E_b k_o^2(\mathbf{b}_t \times \mathbf{k}^r),$$

we obtain

$$\eta_o(\mathbf{H}_b^i + \mathbf{H}_b^r) = -2E_b k_o k_n(\mathbf{b}_t \times \mathbf{n})$$

and, thus, $\mathbf{b}_t\cdot(\mathbf{H}_b^i + \mathbf{H}_b^r) = 0$. On the other hand, we also have

$$\mathbf{a}_t\cdot(\mathbf{E}_b^i + \mathbf{E}_b^r) = E_b\Delta - E_b\Delta = 0.$$

3.5 The boundary impedance and admittance dyadics are connected by (3.8):

$$\overline{\overline{\mathsf{Z}}}_t^{-1} = \mathbf{n}\mathbf{n}{}^\times_\times\overline{\overline{\mathsf{Y}}}_t, \qquad \overline{\overline{\mathsf{Y}}}_t^{-1} = \mathbf{n}\mathbf{n}{}^\times_\times\overline{\overline{\mathsf{Z}}}_t$$

They are inverses of each other if

$$\overline{\overline{\mathsf{Z}}}_t = \mathbf{nn}{}^{\times}_{\times}\overline{\overline{\mathsf{Z}}}_t$$

The decomposition (3.41) of the impedance dyadic

$$\overline{\overline{\mathsf{Z}}}_t = Z_o\overline{\overline{\mathsf{I}}}_t + Z_n\mathbf{n}\times\overline{\overline{\mathsf{I}}} + \overline{\overline{\mathsf{Z}}}_{ta}$$

splits it to the simple-isotropic, antisymmetric PEMC, and symmetric trace-free anisotropic parts.

Due to the dyadic rules $\mathbf{nn}{}^{\times}_{\times}\overline{\overline{\mathsf{I}}} = \overline{\overline{\mathsf{I}}}$ and $\mathbf{nn}{}^{\times}_{\times}(\mathbf{n}\times\overline{\overline{\mathsf{I}}}) = \mathbf{n}\times\overline{\overline{\mathsf{I}}}$, and the following property of the symmetric anisotropic part

$$\begin{aligned}\mathbf{nn}{}^{\times}_{\times}\overline{\overline{\mathsf{Z}}}_{ta} &= \mathbf{nn}{}^{\times}_{\times}\left(Z_1(\mathbf{e}_1\mathbf{e}_1 - \mathbf{e}_2\mathbf{e}_2) + Z_2(\mathbf{e}_1\mathbf{e}_2 + \mathbf{e}_2\mathbf{e}_1)\right)\\ &= Z_1(\mathbf{e}_2\mathbf{e}_2 - \mathbf{e}_1\mathbf{e}_1) + Z_2(-\mathbf{e}_2\mathbf{e}_1 - \mathbf{e}_1\mathbf{e}_2) = -\overline{\overline{\mathsf{Z}}}_{ta}\end{aligned}$$

we can conclude that the relation $\overline{\overline{\mathsf{Z}}}_t = \mathbf{nn}{}^{\times}_{\times}\overline{\overline{\mathsf{Z}}}_t$ is valid only for simple-isotropic and PEMC impedance boundaries or their combinations.

3.6 From

$$\overline{\overline{\mathsf{B}}}_t(\lambda)\cdot\mathbf{x}_t = (\overline{\overline{\mathsf{A}}}_t - \lambda\overline{\overline{\mathsf{C}}}_t)\cdot\mathbf{x}_t = 0, \quad \mathbf{x}_t \neq 0,$$

the 2D dyadic must satisfy

$$\det{}_t\overline{\overline{\mathsf{B}}}_t(\lambda) = \mathbf{nn} : \overline{\overline{\mathsf{B}}}{}_t^{(2)}(\lambda) = 0 \quad\Rightarrow\quad \overline{\overline{\mathsf{B}}}{}_t^{(2)}(\lambda) = 0.$$

Because any 2D dyadic is of the form

$$\overline{\overline{\mathsf{B}}}_t(\lambda) = \mathbf{b}_{1t}\mathbf{c}_{1t} + \mathbf{b}_{2t}\mathbf{c}_{2t},$$

from

$$\overline{\overline{\mathsf{B}}}{}_t^{(2)}(\lambda) = (\mathbf{b}_{1t}\times\mathbf{b}_{2t})(\mathbf{c}_{1t}\times\mathbf{c}_{2t}) = 0$$

and assuming $\mathbf{c}_{1t}\times\mathbf{c}_{2t} \neq 0$, we must have $\mathbf{b}_{1t}\times\mathbf{b}_{2t} = 0$, whence $\overline{\overline{\mathsf{B}}}_t(\lambda)$ must actually be of the form $\mathbf{b}_t\mathbf{c}_t$.

From $\overline{\overline{\mathsf{B}}}_t(\lambda)\cdot\mathbf{x}_t = \mathbf{b}_t\mathbf{c}_t\cdot\mathbf{x}_t = 0$ we conclude that $\mathbf{x}_t$ must be a multiple of $\mathbf{n}\times\mathbf{c}_t$, or, a multiple of

$$\mathbf{a}_t\cdot\mathbf{b}_t(\mathbf{c}_t\times\mathbf{n}) = \mathbf{a}_t\cdot\overline{\overline{\mathsf{B}}}_t(\lambda)\times\mathbf{n}.$$

3.7 From (3.195) and (3.196) we obtain

$$\begin{aligned}\mathbf{k}\times\mathbf{E}_a &= \frac{k_o}{\Delta}\mathbf{k}\times(\mathbf{k}\times\mathbf{a}_t)(\mathbf{b}_t\cdot\eta_o\mathbf{H}) = k_o\eta_o\mathbf{H}_a,\\ \mathbf{k}\times\eta_o\mathbf{H}_a &= \frac{1}{\Delta}\mathbf{k}\times((\mathbf{a}_t\times\mathbf{k})\times\mathbf{k})(\mathbf{b}_t\cdot\eta_o\mathbf{H}) = -k_o\mathbf{E}_a.\end{aligned}$$

From (3.197) and (3.198) we obtain

$$\begin{aligned}\mathbf{k}\times\mathbf{E}_b &= \frac{1}{\Delta}\mathbf{k}\times((\mathbf{b}_t\times\mathbf{k})\times\mathbf{k})(\mathbf{a}_t\cdot\mathbf{E}) = k_o\eta_o\mathbf{H}_b \\ \mathbf{k}\times\eta_o\mathbf{H}_b &= -\frac{k_o}{\Delta}\mathbf{k}\times(\mathbf{k}\times\mathbf{b}_t)(\mathbf{a}_t\cdot\mathbf{E}) = -k_o\mathbf{E}_b.\end{aligned}$$

Since the TE_a fields and TM_b fields satisfy plane-wave equations of their own, they form two independent plane waves.

3.8 Let us consider the problem backwards by representing the vector basis $(\mathbf{e}_1', \mathbf{e}_2')$ in terms of another vector basis $(\mathbf{e}_1, \mathbf{e}_2)$ as

$$\begin{pmatrix}\mathbf{e}_1' \\ \mathbf{e}_2'\end{pmatrix} = \begin{pmatrix}\cos\phi & \sin\phi \\ -\sin\phi & \cos\phi\end{pmatrix}\begin{pmatrix}\mathbf{e}_1 \\ \mathbf{e}_2\end{pmatrix}$$

We can expand

$$\begin{aligned}Z_K'(\mathbf{e}_1'\mathbf{e}_1' - \mathbf{e}_2'\mathbf{e}_2') &= Z_K'(\cos^2\phi - \sin^2\phi)(\mathbf{e}_1\mathbf{e}_1 - \mathbf{e}_2\mathbf{e}_2) \\ &\quad + Z_K'(2\sin\phi\cos\phi)(\mathbf{e}_1\mathbf{e}_2 + \mathbf{e}_2\mathbf{e}_1) \\ &= Z_K\overline{\overline{\mathsf{K}}} + Z_L\overline{\overline{\mathsf{L}}},\end{aligned}$$

from which we have the relations

$$Z_K = Z_K'\cos 2\phi, \qquad Z_L = Z_K'\sin 2\phi.$$

The converse relations are

$$Z_K' = \sqrt{Z_K^2 + Z_L^2}, \qquad \phi = \frac{1}{2}\tan^{-1}(Z_L/Z_K).$$

3.9 Applying $\mathrm{tr}\overline{\overline{\mathsf{K}}} = 0$ and $\det_t\overline{\overline{\mathsf{K}}} = -1$, we can expand

$$\begin{aligned}\Delta &= \det_t(-Z_K\overline{\overline{\mathsf{I}}}_t + \eta_o^2\overline{\overline{\mathsf{I}}}_t) + \eta_o Z_K\mathrm{tr}\overline{\overline{\mathsf{K}}} = -(Z_K^2 - \eta_o^2), \\ \overline{\overline{\mathsf{R}}}_t &= \frac{1}{\Delta}\Big((\Delta - 2\eta_o^2)\overline{\overline{\mathsf{I}}}_t - 2\eta_o Z_K(\overline{\overline{\mathsf{K}}}{}^{\times}_{\times}\mathbf{nn})\Big) \\ &= \frac{1}{Z_K^2 - \eta_o^2}\Big((Z_t^2 + \eta_o^2)\overline{\overline{\mathsf{I}}}_t - 2\eta_o Z_K(\mathbf{e}_1\mathbf{e}_1 - \mathbf{e}_2\mathbf{e}_2)\Big) \\ &= \frac{Z_K - \eta_o}{Z_K + \eta_o}\mathbf{e}_1\mathbf{e}_1 + \frac{Z_K + \eta_o}{Z_K - \eta_o}\mathbf{e}_2\mathbf{e}_2\end{aligned}$$

3.10 A matched incident plane wave must satisfy the impedance conditions (3.1):

$$k_o\eta_o\mathbf{E}_t^i = k_o\eta_o\overline{\overline{\mathsf{Z}}}_t\cdot(\mathbf{n}\times\mathbf{H}^i) = \overline{\overline{\mathsf{Z}}}_t\cdot(\mathbf{n}\times(\mathbf{k}^i\times\mathbf{E}^i) = \overline{\overline{\mathsf{Z}}}_t\cdot(\mathbf{k}_t\mathbf{k}_t/k_n + k_n\overline{\overline{\mathsf{I}}}_t)\cdot\mathbf{E}_t.$$

Assuming $k_n \neq 0$, this has the form $\overline{\overline{\mathsf{A}}}_t\cdot\mathbf{E}_t^i = 0$ with

$$\overline{\overline{\mathsf{A}}}_t = k_o k_n\eta_o\overline{\overline{\mathsf{I}}}_t - \overline{\overline{\mathsf{Z}}}_t\cdot(\mathbf{k}_t\mathbf{k}_t + k_n^2\overline{\overline{\mathsf{I}}}_t).$$

For $\mathbf{E}_t \neq 0$ we must have $\mathrm{det}_t\overline{\overline{\mathsf{A}}}_t = 0$, which yields a condition for the wave vector $\mathbf{k}^i$:

$$\begin{aligned}\mathrm{det}_t\overline{\overline{\mathsf{A}}}_t &= k_o^2k_n^2\eta_o^2 + (\mathrm{det}_t\overline{\overline{\mathsf{Z}}}_t)\mathrm{det}_t(\mathbf{k}_t\mathbf{k}_t + k_n^2\overline{\overline{\mathsf{I}}}_t)\\ &\quad -k_ok_n\eta_o\mathrm{tr}(\overline{\overline{\mathsf{I}}}_t{}^\times_\times(\overline{\overline{\mathsf{Z}}}_t\cdot(\mathbf{k}_t\mathbf{k}_t + k_n 2\overline{\overline{\mathsf{I}}}_t))) = 0.\end{aligned}$$

Here we must substitute

$$\begin{aligned}\mathrm{det}_t(\mathbf{k}_t\mathbf{k}_t + k_n^2\overline{\overline{\mathsf{I}}}_t) &= k_n^4 + k_n^2k_t^2 = k_n^2k_o^2,\\ \mathrm{tr}(\overline{\overline{\mathsf{I}}}_t{}^\times_\times(\overline{\overline{\mathsf{Z}}}_t\cdot(\mathbf{k}_t\mathbf{k}_t + k_n^2\overline{\overline{\mathsf{I}}}_t))) &= \mathrm{tr}(\overline{\overline{\mathsf{Z}}}_t\cdot(\mathbf{k}_t\mathbf{k}_t + k_n^2\overline{\overline{\mathsf{I}}}_t)),\\ &= \overline{\overline{\mathsf{Z}}}_t : \mathbf{k}_t\mathbf{k}_t + k_n^2\mathrm{tr}\overline{\overline{\mathsf{Z}}}_t.\end{aligned}$$

The dispersion equation becomes

$$k_ok_n(\mathrm{det}_t\overline{\overline{\mathsf{Z}}}_t + \eta_o^2) = \eta_o(\overline{\overline{\mathsf{Z}}}_t : \mathbf{k}_t\mathbf{k}_t + k_n^2\mathrm{tr}\overline{\overline{\mathsf{Z}}}_t),$$

where $k_n = -\mathbf{n}\cdot\mathbf{k}^i$. Starting from the reflected wave reverses the sign of k_n, whence the result coincides with (3.72).

3.11 For $\overline{\overline{\mathsf{Z}}}_t = Z_s\overline{\overline{\mathsf{I}}}_t + Z_n\mathbf{n}\times\overline{\overline{\mathsf{I}}}$, the expression (3.51) can be expanded as

$$\begin{aligned}\Delta &= Z_s^2 + Z_n^2 + \eta_o^2 + \frac{\eta_o}{k_ok_n}(Z_s(k_t^2 + 2k_n^2))\\ &= Z_s^2 + Z_n^2 + \eta_o^2 + Z_s\eta_o\frac{k_o^2 + k_n^2}{k_ok_n}\\ &= \frac{1}{k_ok_n}(k_oZ_s + k_n\eta_o)(k_nZ_s + k_o\eta_o) + Z_n^2.\end{aligned}$$

Expanding

$$\begin{aligned}\overline{\overline{\mathsf{J}}}_t\cdot\overline{\overline{\mathsf{Z}}}_t^T\times\mathbf{n} &= Z_s\overline{\overline{\mathsf{J}}}_t\times\mathbf{n} + Z_n\overline{\overline{\mathsf{J}}}_t\\ &= -\frac{Z_s}{k_ok_nk_t^2}(k_o^2(\mathbf{n}\times\mathbf{k})(\mathbf{n}\times\mathbf{k}) + k_n^2\mathbf{k}_t\mathbf{k}_t) + Z_n\overline{\overline{\mathsf{J}}}_t\end{aligned}$$

and substituting in (3.48) we obtain

$$\begin{aligned}\Delta\overline{\overline{\mathsf{R}}}_t &= (\Delta - 2\eta_o^2)\overline{\overline{\mathsf{I}}}_t + 2\eta_o\overline{\overline{\mathsf{J}}}_t\cdot\overline{\overline{\mathsf{Z}}}_t^T\times\mathbf{n}\\ &= \frac{1}{k_ok_nk_t^2}\Big((k_nZ_s - k_o\eta_o)(k_oZ_s + k_n\eta_o)(\mathbf{n}\times\mathbf{k})(\mathbf{n}\times\mathbf{k})\\ &+ (k_oZ_s - k_n\eta_o)(k_nZ_s + k_o\eta_o)\mathbf{k}_t\mathbf{k}_t\Big) + Z_n^2\overline{\overline{\mathsf{I}}}_t + 2\eta_oZ_n\overline{\overline{\mathsf{J}}}_t\end{aligned}$$

3.12 Setting $Z_{dn} = 0$ in (3.236) with $A = D = \cos\varphi$ and $B = -C = \sin\varphi$, yields

$$\begin{aligned}0 &= ACZ_s^2 + (AZ_n - B\eta_o)(CZ_n - D\eta_o))\\ &= -\cos\varphi\sin\varphi(Z_s^2 + Z_n^2 - \eta_o^2) - (\cos^2\varphi - \sin^2\varphi)Z_n\eta_o,\end{aligned}$$

which can be written as a condition for the parameter φ

$$\tan 2\varphi = \frac{2Z_n\eta_o}{Z_s^2 + Z_n^2 - \eta_o^2}.$$

In this case, the transformed impedance becomes

$$\overline{\overline{\mathsf{Z}}}_{dt} = \frac{\eta_o^2 Z_s}{Z_s^2 \sin^2\varphi + (Z_n \sin\varphi + \eta_o\cos\varphi)^2}\overline{\overline{\mathsf{I}}}_t.$$

3.13 (3.142) can be expanded as

$$(-Z_s((\mathbf{n}\times\mathbf{k})(\mathbf{n}\times\mathbf{k}) + k_n^2\overline{\overline{\mathsf{I}}}_t) + Z_n((\mathbf{n}\times\mathbf{k})\mathbf{k}_t + k_n^2\mathbf{n}\times\overline{\overline{\mathsf{I}}})) - \lambda k_o k_n\overline{\overline{\mathsf{I}}}_t)\cdot\mathbf{z}_t = 0$$

Substituting $\mathbf{z}_t = A\mathbf{k}_t + B\mathbf{n}\times\mathbf{k}$ and separating the coefficients of $\mathbf{k}_t$ and $\mathbf{n}\times\mathbf{k}$, we obtain

$$\begin{pmatrix} k_n(Z_s k_n + \lambda k_o) & Z_n k_n^2 \\ -Z_n k_o^2 & k_o(Z_s k_o + \lambda k_n) \end{pmatrix}\begin{pmatrix} A \\ B \end{pmatrix} = \begin{pmatrix} 0 \\ 0 \end{pmatrix}$$

Requiring vanishing of the determinant yields the eigenvalue equation

$$\lambda^2 + \lambda Z_s\frac{k_o^2 + k_n^2}{k_o k_n} + Z_s^2 + Z_n^2 = 0,$$

whose solutions $\lambda_\pm$ can be written as (3.143). From

$$A_\pm = \frac{1}{Z_n k_o}(Z_s k_o + \lambda_\pm k_n)B_\pm$$

we obtain

$$\mathbf{z}_{t\pm} = \frac{B_\pm}{Z_n k_o}((Z_s k_o + \lambda_\pm \eta_o k_n)\mathbf{k}_t + Z_n k_o(\mathbf{n}\times\mathbf{k})).$$

Choosing $B_\pm = Z_n k_o$, this coincides with the expression (3.144).

3.14 Substituting (3.216), (3.217) and (3.218) in the dispersion equation (3.215), we obtain

$$k_o^2(\cos^2\alpha - \sin^2\alpha) = k_t^2(\cos\alpha\cos\varphi + \sin\alpha\sin\varphi)(\cos\alpha\cos\varphi - \sin\alpha\sin\varphi),$$

or,

$$k_o^2\cos 2\varphi = k_t^2\cos(\varphi-\alpha)\cos(\varphi+\alpha)$$

For the lateral wave, $k_t = k_o$ we must have

$$\cos 2\varphi = \cos(\varphi-\alpha)\cos(\varphi+\alpha)$$

or

$$\begin{aligned}\cos^2\varphi - \sin^2\varphi &= \cos^2\varphi\cos^2\alpha - \sin^2\varphi\sin^2\alpha \\ &= \cos^2\varphi - \cos^2\varphi\sin^2\alpha - \sin^2\varphi\sin^2\alpha \\ &= \cos^2\varphi - \sin^2\alpha,\end{aligned}$$

which require $\sin\varphi = \pm\sin\alpha$, or $\varphi = \pm\alpha$.

3.15 By definition, a perfectly anisotropic impedance dyadic $\overline{\overline{\mathsf{Z}}}_{ta}$ satisfies $\overline{\overline{\mathsf{Z}}}_{ta} {}^\times_\times \mathbf{nn} = -\overline{\overline{\mathsf{Z}}}_{ta}$, whence it is both symmetric and trace-free. Expanding

$$(\mathbf{n} \times \overline{\overline{\mathsf{Z}}}_{ta}) {}^\times_\times \mathbf{nn} = -\mathbf{n} \times (\mathbf{n} \times \overline{\overline{\mathsf{Z}}}_{ta} \times \mathbf{n}) = \mathbf{n} \times (\overline{\overline{\mathsf{Z}}}_{ta} {}^\times_\times \mathbf{nn}) = -\mathbf{n} \times \overline{\overline{\mathsf{Z}}}_{ta},$$

we find that $\mathbf{n} \times \overline{\overline{\mathsf{Z}}}_{ta}$ is perfectly anisotropic whenever $\overline{\overline{\mathsf{Z}}}_{ta}$ is perfectly anisotropic. Substituting $\overline{\overline{\mathsf{Z}}}_{ta} = Z_1(\mathbf{e}_1\mathbf{e}_1 - \mathbf{e}_2\mathbf{e}_2) + Z_2(\mathbf{e}_1\mathbf{e}_2 + \mathbf{e}_2\mathbf{e}_1)$, one can check that $\mathbf{e}_3 \times \overline{\overline{\mathsf{Z}}}_{ta}$ is both symmetric and trace free. The dyadic $\overline{\overline{\mathsf{Z}}}_{ta} \times \mathbf{n}$ can be handled similarly.

3.16 Substituting the expansion $\overline{\overline{\mathsf{Z}}}_t = Z_s \overline{\overline{\mathsf{I}}}_t + Z_n \mathbf{n} \times \overline{\overline{\mathsf{I}}} + \overline{\overline{\mathsf{Z}}}_{ta}$ as

$$\mathrm{tr}(\overline{\overline{\mathsf{Z}}}_t \times \mathbf{n}) = Z_s \mathrm{tr}(\overline{\overline{\mathsf{I}}} \times \mathbf{n}) + Z_n \mathrm{tr}(-\overline{\overline{\mathsf{I}}}_t) + \mathrm{tr}(\overline{\overline{\mathsf{Z}}}_{ta} \times \mathbf{n}),$$

applying $\mathrm{tr}(\overline{\overline{\mathsf{I}}} \times \mathbf{n}) = 0$, $\mathrm{tr}\overline{\overline{\mathsf{I}}}_t = 2$, and, because $\overline{\overline{\mathsf{Z}}}_{ta}$ is a symmetric dyadic,

$$\mathrm{tr}(\overline{\overline{\mathsf{Z}}}_{ta} \times \mathbf{n}) = (\overline{\overline{\mathsf{Z}}}_{ta} \cdot (\overline{\overline{\mathsf{I}}} \times \mathbf{n})) : \overline{\overline{\mathsf{I}}}_t = -\overline{\overline{\mathsf{Z}}}_{ta} : (\mathbf{n} \times \overline{\overline{\mathsf{I}}}) = 0,$$

we obtain

$$\mathrm{tr}(\overline{\overline{\mathsf{Z}}}_t \times \mathbf{n}) = -2Z_n.$$

Considering

$$\begin{aligned} \det{}_t \overline{\overline{\mathsf{Z}}}_t &= \mathrm{tr}\overline{\overline{\mathsf{Z}}}_t^{(2)} \\ &= \mathrm{tr}(Z_s^2 \overline{\overline{\mathsf{I}}}_t + Z_n^2 (\mathbf{n} \times \overline{\overline{\mathsf{I}}})^{(2)} + \mathrm{tr}\overline{\overline{\mathsf{Z}}}_{ta}^{(2)}) \\ &\quad + \mathrm{tr}(Z_s Z_n \overline{\overline{\mathsf{I}}}_t {}^\times_\times (\overline{\overline{\mathsf{I}}} \times \mathbf{n}) + Z_s \overline{\overline{\mathsf{I}}}_t {}^\times_\times \overline{\overline{\mathsf{Z}}}_{ta} + Z_n (\overline{\overline{\mathsf{I}}} \times \mathbf{n}) {}^\times_\times \overline{\overline{\mathsf{Z}}}_{ta}), \end{aligned}$$

we can expand

$$\begin{aligned} \det{}_t(\overline{\overline{\mathsf{Z}}}_{ta} \times \mathbf{n}) &= \mathrm{tr}(\overline{\overline{\mathsf{Z}}}_{ta} \times \mathbf{n})^{(2)} = \mathrm{tr}(\overline{\overline{\mathsf{Z}}}_{ta}^{(2)} \cdot (\overline{\overline{\mathsf{I}}} \times \mathbf{n})^{(2)}) = \mathrm{tr}(\overline{\overline{\mathsf{Z}}}_{ta}^{(2)} \cdot \mathbf{nn})) \\ &= \mathrm{tr}(\overline{\overline{\mathsf{Z}}}_{ta}^{(2)}) = \det{}_t \overline{\overline{\mathsf{Z}}}_{ta} \end{aligned}$$

3.17 We can expand the reflection dyadic as

$$\begin{aligned} \overline{\overline{\mathsf{R}}}_t &= \overline{\overline{\mathsf{I}}}_t + \frac{2\eta_o}{\Delta}(\overline{\overline{\mathsf{Z}}}_t \cdot (\mathbf{n} \times \overline{\overline{\mathsf{J}}}_t) - \eta_o \overline{\overline{\mathsf{I}}}_t)^T {}^\times_\times \mathbf{nn}, \\ &= \overline{\overline{\mathsf{I}}}_t + \frac{2\eta_o}{\Delta}((\mathbf{n} \times \overline{\overline{\mathsf{J}}}_t^T \times \mathbf{n}) \cdot \overline{\overline{\mathsf{Z}}}^T \times \mathbf{n} - \eta_o \overline{\overline{\mathsf{I}}}_t), \\ &= (1 - \frac{2\eta_o^2}{\Delta})\overline{\overline{\mathsf{I}}}_t + \frac{2\eta_o}{\Delta}\overline{\overline{\mathsf{J}}}_t \cdot \overline{\overline{\mathsf{Z}}}_t^T \times \mathbf{n}, \end{aligned}$$

The determinant quantity Δ can be expanded as

$$\begin{aligned}
\Delta &= \mathrm{det}_t(\overline{\overline{\mathsf{Z}}}_t\cdot\mathbf{n}\times\overline{\overline{\mathsf{J}}}_t-\eta_o\overline{\overline{\mathsf{I}}}_t)\\
&= \mathbf{nn}:(\overline{\overline{\mathsf{Z}}}_t\cdot\mathbf{n}\times\overline{\overline{\mathsf{J}}}_t-\eta_o\overline{\overline{\mathsf{I}}}_t)^{(2)}\\
&= \mathrm{det}_t\overline{\overline{\mathsf{Z}}}_t+\eta_o^2-\eta_o\mathrm{tr}(\overline{\overline{\mathsf{Z}}}_t\cdot\mathbf{n}\times\overline{\overline{\mathsf{J}}}_t)\\
&= \mathrm{det}_t\overline{\overline{\mathsf{Z}}}_t+\eta_o^2+\frac{\eta_o}{k_ok_n}(\overline{\overline{\mathsf{Z}}}_t:\mathbf{k}_t\mathbf{k}_t+k_n^2\mathrm{tr}\overline{\overline{\mathsf{Z}}}_t).
\end{aligned}$$

3.18 Substituting the dispersion dyadic (3.70) in the dispersion equation (3.71), we obtain

$$\begin{aligned}
\mathrm{det}_t\overline{\overline{\mathsf{D}}}_t(\mathbf{k}) &= \mathrm{tr}\overline{\overline{\mathsf{D}}}_t^{(2)}(\mathbf{k})=\mathbf{nn}:\overline{\overline{\mathsf{D}}}_t^{(2)}(\mathbf{k})\\
&= \mathbf{nn}:((k_ok_n\eta_o)\overline{\overline{\mathsf{I}}}_t+\overline{\overline{\mathsf{Z}}}_t\cdot\mathbf{k}_t\mathbf{k}_t+k_n^2\overline{\overline{\mathsf{Z}}}_t)^{(2)}\\
&= \mathbf{nn}:((k_ok_n\eta_o)^2\overline{\overline{\mathsf{I}}}_t^{(2)}+(\overline{\overline{\mathsf{Z}}}_t\cdot\mathbf{k}_t\mathbf{k}_t)^{(2)}+k_n^4\overline{\overline{\mathsf{Z}}}_t^{(2)}\\
&\quad +k_ok_n\eta_o\overline{\overline{\mathsf{I}}}_t{}^{\times}_{\times}(\overline{\overline{\mathsf{Z}}}_t\cdot\mathbf{k}_t\mathbf{k}_t)\\
&\quad +k_ok_n^3\eta_o\overline{\overline{\mathsf{I}}}_t{}^{\times}_{\times}\overline{\overline{\mathsf{Z}}}_t+k_n^2(\overline{\overline{\mathsf{Z}}}_t\cdot\mathbf{k}_t\mathbf{k}_t){}^{\times}_{\times}\overline{\overline{\mathsf{Z}}}_t)
\end{aligned}$$

We can expand termwise,

$$\mathbf{nn}:((k_ok_n\eta_o)^2\overline{\overline{\mathsf{I}}}_t^{(2)})=(k_ok_n\eta_o)^2$$

$$\mathbf{nn}:(\overline{\overline{\mathsf{Z}}}_t\cdot\mathbf{k}_t\mathbf{k}_t)^{(2)}=\mathbf{n}\cdot((\overline{\overline{\mathsf{Z}}}_t\cdot\mathbf{k}_t)\times(\overline{\overline{\mathsf{Z}}}_t\cdot\mathbf{k}_t))\mathbf{n}\cdot(\mathbf{k}_t\times\mathbf{k}_t)=0$$

$$\mathbf{nn}:(k_n^4\overline{\overline{\mathsf{Z}}}_t^{(2)})=k_n^4\mathrm{det}_t\overline{\overline{\mathsf{Z}}}_t$$

$$\mathbf{nn}:(k_ok_n\eta_o\overline{\overline{\mathsf{I}}}_t{}^{\times}_{\times}(\overline{\overline{\mathsf{Z}}}_t\cdot\mathbf{k}_t\mathbf{k}_t))=k_ok_n\eta_o\overline{\overline{\mathsf{Z}}}_t:\mathbf{k}_t\mathbf{k}_t$$

$$\mathbf{nn}:(k_ok_n^3\eta_o\overline{\overline{\mathsf{I}}}_t{}^{\times}_{\times}\overline{\overline{\mathsf{Z}}}_t)=k_ok_n^3\mathrm{tr}\overline{\overline{\mathsf{Z}}}_t$$

$$\begin{aligned}
\mathbf{nn}:(k_n^2(\overline{\overline{\mathsf{Z}}}_t\cdot\mathbf{k}_t\mathbf{k}_t){}^{\times}_{\times}\overline{\overline{\mathsf{Z}}}_t)) &= -k_n^2\mathbf{n}\cdot((\overline{\overline{\mathsf{Z}}}_t\cdot\mathbf{k}_t)\times\overline{\overline{\mathsf{Z}}}_t\times\mathbf{k}_t)\cdot\mathbf{n}\\
&= k_n^2\mathbf{k}_t\cdot(\overline{\overline{\mathsf{Z}}}_t^T\cdot\mathbf{n}\times\overline{\overline{\mathsf{Z}}}_t)\cdot(\mathbf{k}_t\times\mathbf{n})=k_n^2k_t^2\mathrm{det}_t\overline{\overline{\mathsf{Z}}}_t.
\end{aligned}$$

For the last step we have applied the identity (see Appendix C)

$$\overline{\overline{\mathsf{A}}}_t^T\cdot\mathbf{n}\times\overline{\overline{\mathsf{A}}}_t=\mathrm{det}_t\overline{\overline{\mathsf{A}}}_t(\mathbf{n}\times\overline{\overline{\mathsf{I}}}).$$

Combining the terms we finally obtain the dispersion equation

$$(k_ok_n\eta_o)^2+0+k_n^4\mathrm{det}_t\overline{\overline{\mathsf{Z}}}_t+k_ok_n\eta_o\overline{\overline{\mathsf{Z}}}_t:\mathbf{k}_t\mathbf{k}_t+k_ok_n^3\mathrm{tr}\overline{\overline{\mathsf{Z}}}_t+k_n^2k_t^2\mathrm{det}_t\overline{\overline{\mathsf{Z}}}_t=0.$$

Assuming $k_n\neq 0$ this is reduced a form equivalent to that of (3.72),

$$k_ok_n\eta_o^2+\eta_o\overline{\overline{\mathsf{Z}}}_t:\mathbf{k}_t\mathbf{k}_t+\eta_ok_n^2\mathrm{tr}\overline{\overline{\mathsf{Z}}}_t+k_nk_o\mathrm{det}\overline{\overline{\mathsf{Z}}}_t=0,$$

which also equals $\Delta=0$ with Δ defined by (3.51).

3.19 Assuming that the vectors $\mathbf{a}_{1t}$ and $\mathbf{a}_{2t}$ in (3.17) and (3.18) make a 2D basis, whence

$$\begin{aligned}(\mathbf{a}_{1t}\times\mathbf{a}_{2t})\times\overline{\overline{\mathsf{I}}} &= \mathbf{a}_{2t}\mathbf{a}_{1t}-\mathbf{a}_{1t}\mathbf{a}_{2t} \\ &= (\mathbf{n}\cdot\mathbf{a}_{1t}\times\mathbf{a}_{2t})\mathbf{n}\times\overline{\overline{\mathsf{I}}},\end{aligned}$$

we can express

$$\begin{aligned}\mathbf{I}_t &= -\mathbf{n}\times(\mathbf{n}\times\mathbf{I}) \\ &= \frac{1}{\mathbf{n}\cdot\mathbf{a}_{1t}\times\mathbf{a}_{2t}}\mathbf{n}\times(\mathbf{a}_{1t}\mathbf{a}_{2t}-\mathbf{a}_{2t}\mathbf{a}_{1t}).\end{aligned}$$

Applying (3.17) and (3.18), we obtain

$$\begin{aligned}\mathbf{E}_t &= \frac{1}{\mathbf{n}\cdot\mathbf{a}_{1t}\times\mathbf{a}_{2t}}\mathbf{n}\times(\mathbf{a}_{1t}\mathbf{a}_{2t}-\mathbf{a}_{2t}\mathbf{a}_{1t})\cdot\mathbf{E}_t \\ &= \frac{1}{\mathbf{n}\cdot\mathbf{a}_{1t}\times\mathbf{a}_{2t}}\mathbf{n}\times(-\mathbf{a}_{1t}\mathbf{b}_{2t}+\mathbf{a}_{2t}\mathbf{b}_{1t})\cdot\mathbf{H}_t.\end{aligned}$$

Comparing with (3.1), we can express the impedance dyadic as

$$\overline{\overline{\mathsf{Z}}}_s = -\frac{1}{\mathbf{n}\cdot\mathbf{a}_{1t}\times\mathbf{a}_{2t}}\mathbf{n}\times(-\mathbf{a}_{1t}\mathbf{b}_{2t}+\mathbf{a}_{2t}\mathbf{b}_{1t})\times\mathbf{n}.$$

The condition of reciprocity (3.25) is equal to the condition

$$-\mathbf{a}_{1t}\mathbf{b}_{2t}+\mathbf{a}_{2t}\mathbf{b}_{1t} = -\mathbf{b}_{2t}\mathbf{a}_{1t}+\mathbf{b}_{1t}\mathbf{a}_{2t},$$

or,

$$\mathbf{b}_{2t}\mathbf{a}_{1t}-\mathbf{a}_{1t}\mathbf{b}_{2t} = \mathbf{b}_{1t}\mathbf{a}_{2t}-\mathbf{a}_{2t}\mathbf{b}_{1t},$$

which equals the condition

$$\mathbf{a}_{1t}\times\mathbf{b}_{2t} = \mathbf{a}_{2t}\times\mathbf{b}_{1t}.$$

3.20 Applying the plane-wave relations

$$\eta_o\mathbf{H}^i = -\overline{\overline{\mathsf{J}}}_t\cdot\mathbf{E}^i_t,\quad \eta_o\mathbf{H}^r_t = \overline{\overline{\mathsf{J}}}_t\cdot\mathbf{E}^r_t = R\overline{\overline{\mathsf{J}}}_t\cdot\mathbf{E}^i_t = -R\eta_o\mathbf{H}^i_t,$$

the GSH boundary conditions can be written as

$$\mathbf{a}_t\cdot\mathbf{E}_t = \mathbf{a}_t\cdot(\mathbf{E}^i_t+\mathbf{E}^r_t) = (1+R)\mathbf{a}_t\cdot\mathbf{E}^i_t = 0,$$

$$\mathbf{b}_t\cdot\mathbf{H}_t = \mathbf{b}_t\cdot(\mathbf{H}^i_t+\mathbf{H}^r_t) = (1-R)\mathbf{b}_t\cdot\mathbf{H}^i_t.$$

Expanding $k_o\eta_o\mathbf{b}_t\cdot\mathbf{H}^i_t = k_o\eta_o\mathbf{b}_t\cdot\mathbf{H}^i = \mathbf{b}_t\cdot(\mathbf{k}^i\times\mathbf{E}^i)$ the vector $\mathbf{E}^i$ satisfies the three conditions

$$(1+R)\mathbf{a}_t\cdot\mathbf{E}^i = 0,\quad (1-R)(\mathbf{b}_t\times\mathbf{k}^i)\cdot\mathbf{E}^i = 0,\quad \mathbf{k}^i\cdot\mathbf{E}^i = 0.$$

For $\mathbf{E}^i \neq 0$, the three vectors must satisfy

$$(1-R^2)\mathbf{a}_t \cdot (\mathbf{b}_t \times \mathbf{k}^i) \times \mathbf{k}^i = 0$$

For the general $\mathbf{k}^i$ satisfying $(\mathbf{a}_t \times \mathbf{k}^i)\cdot(\mathbf{b}_t \times \mathbf{k}^i) \neq 0$, we must have $R=1$ or $R=-1$. The two respective cases $\mathbf{E}_t^r = \mathbf{E}_t^i$ and $\mathbf{E}_t^r = -\mathbf{E}_t^i$ correspond to reflection from equivalent PMC and PEC boundaries.

3.21 Expanding

$$\begin{aligned}(\mathbf{a}_t \times (\mathbf{n}\times\mathbf{b}_t))\times \overline{\overline{\mathsf{I}}} &= (\mathbf{a}_t\cdot\mathbf{b}_t)\mathbf{n}\times\overline{\overline{\mathsf{I}}}_t, \\ &= (\mathbf{n}\times\mathbf{b}_t)\mathbf{a}_t - \mathbf{a}_t(\mathbf{n}\times\mathbf{b}_t),\end{aligned}$$

we obtain

$$\begin{aligned}(\mathbf{a}_t\cdot\mathbf{b}_t)\overline{\overline{\mathsf{I}}}_t &= -\mathbf{n}\times((\mathbf{n}\times\mathbf{b}_t)\mathbf{a}_t - \mathbf{a}_t(\mathbf{n}\times\mathbf{b}_t)) \\ &= \mathbf{b}_t\mathbf{a}_t + (\mathbf{n}\times\mathbf{a}_t)(\mathbf{n}\times\mathbf{b}_t).\end{aligned}$$

Applying this to (3.181), we can expand

$$\begin{aligned}\overline{\overline{\mathsf{Z}}}_t/\eta_o &= \delta\mathbf{b}_t\mathbf{a}_t + \frac{1}{\delta}(\mathbf{n}\times\mathbf{a})(\mathbf{n}\times\mathbf{b}) \\ &= \frac{1}{2}(\delta+\frac{1}{\delta})(\mathbf{b}_t\mathbf{a}_t + (\mathbf{n}\times\mathbf{a})(\mathbf{n}\times\mathbf{b})) \\ &\quad +\frac{1}{2}(\delta-\frac{1}{\delta})(\mathbf{b}_t\mathbf{a}_t - (\mathbf{n}\times\mathbf{a})(\mathbf{n}\times\mathbf{b})),\end{aligned}$$

the first term of which yields the simple-isotropic part $Z_s\overline{\overline{\mathsf{I}}}_t$ as

$$\frac{\eta_o}{2}(\delta+\frac{1}{\delta})(\mathbf{b}_t\mathbf{a}_t + (\mathbf{n}\times\mathbf{a}_t)(\mathbf{n}\times\mathbf{b}_t)) = \frac{\eta_o}{2}(\delta+\frac{1}{\delta})(\mathbf{a}_t\cdot\mathbf{b}_t)\overline{\overline{\mathsf{I}}}_t.$$

The second term is trace free and contains the antisymmetric part $Z_n\mathbf{n}\times\overline{\overline{\mathsf{I}}}$ and the anisotropic part $\overline{\overline{\mathsf{Z}}}_{ta}$. Denoting $D=\delta-1/\delta$, they can be identified as

$$\begin{aligned}Z_n\mathbf{n}\times\overline{\overline{\mathsf{I}}} &= \eta_o\frac{D}{2}(\mathbf{b}_t\mathbf{a}_t - (\mathbf{n}\times\mathbf{a}_t)(\mathbf{n}\times\mathbf{b}_t) - \mathbf{a}_t\mathbf{b}_t + (\mathbf{n}\times\mathbf{b}_t)(\mathbf{n}\times\mathbf{a}_t)) \\ &= \eta_o\frac{D}{2}(\mathbf{a}_t\times\mathbf{b}_t + (\mathbf{n}\times\mathbf{a}_t)\times(\mathbf{n}\times\mathbf{b}_t))\times\overline{\overline{\mathsf{I}}}_t \\ &= D(\mathbf{n}\cdot\mathbf{a}_t\times\mathbf{b}_t)\mathbf{n}\times\overline{\overline{\mathsf{I}}}_t, \\ \overline{\overline{\mathsf{Z}}}_{ta} &= \eta_o\frac{D}{2}(\overline{\overline{\mathsf{A}}}_t + \overline{\overline{\mathsf{A}}}_t^T), \quad \overline{\overline{\mathsf{A}}}_t = \mathbf{b}_t\mathbf{a}_t - (\mathbf{n}\times\mathbf{a})(\mathbf{n}\times\mathbf{b}).\end{aligned}$$

3.22 Starting from a representation of the 2D unit dyadic

$$\overline{\overline{\mathsf{I}}}_t = \frac{1}{\mathbf{n}\cdot\mathbf{a}_{1t}\times\mathbf{a}_{2t}}(\mathbf{n}\times\mathbf{a}_{1t}\mathbf{a}_{2t} - \mathbf{n}\times\mathbf{a}_{2t}\mathbf{a}_{1t}),$$

and expanding $\mathbf{E}_t = \overline{\overline{\mathsf{I}}}_t \cdot \mathbf{E}_t$, we obtain the impedance dyadic in the form

$$\overline{\overline{\mathsf{Z}}}_t = Z\mathbf{n}\mathbf{n}{}^{\times}_{\times}(\mathbf{a}_{2t}\mathbf{b}_{1t} - \mathbf{a}_{1t}\mathbf{b}_{2t}), \quad Z = \frac{\eta_o}{\mathbf{n}\cdot\mathbf{a}_{1t}\times\mathbf{a}_{2t}}.$$

The eigendyadic $\overline{\overline{\mathsf{Z}}}_{t+} = Z_s\overline{\overline{\mathsf{I}}}_t + Z_n\mathbf{n}\times\overline{\overline{\mathsf{I}}}$ is obtained as

$$\begin{aligned}\overline{\overline{\mathsf{Z}}}_{t+} &= (1/2)(\overline{\overline{\mathsf{Z}}}_t + \mathbf{n}\mathbf{n}{}^{\times}_{\times}\overline{\overline{\mathsf{Z}}}_t) \\ &= Z(\mathbf{n}\mathbf{n}{}^{\times}_{\times}(\mathbf{a}_{2t}\mathbf{b}_{1t} - \mathbf{a}_{1t}\mathbf{b}_{2t}) + (\mathbf{a}_{2t}\mathbf{b}_{1t} - \mathbf{a}_{1t}\mathbf{b}_{2t})).\end{aligned}$$

Applying the rule

$$\mathbf{n}\mathbf{n}{}^{\times}_{\times}\overline{\overline{\mathsf{A}}}_t = \mathrm{tr}\overline{\overline{\mathsf{A}}}_t\,\overline{\overline{\mathsf{I}}}_t - \overline{\overline{\mathsf{A}}}_t^T,$$

we obtain $\overline{\overline{\mathsf{Z}}}_{t+} = Z_n\mathbf{n}\times\overline{\overline{\mathsf{I}}}_t + Z_s\overline{\overline{\mathsf{I}}}_t$, with

$$Z_s = Z(\mathbf{a}_{2t}\cdot\mathbf{b}_{1t} - \mathbf{a}_{1t}\cdot\mathbf{b}_{2t})$$

and

$$Z_n = Z\mathbf{n}\cdot(\mathbf{b}_{1t}\times\mathbf{a}_{2t} - \mathbf{b}_{2t}\times\mathbf{a}_{1t}).$$

The anisotropic part can be expanded through the same rule as

$$\begin{aligned}\overline{\overline{\mathsf{Z}}}_{ta} &= \overline{\overline{\mathsf{Z}}}_{t-} = (1/2)(\overline{\overline{\mathsf{Z}}}_t + \mathbf{n}\mathbf{n}{}^{\times}_{\times}\overline{\overline{\mathsf{Z}}}_t) \\ &= Z(\mathbf{a}_{1t}\mathbf{b}_{2t} + \mathbf{b}_{2t}\mathbf{a}_{1t} - \mathbf{a}_{2t}\mathbf{b}_{1t} - \mathbf{b}_{1t}\mathbf{a}_{2t} \\ &\quad +(\mathbf{a}_{2t}\cdot\mathbf{b}_{1t} - \mathbf{a}_{1t}\cdot\mathbf{b}_{2t})\overline{\overline{\mathsf{I}}}_t).\end{aligned}$$

3.23 For imaginary surface impedance, $Z_s = jX_s$, the reflection coefficients satisfy

$$|R_{TE}|^2 = |R_{TM}|^2 = 1, \tag{D.1}$$

when $\mathbf{k}^i$ is a real vector. From this it follows that

$$\begin{aligned}|\mathbf{E}_t^r|^2 &= |\mathbf{E}_{TEt}^r + \mathbf{E}_{TMt}^r|^2 \\ &= |R_{TE}\mathbf{E}_{TE_t}^i + R_{TM}\mathbf{E}_{TMt}^i|^2 \\ &= |R_{TE}|^2|\mathbf{E}_{TE_t}^i|^2 + |R_{TM}|^2|\mathbf{E}_{TMt}^i|^2 \\ &= |\mathbf{E}_{TE_t}^i|^2 + |\mathbf{E}_{TMt}^i|^2 = |\mathbf{E}_t^i|^2,\end{aligned}$$

where we have applied (3.81). This verifies that an imaginary impedance surface is lossless.

3.24 The dispersion equation for the general impedance surface (3.72),

$$k_o k_n(\eta_o^2 + \det{}_t\overline{\overline{\mathsf{Z}}}_t) + \eta_o(\overline{\overline{\mathsf{Z}}}_t : \mathbf{k}_t\mathbf{k}_t + k_n^2\mathrm{tr}\overline{\overline{\mathsf{Z}}}_t) = 0$$

for the impedance dyadic $\overline{\overline{\mathsf{Z}}}_t = \eta_o(\mathbf{b}_t\mathbf{a}_t\delta + (\mathbf{n}\times\mathbf{a}_t)(\mathbf{n}\times\mathbf{b}_t)/\delta)$ can be expanded by substituting

$$\mathrm{tr}\overline{\overline{\mathsf{Z}}}_t = \eta_o(\mathbf{a}_t\cdot\mathbf{b}_t)(\delta + 1/\delta),$$

$$\mathrm{det}_t\overline{\overline{\mathsf{Z}}}_t = \eta_o^2\mathbf{n}\cdot(\mathbf{b}_t\times(\mathbf{n}\times\mathbf{a}_t))\mathbf{n}\cdot(\mathbf{a}_t\times(\mathbf{n}\times\mathbf{b}_t)) = \eta_o^2(\mathbf{a}_t\cdot\mathbf{b}_t)^2,$$

$$\overline{\overline{\mathsf{Z}}}_t : \mathbf{k}_t\mathbf{k}_t = \eta_o((\mathbf{a}_t\cdot\mathbf{k}_t)(\mathbf{b}_t\cdot\mathbf{k}_t)\delta + (\mathbf{n}\times\mathbf{a}_t\cdot\mathbf{k}_t)(\mathbf{n}\cdot\mathbf{b}_t\times\mathbf{k}_t)/\delta),$$

whence the resulting equation takes the form

$$k_ok_n(1+(\mathbf{a}_t\cdot\mathbf{b}_t)^2) + ((\mathbf{a}_t\cdot\mathbf{k}_t)(\mathbf{b}_t\cdot\mathbf{k}_t) + k_n^2(\mathbf{a}_t\cdot\mathbf{b}_t))\delta +$$

$$+((\mathbf{n}\times\mathbf{a}_t\cdot\mathbf{k}_t)(\mathbf{n}\cdot\mathbf{b}_t\times\mathbf{k}_t) + k_n^2(\mathbf{a}_t\cdot\mathbf{b}_t))1/\delta = 0$$

For $\delta\to 0$, the last term dominates, whence the dispersion equation is reduced to

$$(\mathbf{n}\times\mathbf{a}_t\cdot\mathbf{k}_t)(\mathbf{n}\cdot\mathbf{b}_t\times\mathbf{k}_t) + k_n^2(\mathbf{a}_t\cdot\mathbf{b}_t) = 0$$

Now we can apply the expansion

$$(\mathbf{a}_t\cdot\mathbf{b}_t)\overline{\overline{\mathsf{I}}}_t = (\mathbf{n}\times\mathbf{a}_t)(\mathbf{n}\times\mathbf{b}_t) + \mathbf{b}_t\mathbf{a}_t,$$

whence the dispersion equation can be written in an equivalent form as

$$(\mathbf{a}_t\cdot\mathbf{b}_t)k_t^2 - (\mathbf{b}_t\cdot\mathbf{k}_t)(\mathbf{a}_t\cdot\mathbf{k}_t) + k_n^2(\mathbf{a}_t\cdot\mathbf{b}_t) = 0.$$

This is equivalent to

$$(\mathbf{a}_t\cdot\mathbf{b}_t)k_o^2 - (\mathbf{a}_t\cdot\mathbf{k}_t)(\mathbf{b}_t\cdot\mathbf{k}_t) = (\mathbf{a}_t\times\mathbf{k})\cdot(\mathbf{b}_t\times\mathbf{k}) = 0,$$

which coincides with (3.215).

4.1 Assuming $\mathbf{k}^i = \mathbf{e}_1k_1 - \mathbf{e}_3k_3$ for the incident wave and $\mathbf{E}^i = \mathbf{e}_2E^i$, the TE electric field is parallel to $\mathbf{e}_2$ in each region. Tangential fields in each region are defined by

- Region $x_3\geq 0$, at $x_3 = 0$,

$$\mathbf{E} = \mathbf{e}_2(E^i + E^r),\quad \omega\mu_o\mathbf{H}_t = (\mathbf{k}^i\times\mathbf{E}^i + \mathbf{k}^r\times\mathbf{E}^r)_t = \mathbf{e}_1(k_3E^i - k_3E^r),$$

- Region $0\geq x_3\geq -d$, with $\beta = \beta_{TE}$,

$$\begin{aligned}\mathbf{E}(x_3) &= \mathbf{e}_2(E^+e^{-j\beta x_3} + E^-e^{j\beta x_3}),\\ \omega\mu_t\mathbf{H}_t(x_3) &= \mathbf{e}_1(\beta E^+e^{-j\beta x_3} - \beta E^-e^{j\beta x_3}),\end{aligned}$$

- Region $-d\geq x_3$, at $x_3 = -d$

$$\mathbf{E} = \mathbf{e}_2E^te^{-jk_od},\qquad \omega\mu_o\mathbf{H}_t^t = \mathbf{e}_1k_3E^te^{-jk_od}.$$

Requiring continuity of the tangential fields at interface $x_3 = -d$, we obtain

$$\begin{aligned}E^te^{-jk_od} &= E^+e^{j\beta d} + E^-e^{-j\beta d},\\ \mu_tk_3E^te^{-jk_od} &= \mu_o\beta(E^+e^{j\beta d} - E^-e^{-j\beta d})\end{aligned}$$

From these we can solve the relations

$$E^- = R'E^+, \quad R' = \frac{\mu_o\beta - \mu_t k_3}{\mu_o\beta + \mu_t k_3} e^{j2\beta d},$$

$$E^t = T'E^+, \quad T' = \frac{2\mu_o\beta}{\mu_o\beta + \mu_t k_3} e^{j\beta d} e^{jk_o d}$$

Continuity at interface $x_3 = 0$ yields

$$E^i + E^r = (1 + R')E^+, \quad k_3\mu_t(E^i - E^r) = \beta\mu_o(1 - R')E^+,$$

from which we obtain $E^r = RE^i$ with

$$\begin{aligned} R &= \frac{k_3\mu_t(1+R') - \beta\mu_o(1-R')}{k_3\mu_t(1+R') + \beta\mu_o(1-R')} \\ &= \frac{j(k_3^2\mu_t^2 - \beta^2\mu_o^2)\sin\beta d}{2\beta k_3\mu_t\mu_o\cos\beta d + j(k_3^2\mu_t^2 + \beta^2\mu_o^2)\sin\beta d}. \end{aligned}$$

From $E^t = TE^i$ with $T = T'(1+R)/(1+R')$ we obtain

$$\begin{aligned} T &= \frac{2k_t\mu_t T'}{k_3\mu_t(1+R') + \beta\mu_o(1-R')} \\ &= \frac{2k_3\beta\mu_t\mu_o e^{-jk_o d}}{2k_3\beta\mu_t\mu_o\cos\beta d + j(k_3^2\mu_t^2 + \beta^2\mu_o^2)\sin\beta d} \end{aligned}$$

for the TE incident field.

4.2 Applying simple duality substitutions [35]

$$\mathbf{E} \to -\mathbf{H}, \quad \mathbf{H} \to -\mathbf{E}, \quad \mu \to \epsilon, \quad \epsilon \to \mu,$$

in which the medium and $\mathbf{k}$ vectors are invariant, we have $m \to e$, whence $\beta_{TE} \to \beta_{TM}$ and $R_{TE} \to R_{TM}$ (4.43) and (4.44) are transformed to

$$R_{TM} = \frac{j(k_3^2\epsilon_t^2 - \beta_{TM}^2\epsilon_o^2)\sin\beta_{TM}d}{2k_3\beta_{TM}\epsilon_t\epsilon_o\cos\beta_{TM}d + j(k_3^2\epsilon_t^2 + \beta_{TM}^2\epsilon_o^2)\sin\beta_{TE}d},$$

$$T_{TM} = \frac{2k_3\beta_{TM}\epsilon_t\epsilon_o e^{-jk_o d}}{2k_3\beta_{TM}\epsilon_t\epsilon_o\cos\beta_{TM}d + j(k_3^2\epsilon_t^2 + \beta_{TM}^2\epsilon_o^2)\sin\beta_{TM}d}.$$

4.3 Applying the duality transformation (1.19) with $\varphi = \pi/2$ as

$$\mathbf{E}_d = \eta_o\mathbf{H}, \quad \mathbf{H}_d = -\mathbf{E}/\eta_o, \quad \mathbf{D}_d = \mathbf{B}/\eta_o, \quad \mathbf{B}_d = -\eta_o\mathbf{D}$$

we obtain

$$\overline{\overline{\epsilon}}_d = \overline{\overline{\mu}}/\eta_o^2, \quad \overline{\overline{\mu}}_d = \overline{\overline{\epsilon}}\eta_o^2, \quad \Rightarrow \quad m_d = e, \quad e_d = m,$$

and $\epsilon_{od} = \mu_o, \quad \mu_{od} = \epsilon_o, \quad \mathbf{k}_{od} = k_o$. The wave vector $\mathbf{k}$ is not changed in the transformation.

Applying these, we have from (4.33) and (4.34)

$$\beta_{TEd}^2 = \omega^2\mu_{td}\epsilon_{td} - \frac{1}{m_d}\mathbf{k}_t\cdot\mathbf{k}_t = \omega^2\epsilon_t\mu_t - \frac{1}{e}\mathbf{k}_t\cdot\mathbf{k}_t = \beta_{TM}^2,$$

and $\beta_{TMd}^2 = \beta_{TE}^2$, whence we can set $\beta_{TEd} = \beta_{TM}$ and $\beta_{TMd} = \beta_{TE}$. The dual of the reflection coefficient R_{TE} in (4.43) can be expanded as

$$\begin{aligned} R_{TEd} &= \frac{j(k_3^2\mu_{td}^2 - \beta_{TEd}^2\mu_{od}^2)}{2k_3\beta_{TEd}\mu_{td}\mu_{od}\cos\beta_{TEd}d + j(k_3^2\mu_{td}^2 + \beta_{TEd}^2\mu_{od}^2)\sin\beta_{TEd}d} \\ &= \frac{j(k_3^2\epsilon_t^2 - \beta_{TM}^2\epsilon_o^2)}{2k_3\beta_{TM}\epsilon_t\epsilon_o\cos\beta_{TM}d + j(k_3^2\epsilon_t^2 + \beta_{TM}^2\epsilon_o^2)\sin\beta_{TM}d}, \end{aligned}$$

and the result can be seen to coincide with R_{TM} in (4.45). The relation $T_{TEd} = T_{TM}$ can be shown similarly.

4.4 Because tangential components of **E** and **H**, and normal components of **B** and **D**, are continuous through the boundary, the plane-wave fields satisfy

$$(\mathbf{H}^r + \mathbf{H}^i)_t = N(\mathbf{E}^r + \mathbf{E}^i)_t$$

$$\mathbf{n}\cdot(\mathbf{D}^r + \mathbf{D}^i) = N\mathbf{n}\cdot(\mathbf{B}^r + \mathbf{B}^i)$$

at the boundary. The last condition can be replaced for the simple-isotropic medium as

$$(\mathbf{n}\times\mathbf{k}_t)\cdot(\mathbf{H}^r + \mathbf{H}^i)_t = -N(\mathbf{n}\times\mathbf{k}_t)\cdot(\mathbf{E}^r + \mathbf{E}^i)_t,$$

while the first condition multiplied by $(\mathbf{n}\times\mathbf{k}_t)\cdot$ yields

$$(\mathbf{n}\times\mathbf{k}_t)\cdot(\mathbf{H}^r + \mathbf{H}^i)_t = N(\mathbf{n}\times\mathbf{k}_t)\cdot(\mathbf{E}^r + \mathbf{E}^i)_t.$$

Comparing these two conditions we conclude that the quantities on both sides must vanish. Thus, we must have

$$(\mathbf{n}\times\mathbf{k}_t)\cdot(\mathbf{H}^r + \mathbf{H}^i)_t = 0, \quad \Rightarrow \quad \mathbf{n}\cdot(\mathbf{D}^r + \mathbf{D}^i) = 0,$$

$$(\mathbf{n}\times\mathbf{k}_t)\cdot(\mathbf{E}^r + \mathbf{E}^i)_t = 0, \quad \Rightarrow \quad \mathbf{n}\cdot(\mathbf{B}^r + \mathbf{B}^i) = 0,$$

which coincide with the conditions of the DB boundary

4.5 Defining the incident and reflected powers as

$$\begin{aligned} P^i &= -\mathbf{n}\cdot\frac{1}{2}\Re\{\mathbf{E}^i\times\mathbf{H}^{i*}\} \\ P^r &= \mathbf{n}\cdot\frac{1}{2}\Re\{\mathbf{E}^r\times\mathbf{H}^{r*}\}, \end{aligned}$$

their difference is

$$P^i - P^r = -\mathbf{n}\cdot\frac{1}{2}\Re\{(\mathbf{E}^i\times\mathbf{H}^{i*} + \mathbf{E}^r\times\mathbf{H}^{r*}\}.$$

Substituting the expressions (4.16) - (4.20) we can expand

$$\mathbf{E}^i \times \mathbf{H}^{i*} + \mathbf{E}^r \times \mathbf{H}^{r*} =$$

$$(\mathbf{E}_+^i + \mathbf{E}_-^i) \times \eta_o(\mathbf{H}_+^i + \mathbf{H}_-^i)^* + (\mathbf{E}_+^r + \mathbf{E}_-^r) \times \eta_o(\mathbf{H}_+^r + \mathbf{H}_-^r)^* =$$

$$= (A_+\mathbf{k}_t + A_-\mathbf{n} \times \mathbf{k}_t) \times (-\frac{k_o}{k_n}A_+^*\mathbf{n} \times \mathbf{k}_t + \frac{k_n}{k_o}A_-^*\mathbf{k}_t)$$

$$+(A_+\mathbf{k}_t - A_-\mathbf{n} \times \mathbf{k}_t) \times (\frac{k_o}{k_n}A_+^*\mathbf{n} \times \mathbf{k}_t + \frac{k_n}{k_o}A_-^*\mathbf{k}_t)$$

$$= -\frac{k_o}{k_n}|A_+|^2\mathbf{k}_t \times (\mathbf{n} \times \mathbf{k}) + \frac{k_n}{k_o}|A_-|^2(\mathbf{n} \times \mathbf{k}) \times \mathbf{k}_t$$

$$+\frac{k_o}{k_n}|A_+|^2\mathbf{k}_t \times (\mathbf{n} \times \mathbf{k}) - \frac{k_n}{k_o}|A_-|^2(\mathbf{n} \times \mathbf{k}) \times \mathbf{k}_t = 0.$$

Because $P^i = P^r$, the power of the wave incident to the DB boundary equals that of the wave reflected from the boundary.

4.6 Let the incidence be in the plane defined by the $\mathbf{e}_1$ and $\mathbf{e}_3$ axes.

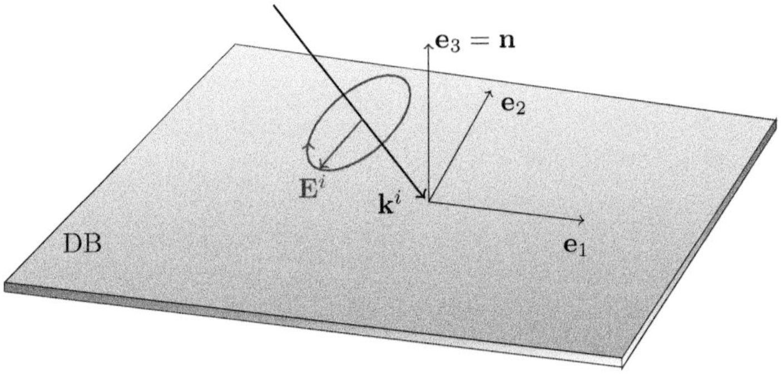

Figure D.2: A right-handed circularly polarized plane wave is incident on a planar DB boundary.

Assuming the incidence angle of 45°, we have

$$\mathbf{k}^i = k_o\frac{\mathbf{e}_1 - \mathbf{e}_3}{\sqrt{2}}$$

and a right-handed circularly polarized electric field is a multiple of the vector $\mathbf{v}^i$ defined by

$$\mathbf{v}^i = \frac{\mathbf{e}_1 + \mathbf{e}_3}{\sqrt{2}} + j\mathbf{e}_2$$

The polarization vector of $\mathbf{v}^i$ is

$$\mathbf{p}(\mathbf{v}^i) = \frac{\mathbf{v}^i \times \mathbf{v}^{i*}}{j\mathbf{v}^i \cdot \mathbf{v}^{i*}} = \frac{\mathbf{e}_1 - \mathbf{e}_3}{\sqrt{2}}$$

In other words, $\mathbf{p}(\mathbf{v}^i)$ has unit amplitude and is parallel to the wave vector $\mathbf{k}^i$ (as is the requirement of a RCP wave). Furthermore, $\mathbf{k}^i \cdot \mathbf{v}^i = 0$.

According to the reflection dyadic (4.14) of the DB boundary, the perpendicular polarization component of the electric field ($\mathbf{e}_2$) changes sign in the reflection while the parallel component ($\mathbf{e}_1$) remains the same. Hence the polarization of the reflected wave has to be

$$\mathbf{v}^r = \frac{\mathbf{e}_1 - \mathbf{e}_3}{\sqrt{2}} - j\mathbf{e}_2$$

where the minus sign of the $\mathbf{e}_3$ component can be written from the requirement that the field be perpendicular to the reflected wave vector $\mathbf{k}^r$:

$$\mathbf{k}^r \cdot \mathbf{v}^r = k_0 \frac{\mathbf{e}_1 + \mathbf{e}_3}{\sqrt{2}} \cdot \left(\frac{\mathbf{e}_1 - \mathbf{e}_3}{\sqrt{2}} - j\mathbf{e}_2 \right) = 0$$

or, alternatively, from the primary DB condition $\mathbf{e}_3 \cdot \mathbf{D} = 0$, which forces the $\mathbf{e}_3$ components of the incident and reflected fields be equal and opposite of each other.

The polarization vector of the reflected wave can be computed:

$$\mathbf{p}(\mathbf{v}^r) = \frac{\mathbf{v}^r \times \mathbf{v}^{r*}}{j\mathbf{v}^r \cdot \mathbf{v}^{r*}} = \frac{\mathbf{e}_1 + \mathbf{e}_3}{\sqrt{2}}$$

Again, this has unit amplitude whence the reflected field is also circularly polarized. And since $\mathbf{k}^r \cdot \mathbf{v}^r > 0$, the reflection is right-handed.

Hence the handedness is retained in reflection from a DB boundary, unlike for PEC and PMC surfaces for which the incident and reflected waves have different handedness.

As to the magnitude, the reflection is total: all power is reflected in this case of oblique incidence. However, at normal incidence, we have the situation of the matched wave. In the case of matched wave, DB is a perfect absorber (see (4.25)).

4.7 The tangential component of the reflection dyadic for the DB boundary (4.14)

$$\overline{\overline{\mathsf{R}}}_t = \frac{1}{\mathbf{k}_t^2}(\mathbf{k}_t\mathbf{k}_t - \mathbf{n}\mathbf{n}{}^\times_\times\mathbf{k}_t\mathbf{k}_t)$$

can be compared with that of the SH boundary, (3.210) with $\mathbf{b}_t = \mathbf{a}_t$, which can be expanded as

$$\begin{aligned}\overline{\overline{\mathsf{R}}}_t &= -\frac{1}{\Delta}((\mathbf{k}^r \times \mathbf{a}_t)(\mathbf{a}_t \times \mathbf{k}^i) + ((\mathbf{a}_t \times \mathbf{k}^r) \times \mathbf{k}^r)\mathbf{a}_t)_t \\ &= -\frac{1}{\Delta}(k_n^2(\mathbf{n} \times \mathbf{a}_t)(\mathbf{n} \times \mathbf{a}_t) + (\mathbf{k}_t(\mathbf{a}_t \cdot \mathbf{k}_t) - k_o^2\mathbf{a}_t)\mathbf{a}_t).\end{aligned}$$

Assuming that the vectors $\mathbf{a}_t$ and $\mathbf{k}_t$ are parallel, whence they satisfy

$$(\mathbf{a}_t \times \mathbf{k}_t) \times \mathbf{k}_t = \mathbf{k}_t(\mathbf{a}_t \cdot \mathbf{k}_t) - \mathbf{k}_t^2\mathbf{a}_t = 0,$$

we can further expand

$$\Delta = -(\mathbf{a}_t \times \mathbf{k}^i)^2 = -\mathbf{a}_t^2 k_o^2 + (\mathbf{a}_t \cdot \mathbf{k}_t)^2 = -\mathbf{a}_t^2 k_o^2 + \mathbf{a}_t^2 \mathbf{k}_t^2 = -\mathbf{a}_t^2 k_n^2$$

and

$$\begin{aligned}
\overline{\overline{\mathsf{R}}}_t &= \frac{1}{\mathbf{a}_t^2 k_n^2}(k_n^2(\mathbf{nn}{}^\times_\times \mathbf{a}_t\mathbf{a}_t) + (\mathbf{k}_t(\mathbf{a}_t \cdot \mathbf{k}_t) - k_o^2 \mathbf{a}_t)\mathbf{a}_t) \\
&= \frac{1}{\mathbf{a}_t^2 k_n^2}(k_n^2(\mathbf{nn}{}^\times_\times \mathbf{a}_t\mathbf{a}_t) + (\mathbf{k}_t^2 \mathbf{a}_t - k_o^2 \mathbf{a}_t)\mathbf{a}_t). \\
&= \frac{1}{\mathbf{a}_t^2}(\mathbf{nn}{}^\times_\times \mathbf{a}_t\mathbf{a}_t - \mathbf{a}_t\mathbf{a}_t) \\
&= \frac{1}{\mathbf{k}_t^2}(\mathbf{nn}{}^\times_\times \mathbf{k}_t\mathbf{k}_t - \mathbf{k}_t\mathbf{k}_t).
\end{aligned}$$

Obviously, the last expression appears similar to (4.14) except for the minus sign. Omitting the sign change, the field reflected from the DB boundary coincides with that reflected from the SH boundary in the case when the wave arrives with $\mathbf{k}_t$ parallel to $\mathbf{a}_t$.

4.8 Starting from the expansion in terms of vector potentials (4.69) and (4.70),

$$\begin{aligned}
\mathbf{E}(\mathbf{r}) &= -\frac{1}{\epsilon_o}\nabla \times (\mathbf{e}_\rho F(\mathbf{r})) + \frac{1}{j\omega\mu_o\epsilon_o}\nabla \times (\nabla \times (\mathbf{e}_\rho A(\mathbf{r}))), \\
\mathbf{H}(\mathbf{r}) &= \frac{1}{\mu_o}\nabla \times (\mathbf{e}_\rho A(\mathbf{r})) + \frac{1}{j\omega\mu_o\epsilon_o}\nabla \times (\nabla \times (\mathbf{e}_\rho F(\mathbf{r}))),
\end{aligned}$$

we can insert them in the Maxwell equations as

$$\begin{aligned}
\nabla \times \mathbf{E} + j\omega\mu_o \mathbf{H} &= \frac{1}{j\omega\mu_o\epsilon_o}\nabla \times (\nabla \times (\nabla \times (\mathbf{e}_\rho A))) + j\omega \nabla \times (\mathbf{e}_\rho A(\mathbf{r})) \\
&= -\frac{1}{j\omega\mu_o\epsilon_o}\nabla \times (\nabla^2 + k_o^2)(\mathbf{e}_\rho A) = 0,
\end{aligned}$$

$$\begin{aligned}
\nabla \times \mathbf{H} - j\omega\epsilon_o \mathbf{E} &= \frac{1}{j\omega\mu_o\epsilon_o}\nabla \times \nabla \times (\nabla \times (\mathbf{e}_\rho F)) + j\omega \nabla \times (\mathbf{e}_\rho F) \\
&= -\frac{1}{j\omega\mu_o\epsilon_o}\nabla \times (\nabla^2 + k_o^2)(\mathbf{e}_\rho F)) = 0.
\end{aligned}$$

Applying differential operator expressions in cylindrical coordinates and taking $\nabla \times \mathbf{e}_\rho = 0$ into account, we can expand

$$\begin{aligned}
\mathbf{e}_\rho \cdot (\nabla \times (\nabla^2 + k_o^2)\mathbf{e}_\rho F) &= -\nabla \cdot (\mathbf{e}_\rho \times (\nabla^2(\mathbf{e}_\rho F))) \\
&= -\nabla \cdot (\mathbf{e}_\rho \times \mathbf{e}_\varphi(\frac{2}{\rho^2}\partial_\varphi F)) \\
&= -\partial_{x_3}(\frac{2}{\rho^2}\partial_\varphi F)) = 0,
\end{aligned}$$

whence the vector potential component must satisfy $\partial_{x_3}\partial_\varphi F = 0$. Similarly, we must also have $\partial_{x_3}\partial_\varphi A = 0$. Obviously, potentials of this kind can only represent fields which are independent of either x_3 or φ coordinate, i.e., constant along the axial or circumferential direction. Because they do not represent the most general fields, the radial components of the vector potentials cannot be applied in general analysis of fields.

4.9 To consider propagation close to $k_o a = x_{11}$, let us restrict to the solutions with the $'+'$ sign in (4.111). For

$$k_c a = x_{11} + \delta, \qquad 0 < \delta \ll x_{11} = 3.832,$$

in the first-order approximation we have

$$k_o a = \sqrt{(x_{11}+\delta)^2 + (\beta a)^2} \approx x_{11} + \delta + \frac{(\beta a)^2}{2x_{11}} \approx x_{11} + \delta.$$

Expanding

$$J_1(k_c a) \approx \delta J_1'(x_{11}),$$

(4.111) becomes

$$\beta a \approx n\delta k_o a / x_{11} \approx n\delta,$$

where n is now 1 or -1. For $n = 1$ $\beta a \approx \delta$ or the dispersion curve has a positive slope at $k_o a = x_{11}$, while for $n = -1$ $\beta a \approx -\delta$ and the slope is negative.

4.10 For the simple-isotropic boundary we can substitute $Z_{TE} = 1/Y_{TM} = Z_s$ in (4.192) and obtain

$$\overline{\overline{\mathsf{R}}}_t = \frac{1}{k_t^2}\left(\frac{k_o Z_s - k_n \eta_o}{k_o Z_s + k_n \eta_o}\mathbf{k}_t\mathbf{k}_t - \frac{k_o \eta_o - k_n Z_s}{k_o \eta_o + k_n Z_s}(\mathbf{n}\times\mathbf{k})(\mathbf{n}\times\mathbf{k})\right),$$

which equals (3.75).

For the PEC boundary $Z_{TE} = 0$ and $1/Y_{TM} = 0$, (4.192) is reduced to $\overline{\overline{\mathsf{R}}}_t = -\overline{\overline{\mathsf{I}}}_t$.

For the PMC boundary $1/Z_{TE} = 0$ and $\overline{\overline{\mathsf{Y}}}_{TM} = 0$, (4.192) is reduced to $\overline{\overline{\mathsf{R}}}_t = \overline{\overline{\mathsf{I}}}_t$.

4.11 For the DB boundary we can substitute $Z_{TE} = Y_{TM} = 0$ in (4.192) and obtain

$$\overline{\overline{\mathsf{R}}}_t = \frac{1}{k_t^2}(\mathbf{k}_t\mathbf{k}_t - (\mathbf{n}\times\mathbf{k})(\mathbf{n}\times\mathbf{k})),$$

which coincides with (4.14).

For the D'B' boundary we can substitute $1/Z_{TE} = 1/Y_{TM} = 0$ in (4.192) and obtain

$$\overline{\overline{\mathsf{R}}}_t = \frac{1}{k_t^2}(-\mathbf{k}_t\mathbf{k}_t + (\mathbf{n}\times\mathbf{k})(\mathbf{n}\times\mathbf{k}))$$

which coincides with (4.149).

4.12 Considering the DB' boundary, the magnetic fields of incident and reflected plane waves satisfy from (4.6)

$$\mathbf{n}\cdot\mathbf{D}=0, \quad \Rightarrow \quad \mathbf{n}\cdot(\mathbf{k}^i\times\mathbf{H}^i+\mathbf{k}^r\times\mathbf{H}^r)=(\mathbf{n}\times\mathbf{k}_t)\cdot(\mathbf{H}_t^i+\mathbf{H}_t^r)=0,$$

and

$$(\mathbf{n}\cdot\nabla)(\mathbf{n}\cdot\mathbf{B})=0 \quad\Rightarrow\quad jk_nH_n^i-jk_nH_n^r=jk_n(H_n^i-H_n^r)=0,$$

whence $H_n^i=H_n^r$ at the boundary. The plane-wave fields satisfy

$$\begin{aligned}\mathbf{k}^i\cdot\mathbf{H}^i &= -k_nH_n^i+\mathbf{k}_t\cdot\mathbf{H}_t^i=0,\\ \mathbf{k}^r\cdot\mathbf{H}^r &= k_nH_n^r+\mathbf{k}_t\cdot\mathbf{H}_t^r=0,\end{aligned}$$

which, when combined, yield

$$\mathbf{k}_t\cdot(\mathbf{H}_t^i+\mathbf{H}_t^r)=k_n(H_n^i-H_n^r)=0.$$

Expanding

$$(\mathbf{k}_t\times(\mathbf{n}\times\mathbf{k}_t))\times\mathbf{H}=(\mathbf{n}\times\mathbf{k}_t)(\mathbf{k}_t\cdot\mathbf{H}_t)-\mathbf{k}_t(\mathbf{n}\times\mathbf{k}_t)\cdot\mathbf{H},$$

with $\mathbf{H}=\mathbf{H}^i+\mathbf{H}^r$, and applying the above properties, we obtain

$$(\mathbf{k}_t\times(\mathbf{n}\times\mathbf{k}_t))\times\mathbf{H}=(\mathbf{k}_t\cdot\mathbf{k}_t)\mathbf{n}\times\mathbf{H}=0.$$

Since this must be valid for any $\mathbf{k}_t$, the total magnetic field must satisfy the PMC condition $\mathbf{n}\times\mathbf{H}=0$ for any plane wave. From linearity, it must be valid for any sum or integral of plane waves, i.e., for any fields.

Similar reasoning is valid for the D'B boundary, which appears equivalent to the PEC boundary.

For smooth non-planar boundaries the above remains valid locally at the boundary surface when, at the reflection point, the surface is replaced by its tangent plane.

5.1 For the DB boundary, defined by

$$\mathbf{a}_1=\mathbf{b}_2=\mathbf{n}, \qquad \mathbf{b}_1=\mathbf{a}_2=0,$$

we have

$$\mathbf{c}_1^{i,r}=-k_o\mathbf{n}, \quad \mathbf{c}_2^{i,r}=\mathbf{k}^{i,r}\times\mathbf{n},$$

$$\mathbf{d}_1^{i,r}=\mathbf{k}^{i,r}\times\mathbf{n}, \quad \mathbf{d}_2^{i,r}=k_o\mathbf{n},$$

$$J^r=-k_o(\mathbf{n}\times(\mathbf{k}_t\times\mathbf{n}))\cdot\mathbf{k}=-k_ok_t^2.$$

The reflection dyadic (5.63) becomes

$$\begin{aligned}\overline{\overline{\mathsf{R}}} &= \frac{-k_o}{J^r}\mathbf{k}^r\times((\mathbf{k}_t\times\mathbf{n})\mathbf{n}-\mathbf{n}(\mathbf{k}_t\times\mathbf{n}))\\ &= \frac{1}{k_t^2}\mathbf{k}^r\times(\mathbf{k}_t\times\overline{\overline{\mathsf{I}}}).\end{aligned}$$

The TE_{c1} wave satisfies

$$\mathbf{d}_1^{i,r} \cdot \mathbf{H}_1^{i,r} = (\mathbf{k}^{i,r} \times \mathbf{n}) \cdot \mathbf{H}_1^{i,r} = 0.$$

For a TE_n wave with $\mathbf{E}^i = E\mathbf{n} \times \mathbf{k}_t$, the reflected field is

$$\mathbf{E}^r = \overline{\overline{\mathsf{R}}} \cdot \mathbf{E}^i = \frac{E}{k_t^2}\mathbf{k}^r \times (\mathbf{k}_t \times (\mathbf{n} \times \mathbf{k}_t)) = E\mathbf{k}_t \times \mathbf{n} = -\mathbf{E}^i,$$

whence the DB boundary acts as a PEC boundary for the TE_n wave.

For a TM_n wave with $\mathbf{H}^i = H\mathbf{n} \times \mathbf{k}_t$ and $\mathbf{H}^r = -\overline{\overline{\mathsf{R}}} \cdot \mathbf{H}^i$, after similar steps, we obtain $\mathbf{H}^r = -\mathbf{H}^i$, which equals the PMC condition.

To find the eigenvalues for the DB boundary we substitute $\mathbf{a}_{1t}' = \mathbf{b}_{2t}' = 0$ and $\mathbf{a}_{2t}' = -\mathbf{b}_{1t}' = \mathbf{n} \times \mathbf{k}_t/k_o$ in (5.98) to obtain

$$\begin{aligned} 0 &= \mathbf{n} \cdot ((1-R)\mathbf{n} \times \mathbf{k} - t \cdot \overline{\overline{\mathsf{J}}}_t) \times ((1+R)\mathbf{n} \times \mathbf{k}) \\ &= (1-R^2)(\mathbf{n} \times \mathbf{k}_t) \cdot \overline{\overline{\mathsf{J}}}_t \cdot \mathbf{k}_t = (k_o k_t^2/k_n)(1-R^2) = 0. \end{aligned}$$

The eigenvalues are $R = 1$ and $R = -1$.

5.2 For the GSH boundary, defined by

$$\mathbf{b}_1 = \mathbf{a}_2 = 0, \quad \mathbf{a}_1 = \mathbf{a}_t, \quad \mathbf{b}_2 = \mathbf{b}_t,$$

we have

$$\mathbf{c}_1^{i,r} = -k_o\mathbf{a}_t, \quad \mathbf{c}_2^{i,r} = \mathbf{k}^{i,r} \times \mathbf{b}_t,$$

$$\begin{aligned} J^r &= -k_o(\mathbf{k}^r \times \mathbf{a}_t) \cdot (\mathbf{k}^r \times \mathbf{b}_t) \\ &= -k_o^3(\mathbf{a}_t \cdot \mathbf{b}_t) + k_o(\mathbf{k}_t \cdot \mathbf{a}_t)(\mathbf{k}_t \cdot \mathbf{b}_t) = J^i. \end{aligned}$$

The reflection dyadic (5.63) can be expanded as

$$\overline{\overline{\mathsf{R}}} = \frac{1}{(\mathbf{k}^r \times \mathbf{a}_t) \cdot (\mathbf{k}^r \times \mathbf{b}_t)}\mathbf{k}^r \times ((\mathbf{k}^r \times \mathbf{b}_t)\mathbf{a}_t - \mathbf{a}_t(\mathbf{k}^i \times \mathbf{b}_t)).$$

Thus, if the incident wave is polarized as TE_a, $\mathbf{E}_a^i = E\mathbf{k}^i \times \mathbf{a}_t$, we have

$$\mathbf{E}_a^r = -E\mathbf{k}^r \times \mathbf{a}_t.$$

Since

$$\begin{aligned} k_o\eta_o(\mathbf{H}_a^i + \mathbf{H}_a^r) &= E(\mathbf{k}^i \times (\mathbf{k}^i \times \mathbf{a}_t) - \mathbf{k}^r \times (\mathbf{k}^r \times \mathbf{a}_t)) \\ &= E(\mathbf{k}^i - \mathbf{k}^r)(\mathbf{k}_t \cdot \mathbf{a}_t) = -2Ek_n\mathbf{n}(\mathbf{k}_t \cdot \mathbf{a}_t), \end{aligned}$$

the TE_a fields satisfy the PMC condition

$$\mathbf{n} \times (\mathbf{H}_a^i + \mathbf{H}_a^r) = 0.$$

On the other hand, if the incident wave is polarized as TM_b, with $\mathbf{E}_b^i = E\mathbf{k}^i \times (\mathbf{k}^i \times \mathbf{b}_t)$, we have

$$\mathbf{E}_b^r = -E\mathbf{k}^r \times (\mathbf{k}^r \times \mathbf{b}_t),$$

whence

$$\mathbf{n} \times (\mathbf{E}_b^i + \mathbf{E}_b^r) = E(\mathbf{k}^i - \mathbf{k}^r)(\mathbf{k}.\mathbf{b}_t) = -2k_n E\mathbf{n}(\mathbf{k}_t \cdot \mathbf{b}_t),$$

and the TE_b fields satisfy the PEC condition

$$\mathbf{n} \times (\mathbf{E}_b^i + \mathbf{E}_b^r) = 0.$$

To find the eigenvalues for the GSH boundary we substitute $\mathbf{a}'_{1t} = \mathbf{a}_t$, $\mathbf{a}'_{2t} = \mathbf{b}'_{1t} = 0$ and $\mathbf{b}'_{2t} = \mathbf{b}_t$ in (5.98), which becomes

$$\begin{aligned} 0 &= \mathbf{n} \cdot ((1+R)\mathbf{a}_t) \times (-(1-R)\mathbf{b}_t \cdot \overline{\overline{\mathsf{J}}}_t) \\ &= -\frac{1-R^2}{k_o k_n}((\mathbf{a}_t \cdot \mathbf{n} \times \mathbf{k}_t)(\mathbf{b}_t \cdot \mathbf{n} \times \mathbf{k}_t) + k_n^2 \mathbf{a}_t \cdot \mathbf{b}_t). \end{aligned}$$

This yields the eigenvalues $R = 1$ and $R = -1$.

5.3 Expanding different terms of (5.98) as

$$\mathbf{n} \cdot ((1+R)\mathbf{a}'_{1t} \times (1+R)\mathbf{a}'_{2t}) = (1+R)^2 \mathbf{n} \cdot (\mathbf{e}_1 \times \mathbf{e}_2) = (1+R)^2$$

$$\mathbf{n} \cdot ((1-R)\mathbf{e}_2(Z_s/\eta_o) \cdot \overline{\overline{\mathsf{J}}}_t) \times ((1-R)\mathbf{e}_1(Z_s/\eta_o) \cdot \overline{\overline{\mathsf{J}}}_t) =$$

$$= (1-R)^2(Z_s/\eta_o)^2 \mathbf{n} \cdot ((\mathbf{e}_2 \times \mathbf{e}_1) \cdot \overline{\overline{\mathsf{J}}}_t^{(2)}) = (1-R)^2(Z_s/\eta_o)^2$$

$$\mathbf{n} \cdot ((1+R)\mathbf{e}_1) \times ((1-R)\mathbf{e}_1(Z_s/\eta_o) \cdot \overline{\overline{\mathsf{J}}}_t) = (1-R^2)(Z_s/\eta_o)\mathbf{e}_1 \cdot \overline{\overline{\mathsf{J}}}_t \cdot \mathbf{e}_2$$

$$\mathbf{n} \cdot ((1-R)\mathbf{e}_2(Z_s/\eta_o) \cdot \overline{\overline{\mathsf{J}}}_t) \times ((1+R)\mathbf{e}_2) = -(1-R^2)(Z_s/\eta_o)\mathbf{e}_2 \cdot \overline{\overline{\mathsf{J}}}_t \cdot \mathbf{e}_1,$$

we can combine them as

$$(1+R)^2 + (1-R)^2(Z_s/\eta_o)^2 + (1-R^2)(Z_s/\eta_o)(\mathbf{e}_1 \cdot \overline{\overline{\mathsf{J}}}_t \cdot \mathbf{e}_2 - \mathbf{e}_2 \cdot \overline{\overline{\mathsf{J}}}_t \cdot \mathbf{e}_1) = 0.$$

Expanding further

$$\mathbf{e}_1 \cdot \overline{\overline{\mathsf{J}}}_t \cdot \mathbf{e}_2 - \mathbf{e}_2 \cdot \overline{\overline{\mathsf{J}}}_t \cdot \mathbf{e}_1 = \frac{1}{k_o k_n}((-k_2 k_2 - k_1 k_1) - 2k_n^2) = -\frac{1}{k_o k_n}(k_o^2 + k_n^2),$$

the equation for R becomes

$$(1+R)^2 + (1-R)^2(Z_s/\eta_o)^2 - (1-R^2)(Z_s/\eta_o)\left(\frac{k_o}{k_n} + \frac{k_n}{k_o}\right) = 0,$$

or,

$$R^2\left(\frac{Z_s}{\eta_o} + \frac{k_o}{k_n}\right)\left(\frac{Z_s}{\eta_o} + \frac{k_n}{k_o}\right) + 2R(1 - \frac{Z_s}{\eta_o}) + \left(\frac{Z_s}{\eta_o} - \frac{k_o}{k_n}\right)\left(\frac{Z_s}{\eta_o} - \frac{k_n}{k_o}\right) = 0,$$

which reduces to the required form.

5.4 Applying

$$\mathbf{c}_1(\mathbf{k}) \times \mathbf{c}_2(\mathbf{k}) = (\alpha\mathbf{n} \times \mathbf{k}_t + k_o\mathbf{a}_t) \times (\mathbf{k} \times \mathbf{a}_t + k_o\alpha\mathbf{n})$$

$$= \alpha\mathbf{k}(\mathbf{n} \times \mathbf{k}_t \cdot \mathbf{a}_t) + k_o\mathbf{k}\,\mathbf{a}_t^2 - k_o(\mathbf{k}_t \cdot \mathbf{a}_t)\mathbf{a}_t + \alpha^2 k_o\mathbf{k}_t + k_o^2\alpha\mathbf{a}_t \times \mathbf{n},$$

we can expand $J(\mathbf{k}) = \mathbf{k} \cdot \mathbf{c}_1(\mathbf{k}) \times \mathbf{c}_2(\mathbf{k})$ as

$$\begin{aligned}
& \alpha k_o^2(\mathbf{n} \times \mathbf{k}_t \cdot \mathbf{a}_t) + k_o^3\mathbf{a}_t^2 - k_o(\mathbf{k}_t \cdot \mathbf{a}_t)^2 + \alpha^2 k_o\mathbf{k}_t^2 + k_o^2\alpha(\mathbf{n} \times \mathbf{k}_t \cdot \mathbf{a}_t) \\
= \; & k_o(2\alpha k_o(\mathbf{n} \times \mathbf{k}_t \cdot \mathbf{a}_t) + (k_n^2 + \mathbf{k}_t^2)\mathbf{a}_t^2 - (\mathbf{k}_t \cdot \mathbf{a}_t)^2 + \alpha^2\mathbf{k}_t^2) \\
= \; & k_o(2\alpha k_o(\mathbf{n} \times \mathbf{k}_t \cdot \mathbf{a}_t) + (\mathbf{k}_t \times \mathbf{a}_t)^2 + k_n^2\mathbf{a}_t^2 + \alpha^2\mathbf{k}_t^2) \\
= \; & k_o((\mathbf{n} \times \mathbf{k}_t \cdot \mathbf{a}_t + \alpha k_o)^2 - \alpha^2 k_o^2 + k_n^2\mathbf{a}_t^2 + \alpha^2\mathbf{k}_t^2) \\
= \; & k_o((\mathbf{n} \times \mathbf{k}_t \cdot \mathbf{a}_t + \alpha k_o)^2 + k_n^2(\mathbf{a}_t^2 - \alpha^2))
\end{aligned}$$

5.5 For the SHDB boundary with $\mathbf{a}_{1t} = -\mathbf{b}_{2t} = \mathbf{a}_t$, $\mathbf{b}_{1t} = \mathbf{a}_{2t} = 0$, $a_{1n} = b_{2n} = 0$ and $b_{1n} = a_{2n} = \alpha$, (5.90) and (5.91) take the form

$$\mathbf{a}'_{1t} = \mathbf{a}_t + (\alpha/k_o)\mathbf{n} \times \mathbf{k}_t, \qquad \mathbf{b}'_{1t} = 0,$$

$$\mathbf{b}'_{2t} = -\mathbf{a}_t + (\alpha/k_o)\mathbf{n} \times \mathbf{k}_t, \qquad \mathbf{a}'_{2t} = 0,$$

whence (5.98) can be expanded as

$$\begin{aligned}
0 &= \mathbf{n} \cdot ((1+R)\mathbf{a}'_{1t}) \times ((1-R)\mathbf{b}'_{2t} \cdot \overline{\overline{\mathsf{J}}}_t) \\
&= (1-R^2)\mathbf{b}'_{2t} \cdot (\overline{\overline{\mathsf{J}}}_t \times \mathbf{n}) \cdot \mathbf{a}'_{1t} \\
&= -\frac{1-R^2}{k_o k_n}\mathbf{b}'_{2t} \cdot ((\mathbf{n} \times \mathbf{k}_t)(\mathbf{n} \times \mathbf{k}_t) + k_n^2\overline{\overline{\mathsf{I}}}_t) \cdot \mathbf{a}'_{1t} \\
&= -\frac{1-R^2}{k_o k_n}(\alpha^2\mathbf{k}_t^2 - (\mathbf{n} \times \mathbf{k}_t \cdot \mathbf{a}_t)^2 - k_n^2\mathbf{a}_t^2).
\end{aligned}$$

Since the expression in brackets is not identically zero, the eigenvalues are $R = +1$ and $R = -1$.

5.6 Applying the relations for the plane-wave fields

$$\mathbf{k} \times \mathbf{E} = k_o\eta_o\mathbf{H}, \quad \mathbf{k} \times \eta_o\mathbf{H} = -k_o\mathbf{E},$$

(5.25) can be expanded as

$$\phi_1 = k_o\mathbf{a}_1 \cdot \mathbf{E} + \mathbf{b}_1 \cdot (\mathbf{k} \times \mathbf{E}) = -(\mathbf{k} \times \mathbf{b}_1 - k_o\mathbf{a}_1) \cdot \mathbf{E},$$

and

$$\phi_1 = \mathbf{a}_1 \cdot (-\mathbf{k} \times \eta_o\mathbf{H}) + k_o\mathbf{b}_1 \cdot \eta_o\mathbf{H} = (\mathbf{k} \times \mathbf{a}_1 + k_o\mathbf{b}_1) \cdot \eta_o\mathbf{H},$$

from which we can identify the vectors $\mathbf{c}_1(\mathbf{k})$ and $\mathbf{d}_1(\mathbf{k})$ as (5.29) and (5.31). The vectors $\mathbf{c}_2(\mathbf{k})$ and $\mathbf{d}_2(\mathbf{k})$ can be found similarly.

We can expand

$$\begin{aligned} J(\mathbf{k}) &= (\mathbf{k} \times \mathbf{d}_1(\mathbf{k})) \cdot \mathbf{d}_2(\mathbf{k}) \\ &= (\mathbf{k}(\mathbf{k} \cdot \mathbf{a}_1) - k_o^2 \mathbf{a}_1 + k_o \mathbf{k} \times \mathbf{b}_1) \cdot (\mathbf{k} \times \mathbf{a}_2 + k_o \mathbf{b}_2) \\ &= k_o^3 (\mathbf{b}_1 \cdot \mathbf{a}_2 - \mathbf{a}_1 \cdot \mathbf{b}_2) + k_o^2 \mathbf{k} \cdot (\mathbf{a}_1 \times \mathbf{a}_2 + \mathbf{b}_1 \times \mathbf{b}_2) \\ &\quad + k_o \mathbf{k}\mathbf{k} : (\mathbf{a}_1 \mathbf{b}_2 - \mathbf{a}_2 \mathbf{b}_1). \end{aligned}$$

The last expression appears invariant to the transformations $\mathbf{a}_1 \to \mathbf{b}_1$, $\mathbf{a}_2 \to \mathbf{b}_2$ and $\mathbf{b}_1 \to -\mathbf{a}_1$, $\mathbf{b}_1 \to -\mathbf{a}_2$ which correspond to $\mathbf{d}_1(\mathbf{k}) \to \mathbf{c}_1(\mathbf{k})$ and $\mathbf{d}_2(\mathbf{k}) \to \mathbf{c}_2(\mathbf{k})$. Thus, the same expression is valid for $\mathbf{k} \cdot \mathbf{c}_1(\mathbf{k}) \times \mathbf{c}_2(\mathbf{k})$ as well.

5.7 Expanding

$$\mathbf{k} \times ((\mathbf{c}_1 \times \mathbf{c}_2) \times \bar{\bar{\mathsf{I}}}) = (\mathbf{c}_1 \times \mathbf{c}_2)\mathbf{k} - \mathbf{k} \cdot (\mathbf{c}_1 \times \mathbf{c}_2)\bar{\bar{\mathsf{I}}}$$

and

$$\mathbf{k} \times ((\mathbf{c}_1 \times \mathbf{c}_2) \times \bar{\bar{\mathsf{I}}}) = \mathbf{k} \times (\mathbf{c}_2 \mathbf{c}_1 - \mathbf{c}_1 \mathbf{c}_2)$$

we obtain

$$(\mathbf{k} \cdot \mathbf{c}_1 \times \mathbf{c}_2)\bar{\bar{\mathsf{I}}} = (\mathbf{c}_1 \times \mathbf{c}_2)\mathbf{k} - \mathbf{k} \times (\mathbf{c}_2 \mathbf{c}_1 - \mathbf{c}_1 \mathbf{c}_2),$$

which yields (5.34). (5.35) is obtained similarly.

5.8 Expanding

$$J(\mathbf{k})(\mathbf{k} \times \mathbf{E}_1 - k_o \eta_o \mathbf{H}_1) = -\mathbf{k} \times (\mathbf{k} \times \mathbf{c}_1 + k_o \mathbf{d}_1)\phi_2$$

and applying (5.29), (5.31), we have

$$\mathbf{k} \times \mathbf{c}_1 + k_o \mathbf{d}_1 = \mathbf{k} \times (\mathbf{k} \times \mathbf{b}_1 - k_o \mathbf{a}_1) + k_o(\mathbf{k} \times \mathbf{a}_1 + k_o \mathbf{b}_1) = \mathbf{k}(\mathbf{k} \cdot \mathbf{b}_1),$$

whence $\mathbf{E}_1, \mathbf{H}_1$ satisfies the plane-wave relation $\mathbf{k} \times \mathbf{E}_1 = k_o \eta_o \mathbf{H}_1$. The other relations can be checked in a similar fashion.

5.9 Assuming $J^i = -k_o(\mathbf{k}^i \times \mathbf{a}_1) \cdot (\mathbf{k}^i \times \mathbf{b}_2) \neq 0$, the vectors $\mathbf{k}^i, k_o \mathbf{a}_1$ and $\mathbf{k}^i \times \mathbf{b}_2$ make a basis. The corresponding reciprocal basis vectors are $-k_o(\mathbf{a}_1 \times (\mathbf{k}^i \times \mathbf{b}_2))/J^i$, $-((\mathbf{k}^i \times \mathbf{b}_2) \times \mathbf{k}^i)/J^i$ and $-k_o(\mathbf{k}^i \times \mathbf{a}_1)/J^i$. Taking into account $\mathbf{k}^i \cdot \mathbf{E}^i = 0$, we can expand

$$\begin{aligned} J^i \mathbf{E}^i &= -k_o((\mathbf{k}^i \times \mathbf{b}_2) \times \mathbf{k}^i)(\mathbf{a}_1 \cdot \mathbf{E}^i) - k_o(\mathbf{k}^i \times \mathbf{a}_1)(\mathbf{k}^i \times \mathbf{b}_2 \cdot \mathbf{E}^i) \\ &= k_o \mathbf{k}^i \times ((\mathbf{k}^i \times \mathbf{b}_2)(\mathbf{a}_1 \cdot \mathbf{E}^i) + k_o \mathbf{a}_1(\mathbf{b}_2 \cdot \eta_o \mathbf{H}^i)), \end{aligned}$$

which yields (5.255). Applying $\mathbf{k}^i \times \mathbf{E}^i = k_o \eta_o \mathbf{H}^i$, the expansion (5.256) can be obtained.

5.10 For (5.261), the plane waves satisfy

$$\begin{aligned}\mathbf{n}\cdot k_o\eta_o(\mathbf{H}^r+\mathbf{H}^i) &= \mathbf{n}\cdot(\mathbf{k}^r\times\mathbf{E}^r+\mathbf{k}^i\times\mathbf{E}^i)\\ &= \mathbf{n}\times\mathbf{k}_t\cdot(\mathbf{E}^r+\mathbf{E}^i)=(\mathbf{n}\times\mathbf{k}_t)\cdot\mathbf{E}=0,\end{aligned}$$

Assuming $\mathbf{a}_{1t}\times(\mathbf{n}\times\mathbf{k}_t)=\mathbf{n}(\mathbf{a}_{1t}\cdot\mathbf{k}_t)\neq 0$, we obtain on one hand

$$(\mathbf{a}_{1t}\times(\mathbf{n}\times\mathbf{k}_t))\times\mathbf{E}=(\mathbf{n}\times\mathbf{k}_t)(\mathbf{a}_{1t}\cdot\mathbf{E})-\mathbf{a}_{1t}(\mathbf{n}\times\mathbf{k}_t)\cdot\mathbf{E}=0$$

and, on the other hand,

$$(\mathbf{a}_{1t}\times(\mathbf{n}\times\mathbf{k}_t))\times\mathbf{E}=(\mathbf{a}_{1t}\cdot\mathbf{k}_t)\mathbf{n}\times\mathbf{E}=0,$$

which equals the condition of the PEC boundary. Since this is a linear condition and valid for any wave vector $\mathbf{k}$ of the plane wave (except when $\mathbf{a}_t\cdot\mathbf{k}_t=0$), it is valid for any fields of the special EH boundary case (5.261). Similarly, (5.262) can be shown to equal the PMC boundary condition $\mathbf{n}\times\mathbf{H}=0$.

5.11 We can expand

$$\begin{aligned}\mathbf{c}_{1d}(\mathbf{k}) &= \mathbf{k}\times\mathbf{b}_{1d}-k_o\mathbf{a}_{1d}\\ &= \mathbf{k}\times(-\mathbf{a}_1\sin\varphi+\mathbf{b}_1\cos\varphi)-k_o(\mathbf{a}_1\cos\varphi+\mathbf{b}_1\sin\varphi)\\ &= \mathbf{c}_1\cos\varphi-\mathbf{d}_1\sin\varphi,\\ \mathbf{d}_{1d}(\mathbf{k}) &= \mathbf{k}\times\mathbf{a}_{1d}+k_o\mathbf{b}_{1d}\\ &= \mathbf{k}\times(\mathbf{a}_1\cos\varphi+\mathbf{b}_1\sin\varphi)+k_o(-\mathbf{a}_1\sin\varphi+\mathbf{b}_1\cos\varphi)\\ &= \mathbf{d}_1\cos\varphi+\mathbf{c}_1\sin\varphi.\end{aligned}$$

$\mathbf{c}_{2d}(\mathbf{k})$ and $\mathbf{d}_{2d}(\mathbf{k})$ obey similar expressions with $1\to 2$. Applying these, we can further expand

$$\begin{aligned}J_d(\mathbf{k}) &= \mathbf{k}\cdot\mathbf{c}_{1d}(\mathbf{k})\times\mathbf{c}_{2d}(\mathbf{k})\\ &= \mathbf{k}\cdot(\mathbf{c}_1\cos\varphi-\mathbf{d}_1\sin\varphi)\times(\mathbf{c}_2\cos\varphi-\mathbf{d}_2\sin\varphi)\\ &= \mathbf{k}\cdot(\mathbf{c}_1\times\mathbf{c}_2\cos^2\varphi+\mathbf{d}_1\times\mathbf{d}_2\sin^2\varphi)+\Delta\\ &= J(\mathbf{k})+\Delta,\\ \Delta &= -\mathbf{k}\cdot(\mathbf{c}_1\times\mathbf{d}_2-\mathbf{c}_2\times\mathbf{d}_1)\sin\varphi\cos\varphi.\end{aligned}$$

Inserting (5.29) - (5.32) we have

$$\mathbf{k}\cdot\mathbf{c}_1\times\mathbf{d}_2=\mathbf{k}\cdot(\mathbf{k}\times\mathbf{b}_1-k_o\mathbf{a}_1)\times(\mathbf{k}\times\mathbf{a}_2+k_o\mathbf{b}_2)$$

$$=k_o\mathbf{k}\cdot(\mathbf{a}_1\mathbf{a}_2+\mathbf{b}_1\mathbf{b}_2)\cdot\mathbf{k}-k_o^2\mathbf{k}\cdot(\mathbf{a}_1\times\mathbf{b}_2+\mathbf{a}_2\times\mathbf{b}_1)-k_o^3(\mathbf{a}_1\cdot\mathbf{a}_2+\mathbf{b}_1\cdot\mathbf{b}_2).$$

$$\mathbf{k}\cdot\mathbf{c}_2\times\mathbf{d}_1=\mathbf{k}\cdot(\mathbf{k}\times\mathbf{b}_2-k_o\mathbf{a}_2)\times(\mathbf{k}\times\mathbf{a}_1+k_o\mathbf{b}_1)$$

$$=k_o\mathbf{k}\cdot(\mathbf{a}_2\mathbf{a}_1+\mathbf{b}_2\mathbf{b}_1)\cdot\mathbf{k}-k_o^2\mathbf{k}\cdot(\mathbf{a}_2\times\mathbf{b}_1+\mathbf{a}_1\times\mathbf{b}_2)-k_o^3(\mathbf{a}_2\cdot\mathbf{a}_1+\mathbf{b}_2\cdot\mathbf{b}_1),$$

whence $\Delta=0$. In conclusion, the function $J(\mathbf{k})$ is invariant in the duality transformation.

5.12 Denoting $\mathbf{k} = \omega\mathbf{p}$, a plane wave with the dependence $\exp(-j\omega\mathbf{p}\cdot\mathbf{r})$ satisfies in any medium the conditions

$$\mathbf{p}\times\mathbf{E} = \mathbf{B}, \qquad \mathbf{p}\times\mathbf{H} = -\mathbf{D}.$$

Substituting fields from (5.165), the latter equation becomes

$$\mathbf{p}\times(\overline{\overline{\beta}}\cdot\mathbf{E} + \overline{\overline{\mu}}^{-1}\cdot\mathbf{B}) = -\overline{\overline{\epsilon}}'\cdot\mathbf{E} - \overline{\overline{\alpha}}\cdot\mathbf{B}.$$

Inserting $\mathbf{B}$ from the first equation, we obtain

$$\begin{aligned}(\mathbf{p}\times\overline{\overline{\beta}} + \overline{\overline{\epsilon}}')\cdot\mathbf{E} &= -(\mathbf{p}\times\overline{\overline{\mu}}^{-1} + \overline{\overline{\alpha}})\cdot\mathbf{B}\\ &= -(\mathbf{p}\times\overline{\overline{\mu}}^{-1} + \overline{\overline{\alpha}})\cdot(\mathbf{p}\times\mathbf{E})\end{aligned}$$

This can be written in the form

$$\overline{\overline{\mathsf{D}}}(\mathbf{p})\cdot\mathbf{E} = 0,$$

in terms of the dispersion dyadic

$$\begin{aligned}\overline{\overline{\mathsf{D}}}(\mathbf{p}) &= \overline{\overline{\alpha}}\times\mathbf{p} + \mathbf{p}\times\overline{\overline{\mu}}^{-1}\times\mathbf{p} + \mathbf{p}\times\overline{\overline{\beta}} + \overline{\overline{\epsilon}}'\\ &= \overline{\overline{\mathsf{A}}}\times\mathbf{p} - \mathrm{tr}\overline{\overline{\mathsf{A}}}\,\overline{\overline{\mathsf{I}}}\times\mathbf{p} + \mathbf{p}\times(\mathbf{g}\times\overline{\overline{\mathsf{I}}})\times\mathbf{p} + \mathbf{p}\times(\overline{\overline{\mathsf{A}}}^T + a\overline{\overline{\mathsf{I}}}) + \mathbf{c}\times\overline{\overline{\mathsf{I}}},\end{aligned}$$

where we have substituted the expressions (5.166) - (5.169). Expanding

$$\mathbf{p}\times(\mathbf{g}\times\overline{\overline{\mathsf{I}}})\times\mathbf{p} = -(\mathbf{p}\cdot\mathbf{g})\overline{\overline{\mathsf{I}}}\times\mathbf{p}$$

$$\overline{\overline{\mathsf{A}}}\times\mathbf{p} + \mathbf{p}\times\overline{\overline{\mathsf{A}}}^T = \overline{\overline{\mathsf{A}}}\times\mathbf{p} - (\overline{\overline{\mathsf{A}}}\times\mathbf{p})^T = (\mathbf{p}\mathrm{tr}\overline{\overline{\mathsf{A}}} - \mathbf{p}\cdot\overline{\overline{\mathsf{A}}})\times\overline{\overline{\mathsf{I}}},$$

the dispersion dyadic takes the form

$$\overline{\overline{\mathsf{D}}}(\mathbf{p}) = -(\mathbf{p}\cdot\overline{\overline{\mathsf{A}}} + (\mathbf{p}\cdot\mathbf{g})\mathbf{p} - a\mathbf{p} - \mathbf{c})\times\overline{\overline{\mathsf{I}}},$$

which is an antisymmetric dyadic. Because $\det\overline{\overline{\mathsf{D}}}(\mathbf{p}) = 0$ for all $\mathbf{p}$, there is no dispersion equation to determine the vector $\mathbf{p}$ and $\mathbf{k}$, and $\mathbf{E}$ must be a multiple of $\mathbf{q}(\mathbf{p})$. Thus, a plane wave with any vector $\mathbf{k}$ is possible in the skewon-axion medium.

5.13 It was shown that the plane-wave field in the skewon-axion medium satisfies $\mathbf{q}(\mathbf{p})\times\mathbf{E} = 0$, whence, applying parameters (5.172) - (5.175), we may set

$$\begin{aligned}\mathbf{E} &= \mathbf{q}(\mathbf{p}) = (\mathbf{g}\cdot\mathbf{p} - a)\mathbf{p} + \mathbf{p}\cdot\overline{\overline{\mathsf{A}}} - \mathbf{c}\\ &= (\mathbf{a}_1\cdot\mathbf{p} - c)(A\mathbf{p} + \mathbf{e}_1).\end{aligned}$$

On the other hand, we have

$$\mathbf{B} = \mathbf{p}\times\mathbf{E} = (\mathbf{a}_1\cdot\mathbf{p} - c)\mathbf{p}\times\mathbf{e}_1.$$

With $\mathbf{e}_3 = \mathbf{e}_1 \times \mathbf{e}_2$, we obtain the relation

$$A\mathbf{e}_3 \cdot \mathbf{B} + \mathbf{e}_2 \cdot \mathbf{E} = (\mathbf{a}_1 \cdot \mathbf{p} - c)(-A\mathbf{p} \cdot \mathbf{e}_2 + A\mathbf{p} \cdot \mathbf{e}_2) = 0,$$

which is one of the SHDB conditions at the interface $\mathbf{e}_3 \cdot \mathbf{r} = 0$.

Expanding

$$\begin{aligned}
\mathbf{H} &= \overline{\overline{\mu}}^{-1} \cdot \mathbf{B} + \overline{\overline{\beta}} \cdot \mathbf{E} \\
&= \mathbf{g} \times \mathbf{B} + (\overline{\overline{\mathsf{A}}}^T + a\overline{\overline{\mathsf{I}}}) \cdot \mathbf{E} \\
&= (A\mathbf{p}\mathbf{a}_1 - A(\mathbf{a}_1 \cdot \mathbf{p})\overline{\overline{\mathsf{I}}} + (B + a)\overline{\overline{\mathsf{I}}} + \mathbf{e}_1\mathbf{a}_1) \cdot \mathbf{q}(\mathbf{p}) \\
&= (\mathbf{a}_1 \cdot \mathbf{p} - c)(B + a - \mathbf{a}_1 \cdot \mathbf{e}_1)(A\mathbf{p} + \mathbf{e}_1) \\
\mathbf{D} &= -\mathbf{p} \times \mathbf{H} = -(\mathbf{a}_1 \cdot \mathbf{p} - c)(B + a - \mathbf{a}_1 \cdot \mathbf{e}_1)(\mathbf{p} \times \mathbf{e}_1)
\end{aligned}$$

These vectors satisfy

$$A\mathbf{e}_3 \cdot \mathbf{D} - \mathbf{e}_2 \cdot \mathbf{H} = (\mathbf{a}_1 \cdot \mathbf{p} - c)(B + a - \mathbf{a}_1 \cdot \mathbf{e}_1)(-A\mathbf{e}_3 \cdot \mathbf{p} \times \mathbf{e}_1 - A\mathbf{e}_2 \cdot \mathbf{p}) = 0,$$

which equals the second of the SHDB conditions (5.178) and (5.179).

5.14 Because $\mathbf{k} \cdot \mathbf{E} = 0$, $\mathbf{c}_1 \cdot \mathbf{E}_1 = 0$ and $\mathbf{c}_2 \cdot \mathbf{E}_2 = 0$, the fields have polarizations $\mathbf{E}_1 \sim \mathbf{k} \times \mathbf{c}_1$ and $\mathbf{E}_2 \sim \mathbf{k} \times \mathbf{c}_2$. Let us expand

$$\begin{aligned}
\mathbf{E}_1 \cdot \mathbf{E}_2 &\sim (\mathbf{k} \times \mathbf{c}_1) \cdot (\mathbf{k} \times \mathbf{c}_2) \\
&= k_o^2(\mathbf{c}_1 \cdot \mathbf{c}_2) - (\mathbf{k} \cdot \mathbf{c}_1)(\mathbf{k} \cdot \mathbf{c}_2) \\
&= k_o^2(\alpha\mathbf{k}_t \times \mathbf{n} - k_o\mathbf{a}_t) \cdot (\mathbf{k} \times \mathbf{a}_t + k_o\alpha\mathbf{n}) - k_o^2(\mathbf{k}_t \cdot \mathbf{a}_t)(\mathbf{k} \cdot \mathbf{n}) \\
&= -k_o^2\alpha(\mathbf{k}_t \times \mathbf{n}) \cdot (\mathbf{k} \times \mathbf{a}_t) - k_o^2\alpha(\mathbf{k}_t \cdot \mathbf{a}_t)(\mathbf{k} \cdot \mathbf{n}),
\end{aligned} \quad \text{(D.2)}$$

the last terms of which can be seen to cancel. Thus, we have $\mathbf{E}_1 \cdot \mathbf{E}_2 = 0$.

Because of $\mathbf{H} \sim \mathbf{k} \times \mathbf{E}$, we have

$$\mathbf{H}_1 \cdot \mathbf{H}_2 \sim (\mathbf{k} \times \mathbf{E}_1) \cdot (\mathbf{k} \times \mathbf{E}_2) = k_o^2(\mathbf{E}_1 \cdot \mathbf{E}_2) - (\mathbf{k} \cdot \mathbf{E}_2)(\mathbf{k} \cdot \mathbf{E}_2).$$

From $\mathbf{k} \cdot \mathbf{E}_1 = \mathbf{k} \cdot \mathbf{E}_2 = 0$ we must have $\mathbf{H}_1 \cdot \mathbf{H}_2 = 0$.

5.15 Because the impedance boundary is the special case of the GBC boundary with the vectors $\mathbf{a}_1 \cdots \mathbf{b}_2$ in (5.8) and (5.9) having no normal components. Setting $a_{1n} = a_{2n} = b_{1n} = b_{2n} = 0$ in the conditions (5.232) and (5.234), the boundary conditions with PEC/PMC equivalence become

$$(B\mathbf{a}_{1t} + D\mathbf{a}_{2t}) \cdot \mathbf{E} = 0, \quad (A\mathbf{b}_{1t} + C\mathbf{b}_{2t}) \cdot \mathbf{H} = 0,$$

which are of the form of (3.180), conditions of the GSH boundary. Thus, most general impedance conditions with PEC/PMC equivalence are the GSH boundary conditions.

5.16 The mixed-impedance (DB/D'B') boundary is defined by the conditions (4.180) and (4.181) as

$$\begin{aligned}(jk_o\eta_o - Z_{TE}\partial_n)(\mathbf{n}\cdot\mathbf{H}) &= 0\\ (jk_o - Y_{TM}\eta_o\partial_n)(\mathbf{n}\cdot\mathbf{E}) &= 0.\end{aligned}$$

Applying $\nabla\cdot\mathbf{E} = \nabla_t\cdot\mathbf{E}_t + \partial_n(\mathbf{n}\cdot\mathbf{E}) = 0$, and similarly for $\nabla\cdot\mathbf{H}$, the conditions (4.182) and (4.183) can be expressed as

$$\begin{aligned}-\mathbf{n}\cdot(\nabla\times\mathbf{E}) + Z_{TE}\nabla_t\cdot\mathbf{H} &= 0\\ \mathbf{n}\cdot(\nabla\times\mathbf{H}) + Y_{TM}\nabla_t\cdot\mathbf{E} &= 0.\end{aligned}$$

These can be rewritten as

$$\begin{aligned}\nabla_t\cdot(\mathbf{n}\times\mathbf{E} + Z_{TE}\mathbf{H}_t) &= 0\\ (\mathbf{n}\times\nabla_t)\cdot(-Z_{TM}\mathbf{H}_t + \mathbf{n}\times\mathbf{E}) &= 0,\end{aligned}$$

where $Z_{TM} = 1/Y_{TM}$. The two bracketed field expressions can now be represented in terms of two scalar functions as

$$\begin{aligned}\mathbf{n}\times\mathbf{E} + Z_{TE}\mathbf{H}_t &= \mathbf{n}\times\nabla_t\phi,\\ -Z_{TM}\mathbf{H}_t + \mathbf{n}\times\mathbf{E} &= \nabla_t\psi,\end{aligned}$$

from which we can solve

$$\begin{aligned}\mathbf{E}_t &= \frac{1}{Z_{TE}+Z_{TM}}(Z_{TM}\nabla_t\phi - Z_{TE}\mathbf{n}\times\nabla_t\psi),\\ \mathbf{H}_t &= \frac{1}{Z_{TE}+Z_{TM}}(\mathbf{n}\times\nabla_t\phi - \nabla_t\psi).\end{aligned}$$

To define the integrand of (5.271), let us expand

$$\begin{aligned}&(Z_{TE}+Z_{TM})^2\mathbf{n}\cdot(\mathbf{E}_t'\times\mathbf{H}_t'' - \mathbf{E}_t''\times\mathbf{H}_t') =\\ =\ &\mathbf{n}\cdot(Z_{TM}\nabla_t\phi' - Z_{TE}\mathbf{n}\times\nabla_t\psi')\times(\mathbf{n}\times\nabla_t\phi' - \nabla_t\psi' b)\\ &-\mathbf{n}\cdot(Z_{TM}\nabla_t\phi' - Z_{TE}\mathbf{n}\times\nabla_t\psi')\times(\mathbf{n}\times\nabla_t\phi'' - \nabla_t\psi'')\\ =\ &-\mathbf{n}\cdot(Z_{TM}\nabla_t\phi'\times\nabla_t\psi'' + Z_{TE}\nabla_t\psi'\times\nabla_t\phi'')\\ &+\mathbf{n}\cdot(Z_{TM}\nabla_t\phi''\times\nabla_t\psi' + Z_{TE}\nabla_t\psi''\times\nabla_t\phi')\\ =\ &-(Z_{TE}+Z_{TM})\mathbf{n}\cdot(\nabla_t\phi'\times\nabla_t\psi'' + \nabla_t\psi'\times\nabla_t\phi''),\end{aligned}$$

whence we finally have

$$\mathbf{n}\cdot(\mathbf{E}_t'\times\mathbf{H}_t'' - \mathbf{E}_t''\times\mathbf{H}_t') = -\frac{1}{Z_{TE}+Z_{TM}}\mathbf{n}\cdot\nabla\times(\phi'\nabla_t\psi'' + \psi'\nabla_t\phi'').$$

Because the integrand of (5.271) is of the form $\mathbf{n}\cdot\nabla_t\times\mathbf{f}_t$, following the reasoning of the DB boundary, the mixed-impedance (DB/D'B') boundary is found to be reciprocal.

6.1 The vectors of (6.20) are valid wave vectors for the isotropic medium because they satisfy

$$\mathbf{k}'\cdot\mathbf{k}' = \mathbf{k}^i\cdot\mathbf{k}^i = k_o^2, \qquad \mathbf{k}''\cdot\mathbf{k}'' = (\mathbf{k}^i\cdot\mathbf{k}^i)^* = k_o^2,$$

At the boundary the condition (6.14) requires that the three wave components satisfy

$$\mathbf{E}_t^i e^{-j\mathbf{k}_t\cdot\mathbf{r}} + \mathbf{E}_t' e^{-j\mathbf{k}_t\cdot\mathbf{r}} + \mathbf{E}_t'' e^{j\mathbf{k}_t^*\cdot\mathbf{r}} =$$

$$= e^{j\Theta}\tan\psi\ \eta_o(\mathbf{H}_t^{i*} e^{j\mathbf{k}_t^*\cdot\mathbf{r}} + \mathbf{H}_t'^* e^{j\mathbf{k}_t^*\cdot\mathbf{r}} + \mathbf{H}_t''^* e^{-j\mathbf{k}_t\cdot\mathbf{r}}),$$

which is split in two parts according to their spatial dependence as

$$\mathbf{E}_t^i e^{-j\mathbf{k}_t\cdot\mathbf{r}} + \mathbf{E}_t' e^{-j\mathbf{k}_t\cdot\mathbf{r}} = e^{j\Theta}\tan\psi\ \eta_o(\mathbf{H}_t''^* e^{-j\mathbf{k}_t\cdot\mathbf{r}}),$$

$$\mathbf{E}_t'' e^{j\mathbf{k}_t^*\cdot\mathbf{r}} = e^{j\Theta}\tan\psi\ \eta_o(\mathbf{H}_t^{i*} e^{j\mathbf{k}_t^*\cdot\mathbf{r}} + \mathbf{H}_t'^* e^{j\mathbf{k}_t^*\cdot\mathbf{r}}).$$

These coincide with (6.21) and (6.22).

6.2 Applying (6.32), (6.37) and (6.38), we can expand

$$\begin{aligned}\mathbf{p}(\mathbf{E}') &= \frac{(\overline{\overline{\mathsf{C}}}\cdot\mathbf{E}^i)\times(\overline{\overline{\mathsf{C}}}\cdot\mathbf{E}^{i*})}{j(\overline{\overline{\mathsf{C}}}\cdot\mathbf{E}^i)\cdot(\overline{\overline{\mathsf{C}}}\cdot\mathbf{E}^{i*})} = \frac{\overline{\overline{\mathsf{C}}}^{(2)}\cdot(\mathbf{E}^i\times\mathbf{E}^{i*})}{j\mathbf{E}^i\cdot(\overline{\overline{\mathsf{C}}}^T\cdot\overline{\overline{\mathsf{C}}})\cdot\mathbf{E}^{i*}} \\ &= -\overline{\overline{\mathsf{C}}}\cdot\frac{\mathbf{E}^i\times\mathbf{E}^{i*}}{j\mathbf{E}^i\cdot\mathbf{E}^{i*}} = -\overline{\overline{\mathsf{C}}}\cdot\mathbf{p}(\mathbf{E}^i).\end{aligned}$$

6.3 Substituting the expressions (6.27) - (6.31), the total tangential electric and magnetic fields become

$$\begin{aligned}\mathbf{E}_t &= \mathbf{E}_t^i + \mathbf{E}_t' + \mathbf{E}_t'' \\ &= (1-\cos 2\psi)\mathbf{E}_t^i - e^{j\Theta}\sin 2\psi\ \overline{\overline{\mathsf{J}}}_t\cdot\mathbf{E}_t^{i*} \\ &= 2\sin^2\psi\ \mathbf{E}_t^i - e^{j\Theta}\sin 2\psi\ \overline{\overline{\mathsf{J}}}_t\cdot\mathbf{E}_t^{i*} \\ \eta_o\mathbf{H}_t &= \eta_o(\mathbf{H}_t^i + \mathbf{H}_t' + \mathbf{H}_t'') \\ &= -\overline{\overline{\mathsf{J}}}_t\cdot(1+\cos 2\psi)\mathbf{E}_t^i + e^{j\Theta}\sin 2\psi\ \mathbf{E}_t^{i*} \\ &= -2\cos^2\psi\ \overline{\overline{\mathsf{J}}}_t\cdot\mathbf{E}_t^i + e^{j\Theta}\sin 2\psi\ \mathbf{E}_t^{i*}.\end{aligned}$$

Expanding

$$\begin{aligned}e^{j\Theta}\tan\psi\ \eta_o\mathbf{H}_t^{i*} &= e^{j\Theta}\tan\psi(-2\cos^2\psi\ \overline{\overline{\mathsf{J}}}_t\cdot\mathbf{E}_t^{i*} + e^{-j\Theta}\sin 2\psi\ \mathbf{E}_t^i) \\ &= -e^{j\Theta}\sin 2\psi\ \overline{\overline{\mathsf{J}}}_t\cdot\mathbf{E}_t^{i*} + 2\sin^2\psi\ \mathbf{E}_t^i,\end{aligned}$$

we find that this equals $\mathbf{E}_t$, whence (6.14) is satisfied.

6.4 Writing the SQ condition as $\cos\psi\ \mathbf{E}_t^* = e^{-j\varphi}\sin\psi\ \mathbf{H}_t$, and multiplying by $\mathbf{E}_t = Z_s\mathbf{n}\times\mathbf{H}_t$, we obtain

$$\cos\psi|\mathbf{E}_t|^2 = Z_s e^{-j\varphi}\sin\psi\ \mathbf{H}_t\cdot(\mathbf{n}\times\mathbf{H}_t) = 0,$$

whence either $\mathbf{E}_t = 0$ at the boundary, which is the PEC condition, or $\cos\psi = 0$, whence $\mathbf{H}_t = 0$ at the boundary, which is the PMC condition.

6.5 Applying the general SQL boundary condition $\mathbf{E}_t = e^{j\Theta}\overline{\overline{\mathsf{X}}}_t\cdot\eta_o\mathbf{H}^*$, with real $\overline{\overline{\mathsf{X}}}_t$ in stead of (6.14), the conditions (6.21) and (6.22) must be replaced by

$$\begin{aligned}\mathbf{E}_t^i + \mathbf{E}_t' &= e^{j\Theta}\overline{\overline{\mathsf{X}}}_t\cdot\eta_o\mathbf{H}_t''^* = e^{j\Theta}\overline{\overline{\mathsf{X}}}_t\cdot\overline{\overline{\mathsf{J}}}_t\cdot\mathbf{E}_t''^*,\\ \mathbf{E}_t'' &= e^{j\Theta}\overline{\overline{\mathsf{X}}}_t\cdot\eta_o(\mathbf{H}_t^{i*}+\mathbf{H}_t'^*) = e^{j\Theta}\overline{\overline{\mathsf{X}}}_t\cdot\overline{\overline{\mathsf{J}}}_t\cdot(-\mathbf{E}_t^{i*}+\mathbf{E}_t'^*).\end{aligned}$$

From these we obtain

$$\mathbf{E}_t^i + \mathbf{E}_t' = \overline{\overline{\mathsf{Y}}}_t\cdot(-\mathbf{E}_t^i+\mathbf{E}_t'),\qquad \overline{\overline{\mathsf{Y}}}_t = (\overline{\overline{\mathsf{X}}}_t\cdot\overline{\overline{\mathsf{J}}}_t)^2,$$

whence $\mathbf{E}_t'$ can be solved in terms of the reflection dyadic $\overline{\overline{\mathsf{R}}}_t'$ as

$$\mathbf{E}_t' = \overline{\overline{\mathsf{R}}}_t'\cdot\mathbf{E}_t^i,\qquad \overline{\overline{\mathsf{R}}}_t' = -(\overline{\overline{\mathsf{I}}}_t - \overline{\overline{\mathsf{Y}}}_t)^{-1}\cdot(\overline{\overline{\mathsf{I}}}_t + \overline{\overline{\mathsf{Y}}}_t)$$

Further, we can write

$$\mathbf{E}_t'' = \overline{\overline{\mathsf{R}}}_t''\cdot\mathbf{E}_t^{i*}.$$

with the reflection dyadic

$$\begin{aligned}\overline{\overline{\mathsf{R}}}_t'' &= -e^{j\Theta}\overline{\overline{\mathsf{X}}}_t\cdot\overline{\overline{\mathsf{J}}}_t\cdot(\overline{\overline{\mathsf{I}}}_t - \overline{\overline{\mathsf{R}}}_t')\\ &= -e^{j\Theta}\overline{\overline{\mathsf{X}}}_t\cdot\overline{\overline{\mathsf{J}}}_t\cdot(\overline{\overline{\mathsf{I}}}_t + (\overline{\overline{\mathsf{I}}}_t - \overline{\overline{\mathsf{Y}}}_t)^{-1}\cdot(\overline{\overline{\mathsf{I}}}_t + \overline{\overline{\mathsf{Y}}}_t)),\end{aligned}$$

For $\overline{\overline{\mathsf{X}}}_t = \tan\psi\mathbf{n}\times\overline{\overline{\mathsf{I}}}$ the conditions (6.15) become valid. The symmetric 2D dyadics $\overline{\overline{\mathsf{X}}}_t\cdot\overline{\overline{\mathsf{J}}}_t$, $\overline{\overline{\mathsf{Y}}}_t$, $\overline{\overline{\mathsf{R}}}_t'$ and $\overline{\overline{\mathsf{R}}}_t''$ can be expanded in terms of eigenvectors $\mathbf{k}_t$ and $\mathbf{n}\times\mathbf{k}_t$, which correspond to TM and TE polarizations, as

$$\overline{\overline{\mathsf{D}}}_t = D_{TE}\frac{(\mathbf{n}\times\mathbf{k})(\mathbf{n}\times\mathbf{k})}{\mathbf{k}_t^2} + D_{TM}\frac{\mathbf{k}_t\mathbf{k}_t}{\mathbf{k}_t^2}.$$

The coefficients in the eigenexpansions can be solved separately from the dyadic equations as

$$(\overline{\overline{\mathsf{X}}}_t\cdot\overline{\overline{\mathsf{J}}}_t)_{TE} = -e^{j\Theta}\tan\psi\frac{k_n}{k_o},\qquad (\overline{\overline{\mathsf{X}}}_t\cdot\overline{\overline{\mathsf{J}}}_t)_{TM} = -e^{j\Theta}\tan\psi\frac{k_o}{k_n},$$

$$Y_{TE} = \tan^2\psi\frac{k_n^2}{k_o^2},\qquad Y_{TM} = \tan^2\psi\frac{k_o^2}{k_n^2},$$

$$(\overline{\overline{\mathsf{I}}}_t + \overline{\overline{\mathsf{Y}}}_t)_{TE} = \frac{k_o^2 + k_n^2\tan^2\psi}{k_o^2},\qquad (\overline{\overline{\mathsf{I}}}_t + \overline{\overline{\mathsf{Y}}}_t)_{TM} = \frac{k_n^2 + k_o^2\tan^2\psi}{k_n^2},$$

$$(\overline{\overline{\mathsf{I}}}_t - \overline{\overline{\mathsf{Y}}}_t)_{TE}^{-1} = \frac{k_o^2}{k_o^2 - k_n^2 \tan^2 \psi}, \quad (\overline{\overline{\mathsf{I}}}_t - \overline{\overline{\mathsf{Y}}}_t)_{TM} = \frac{k_n^2}{k_n^2 - k_o^2 \tan^2 \psi},$$

$$R'_{TE} = -\frac{k_o^2 + k_n^2 \tan^2 \psi}{k_o^2 - k_n^2 \tan^2 \psi} \quad R'_{TM} = -\frac{k_n^2 + k_o^2 \tan^2 \psi}{k_n^2 - k_o^2 \tan^2 \psi}$$

$$R''_{TE} = -2e^{j\Theta} \frac{k_o k_n \tan \psi}{k_o^2 - k_n^2 \tan^2 \psi}, \quad R''_{TM} = -2e^{j\Theta} \frac{k_o k_n \tan \psi}{k_n^2 - k_o^2 \tan^2 \psi}$$

The PEC and PMC special cases correspond to $\psi = 0$ and $\psi = \pi/2$, respectively, whence the reflection coefficients become

$$R'_{TE} = R'_{TM} = -1, \quad R''_{TE} = R''_{TM} = 0, \quad \text{PEC}$$

$$R'_{TE} = R'_{TM} = +1, \quad R''_{TE} = R''_{TM} = 0, \quad \text{PMC}$$

6.6 The polarization vectors of the reflected components corresponding to right-hand circularly-polarized wave $\mathbf{E}^i = \mathbf{E}_t^i = \mathbf{e}_+ E^i$ incident to the isotropic SQL boundary can be found as

$$\mathbf{E}_t^i = \mathbf{e}_+ E^i, \quad \Rightarrow \quad \mathbf{p}(\mathbf{E}_t^i) = \frac{\mathbf{e}_+ \times \mathbf{e}_-}{j\mathbf{e}_+ \cdot \mathbf{e}_-} = \frac{-2j\mathbf{n}}{2j} = -\mathbf{n}.$$

This is right handed because $-\mathbf{n} \cdot \mathbf{p}(\mathbf{E}^i) = 1$.

$$\mathbf{E}'_t = -\cos 2\psi \; \mathbf{e}_+ E^i, \quad \Rightarrow \quad \mathbf{p}(\mathbf{E}'_t) = \mathbf{p}(\mathbf{e}_+) = -\mathbf{n},$$

This is left handed because $\mathbf{n} \cdot \mathbf{p}(\mathbf{E}^i) = -1$.

$$\mathbf{E}''_t = je^{j\Theta} \sin 2\psi \; \mathbf{e}_- E^i, \quad \Rightarrow \quad \mathbf{p}(\mathbf{E}''_t) = \mathbf{p}(\mathbf{e}_-) = \mathbf{n}.$$

This is right handed because $\mathbf{n} \cdot \mathbf{p}(\mathbf{E}''_t) = 1$

Taking into account the properties

$$\mathbf{E}'_t \times \mathbf{E}''^*_t = \mathbf{E}''_t \times \mathbf{E}'^*_t = 0,$$

$$\mathbf{E}'_t \cdot \mathbf{E}''^*_t = \mathbf{E}''_t \cdot \mathbf{E}'^*_t = 0,$$

we can write

$$\begin{aligned} \mathbf{p}(\mathbf{E}'_t + \mathbf{E}''_t) &= \frac{\mathbf{E}'_t \times \mathbf{E}'^*_t + \mathbf{E}''_t \times \mathbf{E}''^*_t}{j(\mathbf{E}'_t \cdot \mathbf{E}''^*_t + \mathbf{E}''_t \cdot \mathbf{E}'^*_t)} \\ &= \frac{-2j\mathbf{n}(\cos^2 2\psi - \sin^2 2\psi)}{2j(\cos^2 2\psi + \sin^2 2\psi)} = -\mathbf{n} \cos 4\psi. \end{aligned}$$

7.1 An electrically small sphere is described by $x \ll 1$, in other words, the diameter of the sphere is small compared with the wavelength of the incident field. The size parameter appears in the argument of the spherical Bessel and Hankel functions in (7.5), and hence it is proper to expand the coefficients into Taylor series:

$$j_1(x) = \frac{x}{3} - \frac{x^3}{30} + O\left(x^4\right),$$
$$h_1^{(2)}(x) = \frac{j}{x^2} + \frac{j}{2} + \frac{x}{3} + O\left(x^{3/2}\right),$$
$$j_2(x) = \frac{x^2}{15} - \frac{x^4}{210} + O\left(x^5\right),$$
$$h_2^{(2)}(x) = \frac{3j}{x^3} + \frac{j}{2x} + O\left(\sqrt{x}\right)\ldots$$

Hence

$$b_{1,\mathrm{PEC}} = \frac{j_1(x)}{h_1^{(2)}(x)} = -\frac{jx^3}{3} + O\left(x^{9/2}\right),$$
$$b_{2,\mathrm{PEC}} = \frac{j_2(x)}{h_2^{(2)}(x)} = -\frac{jx^5}{45} + O\left(x^{11/2}\right).$$

As the coefficients are ratios of the Bessel function over the Hankel one, their dependence on x increases as the order n increases. Therefore, the dominant term in the small-x regime comes from the first-order term in the series in (7.2), and we can neglect terms $b_{2,\mathrm{PEC}}$ and higher.

Likewise for the a_n coefficients:

$$(xj_1(x))' = \frac{2x}{3} - \frac{2x^3}{15} + O\left(x^4\right),$$
$$\left(xh_1^{(2)}(x)\right)' = -\frac{j}{x^2} + \frac{j}{2} + O\left(\sqrt{x}\right).$$

and

$$a_{1,\mathrm{PEC}} = \frac{(xj_1(x))'}{\left(xh_1^{(2)}(x)\right)'} = \frac{2jx^3}{3} + O\left(x^{9/2}\right).$$

This leads to the low-frequency (Rayleigh) approximation for the scattering efficiency:

$$Q_{\mathrm{sca,PEC}} \approx \frac{2}{x^2} \cdot 3\left(\left|a_{1,\mathrm{PEC}}\right|^2 + \left|b_{1,\mathrm{PEC}}\right|^2\right) \approx \frac{10}{3}x^4$$

Note that in the scattered power, the electric dipole coefficient a_1 contributes four times as much as the magnetic dipole b_1.

7.2 Vanishing absorption efficiency is tantamount to the situation that extinction and scattering are equal. Hence, using (7.2) and (7.3), it suffices to prove that

$$\mathrm{Re}\{a_n\} = |a_n|^2 \quad \text{and} \quad \mathrm{Re}\{b_n\} = |b_n|^2.$$

Due to the fact that in all four expansions, the coefficients are expressed in terms of the PEC coefficients (see (7.13) and (7.9)), it remains to prove these conditions for the a_n and b_n coefficients for PEC sphere.

Compute the real part in a straightforward manner, using, instead of Hankel function, the spherical Bessel function of the second kind $y_n(x)$:

$$b_{n,\mathrm{PEC}} = \frac{j_n(x)}{h_n^{(2)}} = \frac{j_n(x)}{j_n(x) - jy_n(x)} = \frac{j_n(x)\,[j_n(x) + jy_n(x)]}{j_n^2(x) + y_n^2(x)}.$$

Functions $j_n(x)$ and $y_n(x)$ are real-valued for real positive arguments x, which leads to

$$\mathrm{Re}\,\{b_{n,\mathrm{PEC}}\} = \frac{j_n^2(x)}{j_n^2(x) + y_n^2(x)} = \left|\frac{j_n(x)}{j_n(x) - jy_n(x)}\right|^2 = |b_{n,\mathrm{PEC}}|^2.$$

As to the coefficients $a_{n,\mathrm{PEC}}$, using for example the recursion formula

$$(xz_n(x))' = xz_{n-1}(x) - nz_n(x)$$

(z_n being either j_n or y_n) distills the derivatives in the numerator and denominator of $a_{n,\mathrm{PEC}}$ into a form for which the same argument applies as in the case for $b_{n,\mathrm{PEC}}$ above.

7.3 Normalize the surface impedance Z_s by the free-space impedance,

$$z_s = r_s + jx_s = \frac{R_s + jX_s}{\eta_o}.$$

For lossless IBC spheres the surface impedance is purely imaginary ($Z_s = jX_s$ and $z_s = jx_s$), and the Mie coefficients (7.19)–(7.20), using normalized impedances, read

$$a_{n,\mathrm{IBC}} = \frac{x\,j_{n-1}(x) - n\,j_n(x) + x_s\,x\,j_n(x)}{x\,h_{n-1}^{(2)}(x) - n\,h_n^{(2)}(x) + x_s\,x\,h_n^{(2)}(x)},$$

$$b_{n,\mathrm{IBC}} = \frac{x\,j_{n-1}(x) - n\,j_n(x) - (1/x_s)\,x\,j_n(x)}{x\,h_{n-1}^{(2)}(x) - n\,h_n^{(2)}(x) - (1/x_s)\,x\,h_n^{(2)}(x)}.$$

From this form, it is obvious that the change $x_s \to -1/x_s$ swaps the coefficients:

$$a_{n,\mathrm{IBC}}(-1/x_s) = b_{n,\mathrm{IBC}}(x_s), \quad \text{and} \quad b_{n,\mathrm{IBC}}(-1/x_s) = a_{n,\mathrm{IBC}}(x_s).$$

And since the scattering efficiency (7.2) contains the sum of squares of the coefficients ($|a_n|^2 + |b_n^2|$), its value is not affected by the change $x_s \to -1/x_s$.

In fact, more can be concluded from the form of the coefficients (7.19)–(7.20). For an arbitrary complex-valued surface impedance Z_s, the relation

$$a_{n,\mathrm{IBC}}(1/z_s) = b_{n,\mathrm{IBC}}(z_s), \quad \text{and} \quad b_{n,\mathrm{IBC}}(1/z_s) = a_{n,\mathrm{IBC}}(z_s)$$

holds. A general Z_s is no longer lossless but its absorption efficiency may be non-zero (positive for lossy spheres, and negative for active ones). The following equalities hold:

$$Q_{\text{sca,IBC}}(x, 1/z_s) = Q_{\text{sca,IBC}}(x, z_s),$$
$$Q_{\text{ext,IBC}}(x, 1/z_s) = Q_{\text{ext,IBC}}(x, z_s),$$
$$Q_{\text{abs,IBC}}(x, 1/z_s) = Q_{\text{abs,IBC}}(x, z_s).$$

7.4 Normalize first the dipole moments into dimensionless quantities:

$$\mathbf{n}_e = \frac{\mathbf{p}_e}{\epsilon_o V E_o}, \qquad \mathbf{n}_m = \frac{\eta_o \mathbf{p}_m}{\mu_o V E_o}.$$

Then the moments read, for excitation $\mathbf{E}^i = E_o \mathbf{u}_x, \mathbf{H}^i = E_o/\eta_o \mathbf{u}_y$ as follows (cf. Eq. (7.11))

$$\begin{pmatrix} \mathbf{n}_e \\ \mathbf{n}_m \end{pmatrix} = \frac{3/2}{1+(M\eta_o)^2} \begin{pmatrix} 2(M\eta_o)^2 - 1 & 3\,M\eta_o \\ 3\,M\eta_o & 2-(M\eta_o)^2 \end{pmatrix} \begin{pmatrix} \mathbf{u}_x \\ \mathbf{u}_y \end{pmatrix}.$$

Using the angle parameter ϑ, defined in (2.31) as

$$\tan\vartheta = M\eta_o,$$

we can rewrite the moments as

$$\begin{pmatrix} \mathbf{n}_e \\ \mathbf{n}_m \end{pmatrix} = \left[\frac{3}{4} \begin{pmatrix} 1 & 0 \\ 0 & 1 \end{pmatrix} + \frac{9}{4} \begin{pmatrix} \cos 2\vartheta & \sin 2\vartheta \\ -\sin 2\vartheta & \cos 2\vartheta \end{pmatrix} \begin{pmatrix} -1 & 0 \\ 0 & 1 \end{pmatrix} \right] \begin{pmatrix} \mathbf{u}_x \\ \mathbf{u}_y \end{pmatrix}$$

with special cases of PEC ($\vartheta = \pi/2$) and PMC ($\vartheta = 0$):

$$\begin{pmatrix} \mathbf{n}_e \\ \mathbf{n}_m \end{pmatrix}_{\text{PEC}} = \begin{pmatrix} 3\mathbf{u}_x \\ -\frac{3}{2}\mathbf{u}_y \end{pmatrix}, \quad \text{and} \quad \begin{pmatrix} \mathbf{n}_e \\ \mathbf{n}_m \end{pmatrix}_{\text{PMC}} = \begin{pmatrix} -\frac{3}{2}\mathbf{u}_x \\ 3\mathbf{u}_y \end{pmatrix}.$$

The magnitudes of the electric and magnetic dipole moments, as functions of ϑ, are

$$|\mathbf{n}_e| = 3\sqrt{\frac{5 - 3\cos 2\vartheta}{8}}, \qquad |\mathbf{n}_m| = 3\sqrt{\frac{5 + 3\cos 2\vartheta}{8}}$$

and the angle γ between the moments obeys

$$\cos\gamma = \frac{\mathbf{n}_e \cdot \mathbf{n}_m}{|\mathbf{n}_e||\mathbf{n}_m|} = \frac{3\sin 2\vartheta}{\sqrt{(5 - 3\cos 2\vartheta)\,(5 + 3\cos 2\vartheta)}}$$

These are plotted in Figure D.3 as functions of ϑ. Even if the magnitudes of the dipole moments vary with the degree of PEMC, the following is valid:

$$|\mathbf{n}_e|^2 + |\mathbf{n}_m|^2 = \frac{45}{4}$$

independently of ϑ.

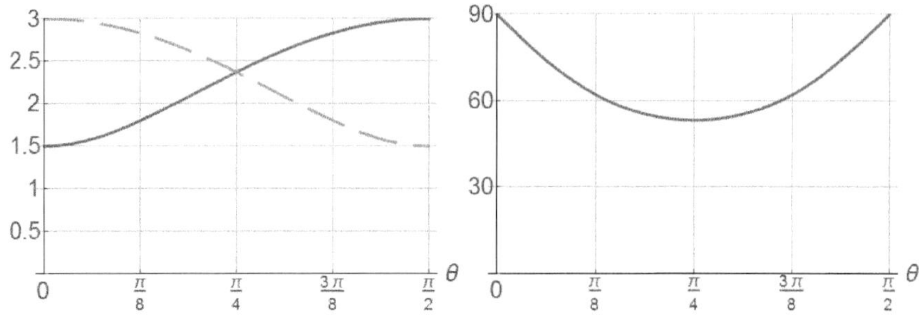

(a) $|\mathbf{n}_e|$ (solid blue) and $|\mathbf{n}_m|$ (dashed orange) (b) γ in degrees

Figure D.3: The magnitudes of the normalized electric and magnetic dipole moments and the angle γ between them as function of the PEMC parameter angle ϑ.

7.5 As can be seen in Figure 7.11, the scattering efficiencies start from zero for $x = 0$ and are low in the Rayleigh regime. For very large scatterers, the scattering efficiencies of lossless scatterers approach the value 2. The maxima have to searched numerically, leading to the following table (note that the PEC and PMC spheres of equal size have the same scattering cross sections):

	$Q_{\text{sca,max}}$	for size parameter x
PEC and PMC	2.29038	1.20878
DB	2.79649	2.98191
D'B'	3.7728	1.12969

7.6 When only the dipole terms ($n = 1$ in (7.2)) are included in the scattering efficiency expression, we capture another low-frequency estimate for Q_{sca} of a PEC sphere. Due to the fact that the first-order Mie coefficients contain x-dependence, the result is a much better approximation compared to the Rayleigh approximation with x^4 dependence. The performance of the two approximations is illustrated in Figure D.4.

In particular, the dipole approximation holds the relative accuracy of one thousand up to spheres of size parameter $x = 0.6988$. If the tolerance is one-percent, the size can be up to $x = 1.083$. (The corresponding figures for the Rayleigh approximation are $x = 0.0646$ for relative accuracy 1‰ and $x = 0.206$ for 1%.)

7.7 The backscattering efficiency Q_b is a function of the electric and magnetic Mie coefficents of the object. For an IBC sphere with $Z_s = \eta_o$, the coffi-

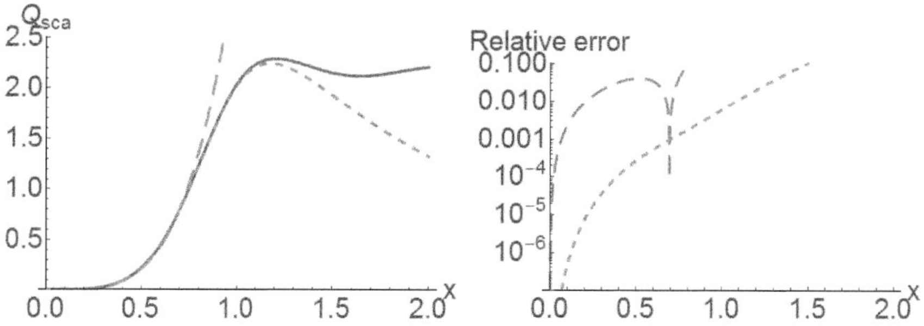

Figure D.4: Scattering efficiency Q_{sca} of a PEC sphere (solid blue), along with the Rayleigh approximation (dashed orange, (7.18)), and the first-term dipoles approximation (dotted green). The relative error of the two approximations on the right-hand side as function of the electrical size of the sphere.

cients (7.19) and (7.20) read

$$a_{n,\text{IBC}} = b_{n,\text{IBC}} = \frac{x\, j_{n-1}(x) - n\, j_n(x) - j\, x\, j_n(x)}{x\, h^{(2)}_{n-1}(x) - n\, h^{(2)}_n(x) - j\, x\, h^{(2)}_n(x)}.$$

Therefore, using (7.6), we have that the backscattering efficiency becomes zero:

$$Q_{\text{b,IBC}} = \frac{1}{x^2}\left|\sum_{n=1}^{\infty}(2n+1)(-1)^n\,(a_{n,\text{IBC}} - b_{n,\text{IBC}})\right|^2 = 0.$$

7.8 Figure 7.18 shows a rather broad minimum for the forward-to-backward scattering ratio in the case of $X_s = -0.1j\eta_o$ around $x \approx 0.36$ for which a closer look gives that the forward scattering efficiency is one thousandth (-30.5 dB) of the backscattering efficiency.

In fact, this minimum can be made even deeper for small spheres of either capacitive or inductive surface reactances. In order to minimize forward scattering for small spheres, Equation (7.7) shows that the sum of the electric and magnetic dipolar ($n = 1$) Mie coefficients should be zero.

Figure D.5 shows the magnitude of the coefficients $a_{1,\text{IBC}}$ and $b_{1,\text{IBC}}$ for a small sphere ($x = 0.1$) as function of its capacitive reactance (for this region, the coefficients are mostly imaginary and the imaginary parts have different sign). Their sum can be seen to vanish as the reactance achieves a value around $-0.025\eta_o$. And for the same condition, the front-to-back ratio reaches a deep minimum (-64 dB), also shown in the figure.

According to the optical theorem [3], the forward scattering amplitude of an object is proportional to its total scattering cross section. Hence a scatterer cannot have a vanishing forward scattering efficiency. Even

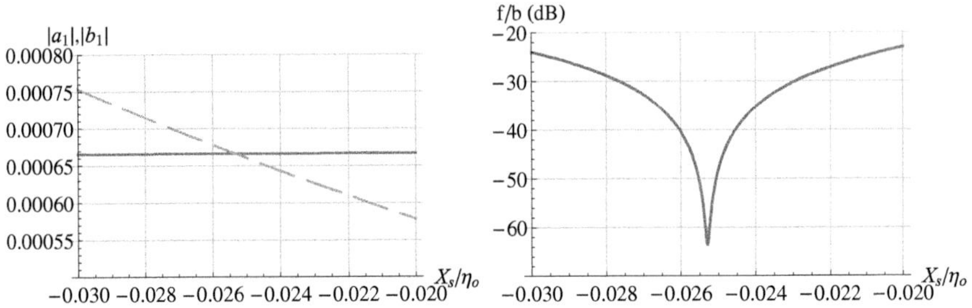

Figure D.5: Left: the absolute values of the first-order Mie coefficients for a lossless IBC sphere with parameter $x = 0.1$ as function of (capacitive) surface reactance X_s (solid blue: $a_{1,\mathrm{IBC}}$, dashed orange: $b_{1,\mathrm{IBC}}$). Right: front-to-back efficiency ratio for the same sphere.

if the dipolar terms were to cancel in the forward scattering efficiency, higher-order multipoles (which are though very small for subwavelength scatterers) will not be matched for the same (x, X_s) combination. However, the minimum exemplified in Figure D.5 can reach very low values for small spheres.

For the example $x = 0.1$ the minimum forward scattering takes place approximately for the reactance value $X_s/\eta_o = -x/4$. This condition becomes increasingly accurate as x decreases. And again, due to the duality correspondence between the electric and magnetic coefficients for the IBC case ($a_{n,\mathrm{IBC}}(X_s/\eta_o) = b_{n,\mathrm{IBC}}(-\eta_o/X_s)$), the same minimum happens in the inductive domain for positive reactance $X_s/\eta_o = 4/x$.

A.1 Expanding

$$\begin{aligned}
\mathbf{a}' \times \mathbf{a}'^* &= (\mathbf{a} \times \mathbf{a}^*)\cos^2\theta + (\mathbf{m} \times \mathbf{a}) \times \mathbf{a}^* \sin\theta\cos\theta \\
&\quad + \mathbf{a}^* \times (\mathbf{m} \times \mathbf{a})\sin\theta\cos\theta + (\mathbf{m} \times \mathbf{a}) \times (\mathbf{m} \times \mathbf{a}^*)\sin^2\theta \\
&= (\mathbf{a} \times \mathbf{a}^*)\cos^2\theta + \mathbf{m}\mathbf{m} \cdot (\mathbf{a} \times \mathbf{a}^*)\sin^2\theta \\
&= j|\mathbf{a}|^2\mathbf{p}(\mathbf{a}),
\end{aligned}$$

and

$$\begin{aligned}
|\mathbf{a}'|^2 &= |\mathbf{a}|^2\cos^2\theta + \mathbf{a} \cdot (\mathbf{m} \times \mathbf{a}^*)\sin\theta\cos\theta \\
&\quad + (\mathbf{m} \times \mathbf{a}) \times \mathbf{a}^* \sin\theta\cos\theta + |\mathbf{a}|^2\sin^2\theta \\
&= |\mathbf{a}|^2,
\end{aligned}$$

we obtain $\mathbf{p}(\mathbf{a}') = \mathbf{p}(\mathbf{a})$.

A.2 In this case we have $\mathbf{m} = \mathbf{n}$. Let us expand

$$\begin{aligned}
\mathbf{a}_t \times \mathbf{a}_t^* &= (\mathbf{b}_t + j\mathbf{n} \times \mathbf{b}_t) \times (\mathbf{b}_t + j\mathbf{n} \times \mathbf{b}_t)^* \\
&= \mathbf{b}_t \times \mathbf{b}_t^* - 2j\mathbf{n}|\mathbf{b}_t|^2 + \mathbf{nn} \cdot \mathbf{b}_t \times \mathbf{b}_t^* \\
&= 2\mathbf{n}(\mathbf{n} \cdot \mathbf{b}_t \times \mathbf{b}_t^* - j|\mathbf{b}_t|^2) \\
&= 2\mathbf{n}(\mathbf{n} \cdot \mathbf{p}(\mathbf{b}_t) - 1)j|\mathbf{b}_t|^2, \\
\mathbf{a}_t \cdot \mathbf{a}_t^* &= (\mathbf{b}_t + j\mathbf{n} \times \mathbf{b}_t) \cdot (\mathbf{b}_t + j\mathbf{n} \times \mathbf{b}_t)^* \\
&= \mathbf{b}_t \cdot \mathbf{b}_t^* + 2j\mathbf{n} \cdot (\mathbf{b}_t \times \mathbf{b}_t^*) + \mathbf{b}_t \cdot \mathbf{b}_t^* \\
&= 2|\mathbf{b}_t|^2 - 2\mathbf{n} \cdot \mathbf{p}(\mathbf{b}_t)|\mathbf{b}_t|^2 \\
&= 2(1 - \mathbf{n} \cdot \mathbf{p}(\mathbf{b}_t))|\mathbf{b}_t|^2.
\end{aligned}$$

Combining these we obtain

$$\mathbf{p}(\mathbf{a}_t) = \frac{2\mathbf{n}(\mathbf{n} \cdot \mathbf{p}(\mathbf{b}_t) - 1)j|\mathbf{b}_t|^2}{j2(1 - \mathbf{n} \cdot \mathbf{p}(\mathbf{b}_t))|\mathbf{b}_t|^2} = -\mathbf{n},$$

whence the polarization of $\mathbf{a}_t$ is circular and left handed with respect to $\mathbf{n}$.

B.1 For any vector $\mathbf{b}$ we can expand

$$(\mathbf{a} \times \overline{\overline{\mathsf{I}}} - \overline{\overline{\mathsf{I}}} \times \mathbf{a}) \cdot \mathbf{b} = \mathbf{a} \times (\overline{\overline{\mathsf{I}}} \cdot \mathbf{b}) - \overline{\overline{\mathsf{I}}} \cdot (\mathbf{a} \times \mathbf{b}) = \mathbf{a} \times \mathbf{b} - \mathbf{a} \times \mathbf{b} = 0,$$

whence the first dyadic in brackets is the null dyadic and, thus,

$$\mathbf{a} \times \overline{\overline{\mathsf{I}}} = \overline{\overline{\mathsf{I}}} \times \mathbf{a}$$

is valid for any vector $\mathbf{a}$.

B.2 Replacing $\overline{\overline{\mathsf{A}}}$ by $\overline{\overline{\mathsf{A}}} + \overline{\overline{\mathsf{B}}}$ in the identity

$$\overline{\overline{\mathsf{A}}}^{(2)} = \overline{\overline{\mathsf{A}}}^{2T} - \mathrm{tr}\overline{\overline{\mathsf{A}}}\,\overline{\overline{\mathsf{A}}}^T + \mathrm{tr}\overline{\overline{\mathsf{A}}}^{(2)}\,\overline{\overline{\mathsf{I}}}$$

we can substitute

$$\begin{aligned}
(\overline{\overline{\mathsf{A}}} + \overline{\overline{\mathsf{B}}})^{(2)} &= \overline{\overline{\mathsf{A}}}^{(2)} + \overline{\overline{\mathsf{A}}}{}^{\times}_{\times}\overline{\overline{\mathsf{B}}} + \overline{\overline{\mathsf{B}}}^{(2)} \\
\mathrm{tr}(\overline{\overline{\mathsf{A}}} + \overline{\overline{\mathsf{B}}}) &= \mathrm{tr}\overline{\overline{\mathsf{A}}} + \mathrm{tr}\overline{\overline{\mathsf{B}}} \\
\mathrm{tr}(\overline{\overline{\mathsf{A}}} + \overline{\overline{\mathsf{B}}})^{(2)} &= \mathrm{tr}\overline{\overline{\mathsf{A}}}^{(2)} + \mathrm{tr}(\overline{\overline{\mathsf{A}}}{}^{\times}_{\times}\overline{\overline{\mathsf{B}}}) + \mathrm{tr}\overline{\overline{\mathsf{B}}}^{(2)} \\
\mathrm{tr}(\overline{\overline{\mathsf{A}}}{}^{\times}_{\times}\overline{\overline{\mathsf{B}}}) &= \overline{\overline{\mathsf{A}}} : (\overline{\overline{\mathsf{B}}}{}^{\times}_{\times}\overline{\overline{\mathsf{I}}}) = \overline{\overline{\mathsf{A}}} : (\mathrm{tr}\overline{\overline{\mathsf{B}}}\,\overline{\overline{\mathsf{I}}} - \overline{\overline{\mathsf{B}}}^T) = \mathrm{tr}\overline{\overline{\mathsf{A}}}\,\mathrm{tr}\overline{\overline{\mathsf{B}}} - \mathrm{tr}(\overline{\overline{\mathsf{A}}} \cdot \overline{\overline{\mathsf{B}}})
\end{aligned}$$

Subtracting the identities involving $\overline{\overline{\mathsf{A}}}$ and $\overline{\overline{\mathsf{B}}}$ the remaining identity becomes

$$\overline{\overline{\mathsf{A}}}{}^{\times}_{\times}\overline{\overline{\mathsf{B}}} = [\mathrm{tr}\overline{\overline{\mathsf{A}}}\,\mathrm{tr}\overline{\overline{\mathsf{B}}} - \mathrm{tr}(\overline{\overline{\mathsf{A}}} \cdot \overline{\overline{\mathsf{B}}})]\overline{\overline{\mathsf{I}}} - \mathrm{tr}\overline{\overline{\mathsf{A}}}\,\overline{\overline{\mathsf{B}}}^T - \mathrm{tr}\overline{\overline{\mathsf{B}}}\,\overline{\overline{\mathsf{A}}}^T + (\overline{\overline{\mathsf{A}}} \cdot \overline{\overline{\mathsf{B}}} + \overline{\overline{\mathsf{B}}} \cdot \overline{\overline{\mathsf{A}}})^T$$

B.3 Using straightforward vector algebra, we can write

$$\begin{aligned}\overline{\overline{\mathsf{A}}}_t {}^\times_\times \mathbf{nn} &= (A_{11}\mathbf{e}_1\mathbf{e}_1 + A_{12}\mathbf{e}_1\mathbf{e}_2 + A_{21}\mathbf{e}_2\mathbf{e}_1 + A_{22}\mathbf{e}_2\mathbf{e}_2) {}^\times_\times \mathbf{nn} \\ &= A_{22}\mathbf{e}_1\mathbf{e}_1 - A_{21}\mathbf{e}_1\mathbf{e}_2 - A_{12}\mathbf{e}_2\mathbf{e}_1 + A_{11}\mathbf{e}_2\mathbf{e}_2,\end{aligned}$$

and

$$\overline{\overline{\mathsf{A}}}_t{}^T = A_{11}\mathbf{e}_1\mathbf{e}_1 + A_{21}\mathbf{e}_1\mathbf{e}_2 + A_{12}\mathbf{e}_2\mathbf{e}_1 + A_{22}\mathbf{e}_2\mathbf{e}_2,$$

whence we obtain

$$\overline{\overline{\mathsf{A}}}_t {}^\times_\times \mathbf{nn} + \overline{\overline{\mathsf{A}}}_t{}^T = (A_{11} + A_{22})(\mathbf{e}_1\mathbf{e}_1 + \mathbf{e}_2\mathbf{e}_2) = \mathrm{tr}\overline{\overline{\mathsf{A}}}_t\, \overline{\overline{\mathsf{I}}}_t.$$

B.4 Inserting $\overline{\overline{\mathsf{A}}} = \sum \mathbf{b}_i\mathbf{c}_i = \sum \mathbf{b}_j\mathbf{c}_j$, we can expand

$$\overline{\overline{\mathsf{A}}}\cdot(\mathbf{a}\times\overline{\overline{\mathsf{A}}}^T) = (\mathbf{c}_1\cdot\mathbf{a}\times\mathbf{c}_2)(\mathbf{b}_1\mathbf{b}_2 - \mathbf{b}_2\mathbf{b}_1)$$

$$+(\mathbf{c}_2\cdot\mathbf{a}\times\mathbf{c}_3)(\mathbf{b}_2\mathbf{b}_3 - \mathbf{b}_3\mathbf{b}_2) + (\mathbf{c}_3\cdot\mathbf{a}\times\mathbf{c}_1)(\mathbf{b}_3\mathbf{b}_1 - \mathbf{b}_1\mathbf{b}_3)$$

$$= -(\mathbf{c}_1\cdot\mathbf{a}\times\mathbf{c}_2)(\mathbf{b}_1\times\mathbf{b}_2)\times\overline{\overline{\mathsf{I}}} - (\mathbf{c}_2\cdot\mathbf{a}\times\mathbf{c}_3)(\mathbf{b}_2\times\mathbf{b}_3)\times\overline{\overline{\mathsf{I}}} - (\mathbf{c}_3\cdot\mathbf{a}\times\mathbf{c}_1)(\mathbf{b}_3\times\mathbf{b}_1)\times\overline{\overline{\mathsf{I}}}$$

$$= \mathbf{a}\cdot((\mathbf{c}_1\times\mathbf{c}_2)(\mathbf{b}_1\times\mathbf{b}_2) + (\mathbf{c}_2\times\mathbf{c}_3)(\mathbf{b}_2\times\mathbf{b}_3) + (\mathbf{c}_3\times\mathbf{c}_1)(\mathbf{b}_3\times\mathbf{b}_1))\times\overline{\overline{\mathsf{I}}}$$

$$= \mathbf{a}\cdot\frac{1}{2}(\overline{\overline{\mathsf{A}}}^T {}^\times_\times \overline{\overline{\mathsf{A}}}^T)\times\overline{\overline{\mathsf{I}}} = \mathbf{a}\cdot\overline{\overline{\mathsf{A}}}^{(2)T}\times\overline{\overline{\mathsf{I}}} = (\overline{\overline{\mathsf{A}}}^{(2)}\mathbf{a})\times\overline{\overline{\mathsf{I}}}.$$

B.5 Applying the rule

$$\overline{\overline{\mathsf{A}}} {}^\times_\times \overline{\overline{\mathsf{B}}} = [(\overline{\overline{\mathsf{A}}}:\overline{\overline{\mathsf{I}}})(\overline{\overline{\mathsf{B}}}:\overline{\overline{\mathsf{I}}}) - \overline{\overline{\mathsf{A}}}:\overline{\overline{\mathsf{B}}}^T]\overline{\overline{\mathsf{I}}} - (\overline{\overline{\mathsf{A}}}:\overline{\overline{\mathsf{I}}})\overline{\overline{\mathsf{B}}}^T - (\overline{\overline{\mathsf{B}}}:\overline{\overline{\mathsf{I}}})\overline{\overline{\mathsf{A}}}^T + [\overline{\overline{\mathsf{A}}}\cdot\overline{\overline{\mathsf{B}}} + \overline{\overline{\mathsf{B}}}\cdot\overline{\overline{\mathsf{A}}}]^T$$

for $\overline{\overline{\mathsf{B}}} = \overline{\overline{\mathsf{A}}}$ we obtain the identity

$$2\overline{\overline{\mathsf{A}}}^{(2)} = [(\mathrm{tr}\overline{\overline{\mathsf{A}}})^2 - \mathrm{tr}\overline{\overline{\mathsf{A}}}^2]\overline{\overline{\mathsf{I}}} - 2\mathrm{tr}\overline{\overline{\mathsf{A}}}\,\overline{\overline{\mathsf{A}}}^T + 2\overline{\overline{\mathsf{A}}}^{2T}.$$

Taking the trace of this, we obtain

$$\mathrm{spm}\overline{\overline{\mathsf{A}}} = \mathrm{tr}\overline{\overline{\mathsf{A}}}^{(2)} = \frac{1}{2}((\mathrm{tr}\overline{\overline{\mathsf{A}}})^2 - \mathrm{tr}\overline{\overline{\mathsf{A}}}^2),$$

whence the above identity yields

$$\overline{\overline{\mathsf{A}}}^{(2)} = \mathrm{spm}\overline{\overline{\mathsf{A}}}\,\overline{\overline{\mathsf{I}}} - \mathrm{tr}\overline{\overline{\mathsf{A}}}\,\overline{\overline{\mathsf{A}}}^T + \overline{\overline{\mathsf{A}}}^{2T}.$$

Substituting this in the rule

$$\overline{\overline{\mathsf{A}}}^{(2)T}\cdot\overline{\overline{\mathsf{A}}} = \det\overline{\overline{\mathsf{A}}}\,\overline{\overline{\mathsf{I}}},$$

yields the Cayley-Hamilton equation

$$\overline{\overline{\mathsf{A}}}^3 - \mathrm{tr}\overline{\overline{\mathsf{A}}}\,\overline{\overline{\mathsf{A}}}^2 + \mathrm{spm}\overline{\overline{\mathsf{A}}}\,\overline{\overline{\mathsf{A}}} - \det\overline{\overline{\mathsf{A}}}\,\overline{\overline{\mathsf{I}}} = 0.$$

B.6 The identity of the previous problem,

$$\overline{\overline{\mathsf{A}}}^{(2)} = \mathrm{spm}\overline{\overline{\mathsf{A}}}\,\overline{\overline{\mathsf{I}}} - \mathrm{tr}\overline{\overline{\mathsf{A}}}\,\overline{\overline{\mathsf{A}}}^T + \overline{\overline{\mathsf{A}}}^{2T},$$

can be written for the 2D dyadic $\overline{\overline{\mathsf{A}}} = \overline{\overline{\mathsf{A}}}_t$ satisfying

$$\overline{\overline{\mathsf{A}}}_t^{(2)} = \mathbf{nn}\ \mathrm{tr}\overline{\overline{\mathsf{A}}}_t^{(2)} = \mathbf{nn}\ \mathrm{spm}\overline{\overline{\mathsf{A}}}_t,$$

in the form

$$\mathrm{spm}\overline{\overline{\mathsf{A}}}_t\ \mathbf{nn} = \mathrm{spm}\overline{\overline{\mathsf{A}}}_t\,\overline{\overline{\mathsf{I}}} - \mathrm{tr}\overline{\overline{\mathsf{A}}}_t\,\overline{\overline{\mathsf{A}}}_t^T + \overline{\overline{\mathsf{A}}}_t^{2T},$$

or, as the 2D Cayley-Hamilton equation

$$\overline{\overline{\mathsf{A}}}_t^2 - \mathrm{tr}\overline{\overline{\mathsf{A}}}_t\,\overline{\overline{\mathsf{A}}}_t + \mathrm{spm}\overline{\overline{\mathsf{A}}}_t\,\overline{\overline{\mathsf{I}}}_t = 0.$$

B.7 Starting from

$$\mathbf{y} = \overline{\overline{\mathsf{A}}}\cdot\mathbf{x} = \alpha\mathbf{x} + \mathbf{a}\times\mathbf{x},$$

operate as

$$\begin{aligned}\mathbf{a}\times\mathbf{y} &= \alpha\mathbf{a}\times\mathbf{x} + \mathbf{a}\times(\mathbf{a}\times\mathbf{x}) \\ &= \alpha\mathbf{a}\times\mathbf{x} + \mathbf{aa}\cdot\mathbf{x} - \mathbf{a}^2\mathbf{x} \\ \mathbf{a}\cdot\mathbf{y} &= \alpha\mathbf{a}\cdot\mathbf{x}\end{aligned}$$

Combine

$$\begin{aligned}\alpha\mathbf{a}\times\mathbf{y} &= \alpha^2\mathbf{a}\times\mathbf{x} + \alpha\mathbf{aa}\cdot\mathbf{x} - \alpha\mathbf{a}^2\mathbf{x} \\ &= \alpha^2(\mathbf{y} - \alpha\mathbf{x}) + \mathbf{aa}\cdot\mathbf{y} - \alpha\mathbf{a}^2\mathbf{x} \\ &= \alpha^2\mathbf{y} + \mathbf{aa}\cdot\mathbf{y} - \alpha(\alpha^2 + \mathbf{a}^2)\mathbf{x}\end{aligned}$$

Solve for $\mathbf{x}$

$$\mathbf{x} = \frac{1}{\alpha(\alpha^2 + \mathbf{a}^2)}(\alpha^2\mathbf{y} + \mathbf{aa}\cdot\mathbf{y} - \alpha\mathbf{a}\times\mathbf{y}) = \overline{\overline{\mathsf{A}}}^{-1}\cdot\mathbf{y},$$

from which the inverse dyadic can be identified as

$$\overline{\overline{\mathsf{A}}}^{-1} = \frac{1}{\alpha(\alpha^2 + \mathbf{a}^2)}(\alpha^2\overline{\overline{\mathsf{I}}} + \mathbf{aa} - \alpha\mathbf{a}\times\overline{\overline{\mathsf{I}}})$$

Verifying

$$\begin{aligned}\overline{\overline{\mathsf{A}}}\cdot\overline{\overline{\mathsf{A}}}^{-1} &= \frac{1}{\alpha(\alpha^2 + \mathbf{a}^2)}(\alpha\overline{\overline{\mathsf{I}}} + \mathbf{a}\times\overline{\overline{\mathsf{I}}})\cdot(\alpha^2\overline{\overline{\mathsf{I}}} + \mathbf{aa} - \alpha\mathbf{a}\times\overline{\overline{\mathsf{I}}}) \\ &= \frac{1}{\alpha(\alpha^2 + \mathbf{a}^2)}(\alpha^3\overline{\overline{\mathsf{I}}} + \alpha\mathbf{aa} - \alpha\mathbf{a}\times(\mathbf{a}\times\overline{\overline{\mathsf{I}}})) = \overline{\overline{\mathsf{I}}}.\end{aligned}$$

The result can be obtained more directly by applying the rule for the dyadic inverse, in terms of which we can expand

$$\begin{aligned}
\overline{\overline{\mathsf{A}}}^{(2)} &= \frac{1}{2}(\alpha\overline{\overline{\mathsf{I}}} + \mathbf{a}\times\overline{\overline{\mathsf{I}}})^{\times}_{\times}(\alpha\overline{\overline{\mathsf{I}}} + \mathbf{a}\times\overline{\overline{\mathsf{I}}}) \\
&= \alpha^2\overline{\overline{\mathsf{I}}} + \alpha(\mathbf{a}\times\overline{\overline{\mathsf{I}}})^{\times}_{\times}\overline{\overline{\mathsf{I}}} + (\mathbf{a}\times\overline{\overline{\mathsf{I}}})^{(2)} \\
&= \alpha^2\overline{\overline{\mathsf{I}}} + \alpha\mathbf{a}\times\overline{\overline{\mathsf{I}}} + \mathbf{a}\mathbf{a}, \\
\mathrm{det}\overline{\overline{\mathsf{A}}} &= \frac{1}{6}(\alpha\overline{\overline{\mathsf{I}}} + \mathbf{a}\times\overline{\overline{\mathsf{I}}})^{\times}_{\times}(\alpha\overline{\overline{\mathsf{I}}} + \mathbf{a}\times\overline{\overline{\mathsf{I}}}) : (\alpha\overline{\overline{\mathsf{I}}} + \mathbf{a}\times\overline{\overline{\mathsf{I}}}) \\
&= \alpha^3 + \alpha\mathrm{tr}(\mathbf{a}\times\overline{\overline{\mathsf{I}}})^{(2)} = \alpha(\alpha^2 + \mathbf{a}^2),
\end{aligned}$$

whence

$$\overline{\overline{\mathsf{A}}}^{-1} = \frac{\overline{\overline{\mathsf{A}}}^{(2)T}}{\mathrm{det}\overline{\overline{\mathsf{A}}}} = \frac{1}{\alpha(\alpha^2 + \mathbf{a}^2)}(\alpha^2\overline{\overline{\mathsf{I}}} - \alpha\mathbf{a}\times\overline{\overline{\mathsf{I}}} + \mathbf{a}\mathbf{a}).$$

Bibliography

[1] K. Achouri and C. Caloz, "Space-wave routing via surface waves using a metasurface system," *Scientific Reports*, Vol. 8:7549, 2018.

[2] C. A. Balanis, *Advanced Engineering Electromagnetics*, New York: Wiley, 1989.

[3] C. F. Bohren and D. R. Huffman, *Absorption and Scattering of Light by Small Particles*, New York: Wiley, 1983.

[4] L. Brillouin, "The scattering cross section of spheres for electromagnetic waves," *J. Appl. Phys.*, Vol. 20, pp. 1110-1125, 1949.

[5] C. Caloz, A. Shahvarpour, D. L Sounas, T. Kodera, B. Gurlek and N. Chamanara, "Practical realization of perfect electromagnetic conductor (PEMC) boundaries using ferrites, magnetless non-reciprocal metamaterials (MNMs) and graphene," *Proc. URSI EMTS*, pp. 652–655, Hiroshima May 2013.

[6] W.C. Chew and T.M. Habashy, "Phase-conjugate mirror and time reversal," *J. Opt. Soc. Am. A*, Vol. 2, no. 6, pp. 808-809, June 1985.

[7] P.J.B. Clarricoats and P.K. Saha, "Propagation and radiation behaviour of corrugated feeds. I. Corrugated waveguide feed," *Proc. Inst. Elec. Eng.*, Vol. 118, pp. 1167-1176, 1971.

[8] R.F. Collin, *Field Theory of Guided Waves*, New York: McGraw-Hill, 1960, Sec. 11.2.

[9] G.A. Deschamps, "The Gaussian beam as a bundle of complex rays," *Electron. Lett.*, Vol. 7, no. 23, pp. 684-685, 1971.

[10] G.A. Deschamps, "Electromagnetics and differential forms," *Proc. IEEE*, Vol. 69, no. 6, pp. 676-696, June 1981.

[11] H. M. El-Maghrabi, A. M. Attiya and E. A. Hashish, "Design of a perfect electromagnetic conductor (PEMC) boundary by using periodic patches," *Prog. Electromag. Res. M*, Vol. 16, pp. 159-169, 2011.

[12] L.B. Felsen and N. Marcuvitz, *Radiation and Scattering of Waves*, Englewood Cliffs, N.J.: Prentice-Hall, 1973, pp. 557-559.

[13] J.W. Gibbs, *Vector Analysis*, 2nd ed., New York: Dover, 1960.

[14] J.W. Gibbs, *Elements of Vector Analysis, Arranged for the Use of Students in Physics.* Not published, privately printed in New Haven, pp. 1-36, 1881, pp. 37-82, 1884. In: *The Scientific Papers of J. Willad Gibbs*, Vol. 2, New York: Dover, 1961, pp. 17–90.

[15] W. H. Greub, *Linear Algebra*, 3rd. ed., Berlin: Springer-Verlag, 1967.

[16] V. Gülzow, "An integral equation method for the time-harmonic Maxwell equations with boundary conditions for the normal components," *J. Integral Equations*, Vol. 1, no. 3, pp. 365-384, 1988.

[17] I. Hänninen, I.V. Lindell, A.H. Sihvola, "Realization of generalized soft-and-hard boundary," *Prog. Electromag. Res.*, Vol. 64, pp. 317-333, 2006.

[18] R. F. Harrington, *Time-Harmonic Electromagnetic Fields*, New York: McGraw-Hill, 1961.

[19] O. Heaviside, *Electrical Papers*, New York: Chelsea, 1970; reprint of the 1st edition, London 1892 (vol. 1, p. 447; Vol. 2, pp. 172-175). The original articles were respectively published in *The Electrician*, 1885, p. 306, and *Philos. Mag.*, August 1886, p. 118.

[20] F.W. Hehl and Yu.N. Obukhov, *Foundations of Classical Electrodynamics*, Boston, MA: Birkhäuser, 2003.

[21] D.J. Hoppe and Y. Rahmat-Samii, *Impedance Boundary Conditions in Electromagnetics*, Washington, D.C.: Taylor and Francis, 1995.

[22] R.G.E. Hutter, *Beam and Wave Electronics in Microwave Tubes*, New York: Van Nostrand, 1960, pp. 220–230.

[23] A. Ishimaru, *Electromagnetic Wave Propagation and Scattering*, Englewood Cliffs, N.J.: Prentice-Hall, 1991.

[24] Ya. Itin, "Covariant boundary conditions in electromagnetism," *ACE 10, 6th Workshop on Adv. Comp. Electromag.*, ETH, Zurich, July 2010.

[25] P.-S. Kildal, "Definition of artificially soft and hard surfaces for electromagnetic waves," *Electron. Lett.*, Vol. 24, pp. 168-170, 1988.

[26] P.-S. Kildal, "Artificially soft and hard surfaces in electromagnetics," *IEEE Trans. Antennas Propagat.*, Vol. 38, no. 10, pp. 1537-1544, Oct. 1990.

[27] P.-S. Kildal, "Fundamental properties of canonical soft and hard surfaces, perfect magnetic conductors and the newly introduced DB surface and their relation to different practical applications including cloaking," *Proc. ICEAA'09*, Torino, Italy, August 2009, pp. 607-610.

[28] J. A. Kong, *Electromagnetic Wave Theory*, 2nd ed., New York: Wiley, 1990.

[29] J.A. Kong, *Electromagnetic Wave Theory*, Cambridge, MA: EMW Publishing, 2005.

[30] G. Kristensson, *Scattering of Electromagnetic Waves by Obstacles*, Edison, NJ, USA: SciTech Publishing, IET, 2016.

[31] E.F. Kuester and D.C. Chang, *Electromagnetic Boundary Problems*, Boca Raton FL: CRC Press, 2016.

[32] M.A. Leontovich, "Methods of solution for problems of electromagnetic wave propagation along the Earth surface," *Bull. Acad. Sci. USSR, Phys. Ser.*, Vol. 8, no. 1, pp. 16-22, 1944 (in Russian).

[33] I. V. Lindell, A. Sihvola, S. Tretyakov and A. Viitanen, *Electromagnetic Waves in Chiral and Bi-Isotropic Media*, Boston: Artech House, 1994.

[34] I.V. Lindell, "Electrostatic image theory for a sphere with impedance surface," *Journal of Physics D: Appl. Phys.*, Vol. 27, pp. 1605-1607, 1994.

[35] I.V. Lindell, *Methods for Electromagnetic Field Analysis*, 2nd ed., New York: Wiley and IEEE Press, 1995.

[36] I.V. Lindell, "Image theory for the soft and hard surface". *IEEE Trans. Antennas Propagat.*, Vol. 43, no. 1, pp. 117-119, January 1995.

[37] I.V. Lindell, A.H. Sihvola and K. Suchy: "Six-vector formalism in electromagnetics of bi-anisotropic media," *J. Electro. Waves Appl.*, Vol. 9, no. 7/8, pp. 887-903, 1995.

[38] I.V. Lindell, P.P. Puska, "Reflection dyadic for the soft and hard surface with application to the depolarizing corner reflector," *IEE Proc. Microw. Ant. Propag.*, Vol. 143, no. 5, pp. 417-421, October 1996.

[39] I.V. Lindell, "Condition for the general ideal boundary," *Microwave Opt. Tech. Lett.*, Vol. 26, no. 1, pp. 61-64, July 2000.

[40] I.V. Lindell, "Image theory for the isotropic ideal boundary," *Microwave Opt. Tech. Lett.*, Vol. 27, no. 1, pp. 68-72, October 2000.

[41] I.V. Lindell, "The ideal boundary and generalized soft-and-hard conditions," *IEE Proc. Microw. Ant. Propag.*, Vol. 147, no. 6, pp. 495-499, December 2000.

[42] I.V. Lindell and F. Olyslager, "Duality in Electromagnetics," *J. Commun. Technology and Electronics*, Vol. 45, supplementary issue 2, pp. 260-268, October–November 2000.

[43] I.V. Lindell, "Generalized soft-and-hard surface," *IEEE Trans. Antennas Propag.*, Vol. 50, no. 7, pp. 926-929, July 2002.

[44] I.V. Lindell, *Differential Forms in Electromagnetics*, New York: Wiley, 2004.

[45] I.V. Lindell, A.H. Sihvola, "Perfect electromagnetic conductor," *J. Electro. Waves Appl.*, Vol. 19, no. 7, pp. 861-869, 2005.

[46] I.V. Lindell, A.H. Sihvola, "Transformation method for problems involving perfect electromagnetic conductor (PEMC) structures," *IEEE Trans. Antennas Propag.*, Vol. 53, no. 9, pp. 3005-3011, September 2005.

[47] I.V. Lindell, A.H. Sihvola, "Realization of the PEMC boundary," *IEEE Trans. Antennas Propag.*, Vol. 53, no. 9, pp. 3012-3018, September 2005.

[48] I.V. Lindell, A. Sihvola, "Electromagnetostatic image theory for the PEMC sphere," *IEE Proc. Sci. Meas. Tech.*, Vol. 153, no. 3, pp. 120-124, May 2006.

[49] I.V. Lindell, A.H. Sihvola, "The PEMC resonator," *J. Electro. Waves Appl.*, Vol. 20, no. 7, pp. 849-859, 2006.

[50] I.V. Lindell, A.H. Sihvola, "Losses in the PEMC boundary," *IEEE Trans. Antennas Propag.*, Vol. 54, no. 9, pp. 2553-2558, September 2006.

[51] I.V. Lindell, A.H. Sihvola, "Realization of impedance boundary," *IEEE Trans. Antennas Propag.*, Vol. 54, no. 12, pp. 3669-3676, December 2006.

[52] I.V. Lindell, A.H. Sihvola, I. Hänninen, "Perfectly anisotropic impedance boundary," *IET Microw. Antennas Propag.*, Vol. 1, no. 3, pp. 561-566, June 2007.

[53] I.V. Lindell, A.H. Sihvola, "Reflection and transmission of waves at the interface of perfect electromagnetic conductor (PEMC)," *PIER B*, Vol. 5, pp. 169-183, 2008.

[54] I.V. Lindell, A.H. Sihvola, "Electromagnetic DB boundary," *Proc. XXXI Finnish URSI Convention*, Espoo, October 2008, pp. 81-82.

[55] I.V. Lindell, A.H. Sihvola, "DB boundary as isotropic soft surface," *Proc. Asian Pacific Microwave Conference*, Hong Kong, December 2008 (4 pages).

[56] I.V. Lindell, A. Sihvola, "Zero axial parameter (ZAP) sheet," *Prog. Electromag. Res.*, Vol. 89, pp. 213-224, 2009.

[57] I.V. Lindell, A. Sihvola, "Spherical resonator with DB-boundary conditions," *Prog. Electromag. Res. Lett.*, Vol. 6, pp. 131-137, 2009.

[58] I.V. Lindell, A. Sihvola, "Electromagnetic boundary condition and its realization with anisotropic metamaterial," *Phys. Rev. E*, Vol. 79, no. 2, 026604 (7 pages), 2009.

[59] I.V. Lindell, A. Sihvola, "Uniaxial IB-medium interface and novel boundary conditions," *IEEE Trans. Antennas Propag.*, Vol. 57, no. 3, pp. 694-700, March 2009.

[60] I.V. Lindell, A. Sihvola, "Electromagnetic boundary conditions defined in terms of normal field components," *IEEE Trans. Antennas Propag.*, Vol. 58, no. 4, pp. 1128-1135, April 2010.

[61] I.V. Lindell, H. Wallén, A. Sihvola, "General electromagnetic boundary conditions involving normal field components," *IEEE Antennas Wireless Propag. Lett.*, Vol. 8, pp. 877-880, August 11, 2009.

[62] I.V. Lindell, A. Sihvola, P. Ylä-Oijala, H. Wallén, "Zero backscattering from self-dual objects of finite size," *IEEE Trans. Antennas Propag.*, Vol. 57, no. 9, pp. 2725-2731, September 2009.

[63] I.V. Lindell, A. Sihvola, "Circular waveguide with DB-boundary conditions," *IEEE Trans. Microwave Theory Tech.*, Vol. 58, no. 4, pp. 903-909, April 2010.

[64] I.V. Lindell, A. Sihvola, L. Bergamin, A. Favaro, "Realization of the D'B' boundary condition," *IEEE Antennas Wireless Propag. Lett.*, Vol. 10, pp. 643-646, July 5, 2011.

[65] I.V. Lindell, J. Markkanen, A. Sihvola, P. Ylä-Oijala, "Realization of spherical D'B' boundary by a layer of wave-guiding medium;" *Metamaterials*, Vol. 5, pp. 149-154, 2011.

[66] I.V. Lindell, A. Sihvola, "Simple skewon medium realization of DB boundary condition," *Prog. Electromag. Res. Letters*, Vol. 30, pp. 29-39, 2012.

[67] I.V. Lindell, A. Sihvola, "Skewon-axion medium and soft-and-hard/DB boundary condition," `arXiv:1201.4738v1 [physics.class-ph] 20 Jan 2012`.

[68] I.V. Lindell, A. Sihvola, "Soft-and-hard/D'B' boundary conditions and their realization by electromagnetic media," *IEEE Trans. Antennas Propag.*, Vol. 61, no. 1, pp. 478-482, January 2013.

[69] I.V. Lindell, A. Sihvola, "Soft-and-hard/DB boundary conditions realized by a skewon-axion medium," *IEEE Trans. Antennas Propag.*, Vol. 61, no. 2, pp. 768-774, February 2013.

[70] I.V. Lindell, A. Sihvola, "SHDB Boundary Conditions Realized by Pseudochiral Media," *IEEE Antennas Wireless Propag. Lett.*, Vol. 12, pp. 591-594, 2013.

[71] I.V. Lindell, A. Sihvola, "Surface waves on SHDB boundary," *IEEE Antennas Wireless Propag. Lett.*, Vol. 13, pp. 1027-1030, 2014.

[72] I.V. Lindell, *Multiforms. Dyadics, and Electromagnetic Media*, Hoboken, NJ: IEEE Press and Wiley, 2015.

[73] I.V. Lindell, A. Sihvola, "Electromagnetic boundaries with PEC/PMC equivalence," *Prog. Electromag. Res. Lett.*, Vol. 61, pp. 119-123, 2016.

[74] I.V. Lindell, A. Sihvola, "Generalized Soft-and-Hard/DB Boundary" *IEEE Trans. Antennas Propag.*, Vol. 65, no. 1, pp. 226-233, January 2017.

[75] I.V. Lindell, A. Sihvola, "Electromagnetic wave reflection from boundaries defined by general linear and local conditions," *IEEE Trans. Antennas Propag.*, Vol. 65, no. 9, pp. 4656-4663, September 2017.

[76] F. Liu, S. Xiao, A. Sihvola and J. Li, "Perfect co-circular polarization reflector: A class of reciprocal perfect conductors with total co-circular polarization reflector," *IEEE Trans. Antennas Propag.*, Vol. 62, no. 12, pp. 6274-6281, 2014.

[77] L. Lorenz, "Lysbevægelser i og uden for en af plane Lysbolger belyst Kugle", *Kongelige Danske Videnskabernes Selskabs Skrifter, (Denmark)*, 6:2-62, 1890.

[78] S. Maci, G. Minatti, M. Casaletti and M. Bosilvejac, "Metasurfing: Addressing waves on impenetrable metasurfaces," *IEEE Antennas and Wireless Propagation Letters*, Vol. 10, pp. 1499-1502, 2011.

[79] J.C. Maxwell, *Treatise on Electricity and Magnetism*, 3rd ed., Oxford: Clarendon Press, 1904 (1st ed. 1873).

[80] J. C. Maxwell Garnett, "Colours in metal glasses and metal films," *Transactions of the Royal Society (London)*, Vol. CCIII, pp. 385-420, 1904.

[81] G. Mie, "Beiträge zur Optik trüber Medien, speziell kolloidaler Metallösungen", *Annalen der Physik*, 25:377-445, 1908.

[82] G. W. Milton. *The Theory of Composites*, Cambridge University Press, 2002.

[83] C. Paiva and S. Matos, "Minkowskian isotropic media and the perfect electromagnetic conductor medium," *IEEE Trans. Antennas Propagat.*, Vol. 60, no. 7, pp. 3231-3245, 2012.

[84] G. Pelosi and P.Y. Ufimtsev, "The impedance-boundary condition," *IEEE Antennas Propag. Mag.*, Vol. 38, pp. 31-35, 1996.

[85] R. Picard, "Zur Lösungstheorie der zeitunabhängigen Maxwellschen Gleichungen mit der Randbedingung $\mathbf{n}\cdot\mathbf{B} = \mathbf{n}\cdot\mathbf{D} = 0$ in anisotropen inhomogenen Medien", *Manuscr. Math.*, Vol. 13, pp. 37-52, 1974.

[86] R. Picard, "Ein Randwertproblem für die zeitunabhängigen Maxwellschen Gleichungen mit der Randbedingungen $\mathbf{n}\cdot\epsilon\mathbf{E} = \mathbf{n}\cdot\mu\mathbf{H} = 0$ in beschränkten Gebieten Beliebigen Zusammenhangs," *Appl. Anal.*, Vol. 6, pp. 207-221, 1977.

[87] V. H. Rumsey, "Some new forms of Huygens' principle," *IRE Trans. Antennas Propag., Special Supplement*, Vol. 7, pp. S103-S116, December 1959.

[88] T.B.A. Senior and J.L. Volakis, *Approximate Boundary Conditions in Electromagnetics*, London, U.K.: IEE, 1995.

[89] A. Serdyukov, I. Semchenko, S. Tretyakov and A. Sihvola, *Electromagnetics of Bi-Anisotropic Materials. Theory and Applications.*, Amsterdam: Gordon and Breach, 2001.

[90] A. Shahvarpour, T. Kodera, A. Parsa and C. Caloz, "Arbitrary electromagnetic conductor boundaries using Faraday rotation in a grounded ferrite slab," *IEEE Trans. Microwave Theory Tech.*, Vol. 58, no. 11, pp. 2781-2793, 2010.

[91] A.N. Shchukin, *Propagation of Radio Waves*, Moscow: Svyazizdat, 1940 (in Russian).

[92] D. Sievenpiper, L. Zhang, R.F. Jimenez Broas, N.G. Alexopoulos and E. Yablonovitch, "High-impedance electromagnetic surfaces with a forbidden frequency band," *Trans. IEEE Microwave Theory Tech.*, Vol. 47, no. 11, pp. 2059-2074, November 1999.

[93] A. Sihvola, *Electromagnetic Mixing Formulas and Applications*, Vol. 47 of Electromagnetic Waves Series, IEE/IET Publishing, London, 1999.

[94] A. Sihvola and I.V. Lindell, "Perfect electromagnetic conductor as building block for complex materials," *Electromagnetics*, Vol. 26, no. 3–4, pp. 279-287, April–June 2006.

[95] A. Sihvola, P. Ylä-Oijala, I.V. Lindell, "Scattering by perfect electromagnetic conductor (PEMC) spheres: surface integral equation approach," *ACES Journal*, Vol. 22, no. 2, pp. 236-249, July 2007.

[96] A. Sihvola, P. Ylä-Oijala and I.V. Lindell, "Scattering by PEMC (Perfect Electromagnetic Conductor) spheres using surface integral equation approach", *ACES, Applied Computational Electromagnetics Society Journal*, Vol. 22, no. 2, pp. 236-249, July 2007.

[97] A. Sihvola, I. V. Lindell, "Perfect electromagnetic conductor medium," *Ann. Phys.* (Berlin) Vol. 17, pp. 787-802, September/October 2008.

[98] A. Sihvola, H. Wallén, M. Taskinen, P. Ylä-Oijala, H. Kettunen, I.V. Lindell, "Scattering by DB spheres," *IEEE Antennas Wireless Propag. Lett.*, Vol. 8, pp. 542-545, June 24, 2009.

[99] A. Sihvola, I.V. Lindell, "Bianisotropic materials and PEMC," Chapter 26 in *Metamaterials Handbook, Theory and Phenomena of Metamaterials*, Boca Raton: CRC Press, pp. 26.1–26.7, 2009.

[100] A. Sihvola, J. Qi, I.V. Lindell, "Bridging the gap between plasmonics and Zenneck waves," *IEEE Antennas Propag. Mag.*, Vol. 52, no. 1, pp. 125-136, February 2010.

[101] A. Sihvola, I.V. Lindell, H. Wallén, P. Ylä-Oijala, "Material realizations of perfect electric conductor objects," *ACES Journal*, Vol. 25, no. 12, pp. 1007-1016, 2010.

[102] A. Sihvola, D.C. Tzarouchis, P. Ylä-Oijala, H. Wallén, B.B. Kong, "Resonances in small scatterers with impedance boundary," *Physical Review B*, Vol. 98, 235417, 2018.

[103] A. Sommerfeld, *Partielle Differentialgleichungen der Physik*, Band VI, 6th ed., Leipzig: Akademische Verlagsgesellschaft Geest & Portig, 1966, Section 32.

[104] J.A. Stratton, *Electromagnetic Theory*, New York: McGraw-Hill, 1941, p. 487.

[105] M. Taskinen and P. Ylä-Oijala, "Current and charge integral equation formulation" *IEEE Trans. Antennas Propag.*, Vol. 54, no. 1, pp. 58-67, January 2006.

[106] N. Tedeschi, F. Frezza and A. Sihvola, "On the perfectly matched layer and the DB boundary condition," *J. of the Optical Society of America A*, Vol. 30, no. 10, pp. 1941-1946, October 2013.

[107] N. Tedeschi, F. Frezza and A. Sihvola, "Reflection and transmission at the interface with an electric - magnetic uniaxial medium with application to boundary conditions," *IEEE Trans. Antennas Propagat.*, Vol. 61, no. 11, pp. 5666-5675, November 2013.

[108] S. A. Tretyakov, A. H. Sihvola, A. A. Sochava, and C. R. Simovski, " Magnetoelectric interactions in bi-anisotropic media," *J. Electromag. Waves Appl.*, Vol. 12, no. 4, pp. 481-497, 1998. "Correction," *ibid.*, Vol. 13, no. 2, p. 225, 1999.

[109] J. Van Bladel, *Electromagnetic Fields*, 2nd ed., Piscataway, NJ: Wiley and IEEE Press, 2007. Appendix 4: Dyadic Analysis.

[110] V. H. Weston, "Theory of absorbers in scattering," *IEEE Trans. Antennas Propag.*, Vol. 11, pp. 578-584, September 1963.

[111] A.J. Viitanen, S.A. Tretyakov, I.V. Lindell, "On the realization of the generalized soft-and-hard surface," *Radio Sci.*, Vol. 35, no. 6, pp. 1257-1264, November–December 2000.

[112] H. Wallén, I.V. Lindell, A. Sihvola, "Mixed-impedance boundary conditions," *IEEE Trans. Antennas Propag.*, Vol. 59, no. 5, pp. 1580-1586, May 2011. Note that Equation (12) is incorrect and should be replaced by Equation (13).

[113] H. Wallén, H. Kettunen and A. Sihvola, "Anomalous absorption, plasmonic resonances, and invisibility of radially anisotropic spheres," *Radio Science*, Vol. 50, no. 1, pp. 18-28, January 2015.

[114] R. Weder, "The boundary conditions for point transformed electromagnetic invisible cloaks," *J. Phys. A*, Vol. 41, 415401 (17 pages), 2008.

[115] W.J. Wiscombe, "Improved Mie scattering algorithm," *Applied Optics*, Vol. 19, no. 9, pp. 1505-1509, May 1980.

[116] A. Yaghjian and S. Maci, "Alternative derivation of electromagnetic cloaks and concentrators," *New J. Phys.*, Vol. 10, 115022, 2008; "Corrigendum", *ibid*, Vol. 11, 039802, 2009.

[117] A. Yaghjian, "Extreme electromagnetic boundary conditions and their manifestation at the inner surfaces of spherical and cylindrical cloaks," *Metamaterials*, Vol. 4, pp. 70-76, 2010.

[118] K. S. Yee, "Uniqueness theorem for an exterior electromagnetic field," *SIAM J. Appl. Math.*, Vol. 18, no. 1, pp. 77-83, 1970.

[119] C. Yeh, "Boundary conditions in electromagnetics," *Phys. Rev.*, Vol. E 48, pp. 1426-1427, 1993.

[120] D. Zaluski, D. Muha and S. Hrabar, "DB boundary based on resonant metamaterial inclusions," Proc. *Metamaterials 2011*, Barcelona, October 2011, pp. 820-822.

[121] D. Zaluski, S. Hrabar and D. Muha, "Practical realization of DB metasurface," *Appl. Phys. Lett.*, Vol. 104, 234106, 2014.

[122] B. H. Zhang, Chen, B.-I. Wu and J. A. Kong, "Extraordinary surface voltage effect in the invisibility cloak with an active device inside," *Phys. Rev. Lett.*, Vol. 100, 063904 (4 pages), February 15, 2008.

[123] Y. Zhao and A. Alù, "Manipulating light polarization with ultrathin plasmonic metasurfaces," *Physical Review B*, Vol. 84, 205428, 2011.

[124] Y. Zhao, N. Engheta and A. Alù, "Homogenization of plasmonic metasurfaces modeled as transmission-line loads," *Metamaterials*, Vol. 5, pp. 90-96, 2011.

Index

IEEE PRESS SERIES ON ELECTROMAGNETIC WAVE THEORY

Field Theory of Guided Waves, Second Edition
Robert E. Collin

Field Computation by Moment Methods
Roger F. Harrington

Radiation and Scattering of Waves
Leopold B. Felsen, Nathan Marcuvitz

Methods in Electromagnetic Wave Propagation, Second Edition
D. S. J. Jones

Mathematical Foundations for Electromagnetic Theory
Donald G. Dudley

The Transmission-Line Modeling Method: TLM
Christos Christopoulos

The Plane Wave Spectrum Representation of Electromagnetic Fields, Revised Edition
P. C. Clemmow

General Vector and Dyadic Analysis: Applied Mathematics in Field Theory
Chen-To Tai

Computational Methods for Electromagnetics
Andrew F. Peterson

Plane-Wave Theory of Time-Domain Fields: Near-Field Scanning Applications
Thorkild B. Hansen

Foundations for Microwave Engineering, Second Edition
Robert Collin

Time-Harmonic Electromagnetic Fields
Robert F. Harrington

Antenna Theory & Design, Revised Edition
Robert S. Elliott

Differential Forms in Electromagnetics
Ismo V. Lindell

Conformal Array Antenna Theory and Design
Lars Josefsson, Patrik Persson

Multigrid Finite Element Methods for Electromagnetic Field Modeling
Yu Zhu, Andreas C. Cangellaris

Electromagnetic Theory
Julius Adams Stratton

Electromagnetic Fields, Second Edition
Jean G. Van Bladel

Electromagnetic Fields in Cavities: Deterministic and Statistical Theories
David A. Hill

Discontinuities in the Electromagnetic Field
M. Mithat Idemen

Understanding Geometric Algebra for Electromagnetic Theory
John W. Arthur

The Power and Beauty of Electromagnetic Theory
Frederic R. Morgenthaler

Modern Lens Antennas for Communications Engineering
John Thornton, Kao-Cheng

Electromagnetic Modeling and Simulation
Levent Sevgi

Multiforms, Dyadics, and Electromagnetic Media
Ismo V. Lindell

Low-Profile Natural and Metamaterial Antennas: Analysis Methods and Applications
Hisamatsu Nakano

From ER to E.T.: How Electromagnetic Technologies Are Changing Our Lives
Rajeev Bansal

Electromagnetic Wave Propagation, Radiation, and Scattering: From Fundamentals to Applications, Second Edition
Akira Ishimaru

Time-Domain Electromagnetic Reciprocity in Antenna Modeling
Martin Štumpf

Boundary Conditions in Electromagnetics
Ismo V. Lindell and Ari Sihvola